MAGNETIC FIELDS OF CELESTIAL BODIES

ASTROPHYSICS AND SPACE SCIENCE LIBRARY

VOLUME 198

MAGNETIC FIELDS OF CELESTIAL BODIES

by

YE SHI-HUI

Purple Mountain Observatory and
Yunnan Astronomical Observatory,
Academia Sinica

SPRINGER SCIENCE+BUSINESS MEDIA, B.V.

A C.I.P. Catalogue record for this book is available from the Library of Congress

ISBN 978-94-010-4407-3 ISBN 978-94-011-0944-4 (eBook)
DOI 10.1007/978-94-011-0944-4

Printed on acid-free paper

Dedicated to the memory of my late Soviet advisor,

Academician Andrei Borisovich Severny
(1913 - 1987)

Contents

Preface

Magnetism is one of the basic properties of matter. Mankind has travelled a long road in discovering and utilizing magnetism, and in this respect the ancient Chinese people have made outstanding contributions. In the book 'Lu's Spring and Autumn', written near the end of the Warring States Period, i.e. in the third century B.C., there is a statement on the "attraction of iron by lodestones". So at that time it was known that magnets can attract ferromagnetic material. At the beginning of the first century A.D., viz. in the early years of the East Hang Dynasty, the famous scholar Wang Chong wrote in his masterpiece 'Len Hen' that the handle of a magnetic dipper pointed to the south. It was thus discovered at the time that magnets can point to the poles of the geomagnetic field. At the beginning of the twelfth century, during the reign of Emperor Hui of the Sung Dynasty, in the two books written by Zhu Yo and Xu Jin, respectively, there are descriptions of the compass used in navigation. This tells us that the application of compasses was rather widespread at that time. The distinguished scientist Sen Go (1031–1085) discovered the declination of the terrestrial magnetic field. This is four hundred and more years earlier than its discovery by Christopher Columbus in 1492 during his voyage across the Atlantic Ocean. Such facts as these manifest the important contributions of ancient China to global civilization.

Since the seventeenth century the study of terrestrial magnetisn has become a mature branch of natural science. However, only at the beginning of the twentieth century did mankind start to investigate the magnetic fields in the vastness of the universe. In the last eighty and more years, thanks to the efforts of astronomers of several generations, tremendous progress has been achieved. Nowadays we are profoundly aware of the fact that magnetic lines of force pervade everywhere in the Sun, Moon, stars, galaxies, as well as in the immense cosmic space. Numerous astronomical phenomena, such as sunspots, flares, solar wind, cosmic rays and the polarization of starlight, are closely connected with magnetism. Therefore, research on the magnetic fields of celestial bodies has become an important part of contemporary astrophysics. The author has compiled this monograph in order to introduce this branch of knowledge to the readers, in particular to scientific workers as well as the postgraduate and undergraduate students of astronomy, space physics, geophysics and other related sciences. The book presents a comprehensive account of the methods of observation, the results of data analysis and its interpretation, as well as several selected topics of theoretical research. What we seek to do is chiefly to clarify some basic concepts and principles and to interpret a series of important phenomena so that readers may acquire an overall and clear understanding of the present status of the study of cosmic magnetic fields. In this regard the author has tried to do his best, but there are unavoidably inexactitudes, slips and mistakes. He sincerely wishes that readers would be so kind as to send their comments to him, as these would be greatly valued.

This book was originally written in Chinese and published by the Science Press in Beijing in 1978. Thanks to the enthusiastic encouragement and suggestion of Professors C. de Jager and S.-I. Akasofu, in 1990 the author undertook to rewrite and publish this monograph in English, in order that it might be available to the wide international astronomical community. Because the study of celestial magnetism has achieved very rapid progress in the last decade, a significant portion of the original book has become out of date. Being aware of this, the author has rewritten the text from the very beginning to the end. Although the framework remains basically the same, the content has been greatly enriched and changed. Thus the English edition is practically a new book, which has little connection with the Chinese one. The author hopes that this thoroughly revised work will be of some interest and use to the reader. If this wish should become reality, the author would like to credit this to the good advice and kind concern of Professors de Jager and Akasofu and hereby expresses to them his hearty gratitude.

Finally, the author's sincere thanks are also due to the responsible editor, Dr. M. Kolman, and the staff of Kluwer Academic Publishers for the pleasant cooperation in editing and publishing this book.

April, 1993
Ye Shi-hui

Acknowledgements

In the preparation of this monograph I have much benefited by the valuable help and kind assistance of many friends, both in China and abroad. It is a great pleasure for me here to mention their names, affiliations and contributions, with my gratitude. Professor Ai Guoxiang of the Beijing Observatory and Professor W.C. Livingston of the National Solar Observatory at Kitt Peak have provided me with their excellent originals of Figures 3.16, 3.17, 3.18 and Figures 3.10, 3.11, respectively.

The following colleagues have supplied me with their offprints, dissertations and books, and/or had enlightening discussions with me: Professor Ai Guoxiang, Dr. Li Wei and Professor Lin Yuanzhang (Beijing Observatory), Professors S.I. Gopasyuk and D.N. Rachkovskii (Crimean Astrophysical Observatory), Professor V.M. Grigoriev (Sayan Observatory), Professor E. Hiei (National Astronomical Observatory of Japan), Professor V.L. Khokhlova (Institute of Astronomy, Russian Academy of Sciences), Professor I.S. Kim (Sternberg Institute of Astronomy of Moscow University), Professor M. Makita (Kwasan and Hida Observatories), Professor V.N. Obridko (IZMIRAN, Moscow), Dr. Qu Zhongquan (Yunnan Observatory), Professor J.O. Stenflo (Institute of Astronomy, Zürich), Professors Wang Zhen-yi and You Jiang-qi (Purple Mountain Observatory), Professor H. Wöhl (Kiepenheuer Institut für Sonnenphysik), Professor Wu Lin-xiang (Beijing University), Professor H. Yoshimura (Tokyo University), Professor Zhang Zhen-da (Nanjing University), Professor H. Zirin and Dr. Haimin Wang (Big Bear Solar Observatory).

The following scientists have kindly permitted me to reproduce in my book their published illustrations and some of them have sent me high-quality figures: Professor R. Beck (Max Planck Institut für Radioastronomie), Figure 10.2; Professor R.J. Bray (CSIRO, Australia), Figures 2.6, 3.1, 3.3, 3.9 and 4.1; Professor V. Bumba (Ondrejov Observatory), Figure 4.17; Professor M.J. Hagyard (Marshall Space Flight Center), Figures 3.13, 3.14 and 4.10; Professor A.P. Hatzes (University of Texas), Figure 9.6; Professor T.E. Holzer (High Altitude Observatory), Figure 8.4; Professor H. Kawakami (National Astronomical Observatory of Japan), Figure 6.8; Professor V.L. Khokhlova, Figures 9.4 and 9.5; Professor Lin Yuanzhang, Figure 5.18; Professor M. Makita, Figures 3.19 and 3.20; Dr. S.H.B. Livi (Universidade Federal, Brazil), Figure 5.4; Professor V.N. Obridko, Figure 4.5; Professor R.S. Ronan (Stanford University), Figures 3.23 and 3.24; Professor D.M. Rust (Johns Hopkins University), Figures 3.21 and 3.22; Professor G.D. Schmidt (Steward Observatory), Figure 9.8; Professor Y. Sofue (Tokyo University), Figure 10.6; Professor S.K. Solanki (Institute of Astronomy, Zürich), Figures 6.1 and 6.2; Professor J.O. Stenflo, Figures 2.10, 5.1, 5.2 and 5.14; Professor J.P. Vallée (Herzberg Institute), Figure 10.4; Dr. Yang Wei-min (Yunnan Observatory), Figure 3.35; Professor H. Yoshimura, Figure 11.4.

To all the persons mentioned above and still some others, who have kindly assisted me in the writing of this book, I would like to express my hearty indebtedness.

Ye Shi-hui

CHAPTER 1

Introduction

It appears that the radical element responsible for the continuing thread of cosmic unrest is the magnetic field.

E.N. Parker,

Cosmical Magnetic Fields (1979).

1.1 A Brief Historical Account

The study of magnetic fields outside the Earth commenced in the beginning of the twentieth century. Not long before this, Bigelow (1889) suggested that the Sun might possess a magnetic field like that of the Earth. His speculation was based upon the similarity of the fine coronal plumes near the polar regions, photographed during solar total eclipses, to the field lines of a magnetized sphere. But the first formal and reliable evidence of the existence of solar magnetism was obtained only almost twenty years later. In 1908 the distinguished American astronomer and the father of solar physics, G.E. Hale (1868–1938), discovered the *Zeeman effect of spectral lines of sunspots* and confirmed that they have magnetic fields of several thousand gauss. This is the very beginning of the study of the magnetism of heavenly objects. Ever since that time the research on solar magnetic fields has recorded outstanding achievements and up to the present day it occupies a leading place in the investigation of cosmic magnetism.

In 1947 H.W. Babcock used the 100-inch reflector of the Mt. Wilson Observatory to measure the magnetic field of Virgo 78 and found its strength to be 1500 G. This was the first magnetic observation of stars besides the Sun and appears to be the second milestone in the exploration of celestial magnetic fields.

The third milestone may be considered to be the observation of the *interstellar magnetic field* by means of the Zeeman effect of the 21 cm hydrogen line. Such a possibility was pointed out by J.G. Bolton and J.P. Wild in 1957, but actual measurements were performed only ten years later.

Since the beginning of the space era in 1957, the study of astronomical magnetic fields has been once again proposed. Thanks to the successful launching of many artificial satellites, orbiting observatories and other spacecraft, man can make direct and in situ measurements of the magnetic fields of the Moon, the planets and interplanetary space. The discovery of the sectoral structure of the heliospheric magnetic field in the early sixties is a celebrated achievement and has important practical significance.

So the scope and content of the study of cosmic magnetism has once again become wider and more abundant. The large amount of observational material provides a sound basis for theoretical research. Since the fifties of this century much progress has been achieved in this respect. It seems to be sufficient to mention merely the theory of spectral line formation in magnetic fields, the mechanism of solar activity, the acceleration of cosmic rays and the origin of cosmic magnetism. All these have become new and important constituents of contemporary astrophysics. In order to review and exchange results and opinions of scientific research, many international conferences have been held. Among them are the meeting on “Electromagnetic phenomena in cosmic physics” (Lehnert, 1958), that on “Stellar and solar magnetic fields” (Lüst, 1965) and that on “Solar magnetic fields” (Howard, 1971). For others see, e.g., Cimino (1966), Ferris and Goldsworthy (1971), Athay *et al.* (1972), Canuto (1975), Stenflo (1983), Schröter *et al.* (1987), Beck and Gräve (1987), Beck *et al.* (1990), November (1991) and Zirin *et al.* (1993).

In Table 1.1 we list a series of important events in the investigation of cosmic magnetism. It may give readers a rough impression of the main achievements in this field of research.

1.2 Significance of the Study of Cosmic Magnetic Fields

Why is it necessary to study the magnetic fields of celestial bodies? This question can be answered from the following three aspects.

First of all, it has to be pointed out that the overwhelming majority of heavenly bodies, including the Sun, all other stars, galaxies and interstellar matter, are composed of plasmas, i.e., ionized gases with

TABLE 1.1
Chronology of cosmic magnetic field study

Year	Contents	References
1908	magnetic fields of sunspots	Hale (1908), Hale and Nicholson (1938)
1912	general magnetic field of the Sun	Hale (1913), Hale *et al.* (1918)
1919–	dynamo theory of generation of magnetic fields	Larmor (1919), Cowling (1934)
1945	theory of fossil magnetic fields	Cowling (1945)
1947	magnetic fields of early-type stars	Babcock (1947)
1950	oblique rotator model of magnetic stars	Stibbs (1950)
1953	solar photoelectric magnetograph	Babcock (1953)
1956–1958	inverse Zeeman effect and theory of spectral line formation in magnetic fields	Unno (1956) Stepanov (1958a, b)
1957–1958	first discovery of reversal of solar general magnetic field	Babcock (1959)
1957, 1968	interstellar magnetic field observed with 21 cm hydrogen line	Bolton and Wild (1957), Verschuur (1968, 1971)
1958–	solar local magnetic fields and neutral-point theory of flares	Severny (1958a, b; 1960b)
1960	direction of sunspot magnetic fields	Treanor (1960)
1961	magnetic fields of solar prominences	Zirin and Severny (1961)
1961	theory of solar magnetic cycle	Babcock (1961)
1962	solar vector magnetograph	Stepanov and Severny (1962)
1962	magnetic field of Venus	Smith (1967)
1965	magnetic field of Mars	Dolginov *et al.* (1972)
1965	Hanle effect of solar prominences and corona	Hyder (1965a, b)
1966	magnetic fields of Magellanic Clouds	Schmidt (1970)
1968	sector structure of interplanetary magnetic fields	Wilcox (1968)
1968–	diagnostics of solar magnetic flux tubes	Stenflo (1968, 1973)
1969	global magnetic field of the Sun	Severny (1969)
1970	magnetographic observation of stellar magnetic fields	Severny (1970)
1970	laboratory study of magnetism of lunar rocks	Doell *et al.* (1970)
1971	videomagnetograph	Ramsay *et al.* (1971)
1971	interstellar magnetic field observed with maser radiation	Litvak (1971)
1971	superstrong magnetic fields of white dwarfs	Angel and Landstreet (1971)
1974	sunspot cooling by propagation of Alfvén waves	Parker (1974a, b; 1975)
1974	stokesmeter	Baur *et al.* (1974)
1980	magnetic fields of late-type stars	Robinson (1980)
1981	interstellar magnetic field observed with radio spectral lines	Goldreich and Kylafis (1981)
1982	theory of magnetic reconnection	Priest (1982)
1985	Stokes polarimetry	Stenflo (1985b)
1986	solar background magnetic field	Zirin (1986)
1989	internal magnetic field of the Sun	Dudorov *et al.* (1989, 1990)
1990	magnetic fields of galaxies	Kim and Kronberg (1990)
1992	intrinsically weak solar magnetic fields	Stenflo (1992a)

freely moving ions and electrons. (Our solid Earth, the other planets as well as their natural satellites may be said to be rare exceptions. They occupy only a negligible part of the total mass of the solar system.) Because the plasma consists of charged particles, its physical state and motion are closely connected with the magnetic field. Moreover, the role played by a magnetic field on the cosmic scale is quite different from that in a terrestrial laboratory. As will be explained in the following section, the larger the geometrical scale, the stronger is the coupling of electromagnetic phenomena and fluid motions. Therefore, cosmical plasmas are profoundly influenced and predominantly controlled by magnetic fields.

It is still more important to note that almost all kinds of active phenomena in the universe are due to the existence, changes, reconnection and disturbances of magnetic fields. The remark of Parker (1979a) quoted at the beginning of this chapter may impress the reader. If there were no magnetic fields, the heavenly objects would be quiet, monotonous and uninteresting, and only thermal evolution and nuclear fusion could initiate their changes. Indeed, all such active features, such as sunspots, solar flares, prominences, radio bursts and active galactic nuclei, can come into existence solely due to the action of magnetic fields. Thus it is impossible to understand the nature of many astrophysical phenomena without studying the profound and pervasive effect of magnetic fields.

Secondly, strong magnetism may have a significant influence on the radiation of celestial bodies. The spectra of sunspots provide a good example. Intense magnetic fields of several thousand gauss can cause further broadening of spectral lines, an increase in their equivalent widths and deformation of the curve of growth of spots. Such effects may be more remarkable for magnetic stars. Everybody knows that astronomers study heavenly objects mainly by analyzing their radiation. So, in order to obtain correct information, the effects of magnetic fields on radiation must be taken into account.

The factor of the magnetic field should be considered not merely in the analysis of radiation. As will be stated later in this book, the magnetic field strength in the interior of ordinary stars like the Sun may attain 10^5–10^7 G. So the magnetic pressure of such strong fields cannot be overlooked in the construction of stellar models. Moreover, the formation, stability, rotation and evolution of stars and galaxies are to various degrees connected with their respective magnetic fields.

Thirdly, the solar wind gives rise to the expansion of the solar magnetic field into interplanetary space and compresses the magnetic fields of the Earth, some planets and even, possibly, certain satellites into the limited regions of magnetospheres. This is an important situation which has to be considered in the investigation of the magnetic fields of the Earth and other planets, as well as their ionospheres. Besides, this may also have some relation with interplanetary navigation.

Furthermore, it is interesting to note that astronomical observations may be utilized to test some new physical theories. The justification of Einstein's general theory of relativity with the bending of stellar light during solar total eclipses is a well-known example. Such a role can also be played by celestial magnetism. In recent years the observations of the exceedingly intense magnetic fields of white dwarfs (10^5–10^7 G) and pulsars (10^{12}–10^{14} G) have given a preliminary verification of the theory of radiation in superstrong magnetic fields. This will be discussed in detail in Chapter 9.

1.3 Electromagnetic Properties of Celestial Bodies

1.3.1 Electrical Conductivity

Before our discussion of cosmic magnetism it is necessary to talk about the electromagnetic nature of heavenly objects. This is because only thus can we judge whether some magnetohydrodynamic phenomena, such as magnetic coupling, pinch effect and Alfvén waves, may exist in astronomic plasmas. Of course, the properties of various heavenly bodies can be quite different and their characteristic quantities rather numerous. However, generally speaking, the most important physical parameter which can represent electromagnetic properties is taken to be the *electrical conductivity*. Its mathematical expression is

$$\sigma = \frac{\gamma_c 4(2kT)^{3/2}}{\pi^{3/2} m^{1/2} Z e^2 L}, \tag{1.1}$$

where γ_c is a coefficient which varies with the atomic number Z. When $Z = 1, 2$ or approaches ∞, $\gamma_c = 0.58, 0.68$ or 1, respectively. $L = 2\ln(D/P_0)$, where D is the Debye radius and P_0 is the mean distance between two particles. The others are commonly used symbols. It is easy to see from the above expression that σ is mainly a function of temperature. Adopting some data of Pikel'ner (1966), we have compiled Table 1.2. It shows that not only the Sun and interstellar gas clouds, but also the Earth's ionosphere have very large values of σ. For comparison, we like to say that the electrical conductivities of common low-temperature neutral gases are only about 10^{-3}. Even for copper,

TABLE 1.2
Electrical conductivity of the Sun, interstellar gas clouds and the Earth's ionosphere

	electron density (cm^{-3})	atom density (cm^{-3})	temperature (K)	collision cross-section (cm^2)	mean free path (cm)	electrical conductivity
Sun						
photosphere	3×10^{13}	10^{17}	5800	10^{-12}	3×10^{-2}	10^{13}
chromosphere	10^{10}	10^{10}	6000	10^{-12}	60	6×10^{12}
corona	10^{8}		10^{6}	10^{-16}	10^{8}	7×10^{15}
Interstellar gas clouds						
H II	10	1	10,000	10^{-12}	10^{11}	6×10^{12}
H I	10^{-3}	10	100	10^{-8}	10^{11}	7×10^{9}
Ionosphere						
F2-region	10^{6}	2×10^{8}	1000	10^{-10}	10^{4}	3×10^{11}
E-region	10^{5}	10^{13}	300	10^{-9}	10^{4}	2×10^{10}

which is a good conductor and is used for making electric wires, the value of σ is no more than 5×10^{17}.

1.3.2 MAGNETIC COUPLING

In the investigation of astronomic magnetism we often face a problem of important significance, i.e., the *coupling of electromagnetic processes and fluid motions*. Sometimes phenomena of these two categories should be treated together, and separately at other times. The criterion is the following inequality:

$$lB\sigma/\sqrt{\rho} > 1. \tag{1.2}$$

This condition tells us that the bigger the object of study (i.e. the larger its linear scale l), the stronger is the magnetic field strength (B), the higher the electrical conductivity (σ) and the lower the density (ρ); it is then more likely that the above inequality is valid. In such cases the magnetic field is coupled with fluid motion and so they must be treated together. For the overwhelming majority of celestial bodies, criterion (1.2) is satisfied. For instance, the solar atmosphere has

$$lB\sigma/\sqrt{\rho} \simeq 3 \times 10^{6} \gg 1.$$

Therefore, the condition of magnetic coupling is valid in almost all celestial objects. However, for the lower layers of planetary atmospheres as well as for the Earth's oceans and lakes, the case is the opposite.

By consideration of energy it is also possible to estimate how large the influence of magnetic field is on the physical state of an astrophysical plasma. When the magnetic energy per unit volume is far larger than the kinetic energy of thermal motions, viz.,

$$\frac{B^2}{8\pi} \gg \frac{1}{2}\rho v^2, \tag{1.3}$$

the magnetic field dominates over plasma motions. In this case the plasma is said to be "frozen" in the magnetic field and matter moves along magnetic field lines. This is just like a string of beads which can move only along the connecting wire. Such a situation exists in the solar chromosphere and corona and also in the upper layer of the photosphere. Due to this reason, the solar monochromatic images and magnetic maps are similar and exhibit almost one-to-one correspondence in details.

1.4 Present Status of the Study of Cosmic Magnetic Fields

Since the beginning of this century the investigation of the magnetism of celestial bodies has seen many brilliant achievements, of which some were mentioned in Section 1.1. As a brief summary, we can say that much progress has been achieved in the following four aspects. (1) A large number of instruments have been built and a whole set of methods of analyses of observational data have been established. (2) The magnetic fields of a series of celestial objects have been discovered, and their strengths, models, regularities of variations, etc. have been discovered. (3) The role which the magnetic field plays in many astrophysical processes, such as the birth of sunspots, outburst of flares, formation of solar activity cycle and acceleration of cosmic ray particles, has been elaborated. (4) There is a good beginning in the theoretical investigation of the origin of cosmic magnetism, the formation of spectral lines in magnetic fields, the effect of the magnetic field on the structure and evolution of stars and galaxies and some other topics. All these are gratifying achievements in mankind's investigation of the immense universe.

However, it has to be realized that at the present stage the study of cosmic magnetism has the following shortcomings and limitations. (1) With the exception of the Sun, the magnetic measurements of heavenly objects are rather scarce and crude. Only a few hundred magnetic stars have been discovered and this number is too small to be counted in the stellar world. Besides, the existing observations relate only to their mean magnetic field strengths and we know very little about the fine structures. (2) Even for the Sun, magnetic observations can be made directly only the photosphere. The chromospheric and coronal fields are studied by indirect methods and so it is difficult to get accurate information. As for the magnetic field of the solar interior, only in recent years has it become possible to estimate its strength by analyzing the g-mode oscillations of the surface layers. (3) Only for the Moon and a few planets is it possible to perform explorations by means of space technology on our own initiative. All other observations of celestial magnetism are carried out in a passive way. In other words, we have to obey the demands of Nature. As for simulation experiments, they are still in their infancy. (4) Generally speaking, theoretical research is performed fragmentarily on individual objects and phenomena, and the level of reasoning is not high enough to form a systematic and deep understanding of the cosmic magnetism.

The above questions arise due to both technical and theoretical difficulties, which will be discussed in the following two chapters. Here we want to emphasize only the following. At the present stage most measurements of celestial magnetic fields are based on the Zeeman effect of magneto-sensitive spectral lines. Because spectral lines are formed in stellar atmospheres through a very complex process of radiative transfer, a correct interpretation of the magnetographic observational material is concerned with the mechanisms of line re-radiation, the inhomogeneity of the atmosphere, the deviations from local thermodynamic equilibrium, etc. Although much progress has been achieved in recent decades in constructing and improving this theory, it is still far from being satisfactory. All the above implies that much still remains to be done in the study of astronomical magnetic fields.

CHAPTER 2

Principles of Measurement of Magnetic Fields of Celestial Bodies

> **Thanks to Zeeman's discovery of the effect of magnetism on radiation, it appeared that the detection of such a magnetic field should offer no great difficulty, provided that it were sufficiently intense.**
>
> G.E. Hale,
>
> On the possible existence of a magnetic field in sunspots,
>
> *Astrophys.J.* **28** (1908), 315.

Magnetic lines of force pervade all parts of the universe. From the Earth to the Sun and from stars to galaxies, thousands of millions of heavenly objects possess magnetic fields, which may differ significantly in magnitude and other characteristics. Measurements of the strengths, directions, distributions and variations of so many magnetic fields have become one of the main tasks of contemporary astrophysical observations. After several decades of hard work, astronomers have mastered a series of methods for such measurements. For meteorites, moonrock, solar wind particles and some other samples of astronomical substance which may be captured directly, the magnetic field can be measured in the terrestrial laboratory. For the Moon and some planets, magnetic fields can be observed *in situ* by means of space probes. However, for the overwhelming majority of celestial objects, magnetic fields are measured only indirectly, e.g. through their influence on the electromagnetic radiation. This influence may be exhibited in the *Zeeman effect*, *Paschen-Back effect*, *Hanle effect*, *Faraday rotation*, etc. Among them the Zeeman effect is the most important. At present, the bulk of observational data of cosmical magnetic fields has been obtained by means of this effect.

Apart from the influence of magnetic fields on radiation, celestial magnetism can also be studied by its action on the morphological

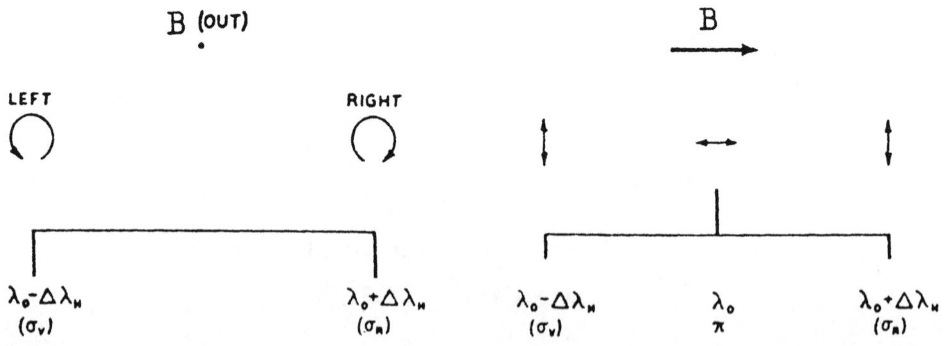

Fig. 2.1. Normal Zeeman effect. (a, *Left*) Longitudinal field. (b, *Right*) Transverse field.

changes of plasmas (such as rapid disturbances of the solar surface with velocities larger than that of sound and the pulsations of prominences) and by certain theoretical considerations (e.g. the equipartition of magnetic and kinetic energies). In this chapter we present a comprehensive description of the principles of all the methods of the above-mentioned categories.

2.1 Zeeman Effect

2.1.1 Basic Principles

In 1896 the Dutch physicist Zeeman (1897) discovered that the yellow D lines of incandescent sodium placed in a magnetic field are split into components and that the state of splitting depends on the direction of the field. If one observes along magnetic lines of force (the longitudinal case), a spectral line is divided into two components, called σ_V and σ_R, which have wavelengths a little shorter and longer, respectively, than that of the original line λ_0. In other words, one component is located on the violet side of the original line and the other on the red side (see Figure 2.1a). The wavelength shift with respect to λ_0 is $\Delta\lambda_B$[1], which is the magnitude of the *Zeeman splitting*. When observation is made in a direction perpendicular to magnetic lines of force (the transverse case), one line is split into three components. The central one (π) has the original wavelength λ_0 and the other two: $\lambda_0 - \Delta\lambda_B$ and $\lambda_0 + \Delta\lambda_B$. The mathematical expression for $\Delta\lambda_B$ is

[1]Sometimes denoted by $\Delta\lambda_H$

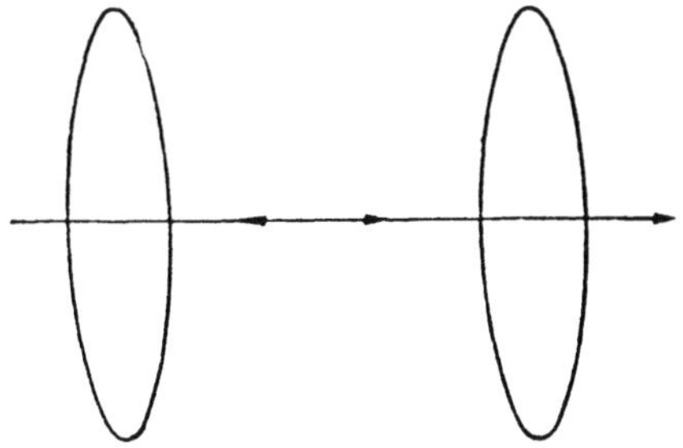

Fig. 2.2. Classical interpretation of normal Zeeman effect.

$$\Delta\lambda_B = 4.67 \times 10^{-5} g\lambda^2 B, \tag{2.1}$$

where g is the *Landé splitting factor*, B is the magnetic field strength and the other symbols have their common meanings. Both λ and $\Delta\lambda_B$ are in units of centimeters.

Soon after the discovery of the Zeeman effect, H.A. Lorentz pointed out that the radiation of every component should be polarized and this was justified later by Zeeman himself. To be precise, in the longitudinal case both σ_V and σ_R are circularly polarized. Moreover, one of them exhibits left rotation and the other right rotation. For the transverse effect, all three components are linearly polarized, but the direction of polarization of π is perpendicular to that of σ_V and σ_R. Besides, the Zeeman components are not equal in brightness and their relative intensities are expressed in Figure 2.1 by the lengths of the straight lines which represent them. What has been stated above is called the normal Zeeman effect and it can be explained as follows, using Lorentz's theory.

According to the classical theory the motion of a harmonic oscillator may be decomposed into a linear vibration plus two circular movements (Figure 2.2). Looking along the magnetic line of force, one cannot see the linear vibration. In this case one circular motion gets energy from the magnetic field and the other imparts energy to it. So the radiations corresponding to them have wavelengths which are shifted to the violet and red sides, respectively. In the transverse case, all the motions are visible. Moreover, circular motions also become linear vibrations. Thus in this case all the three Zeeman components are linearly polarized. Besides, with the classical theory we can calculate the magnitude of Zeeman splitting. In the frequency system it is

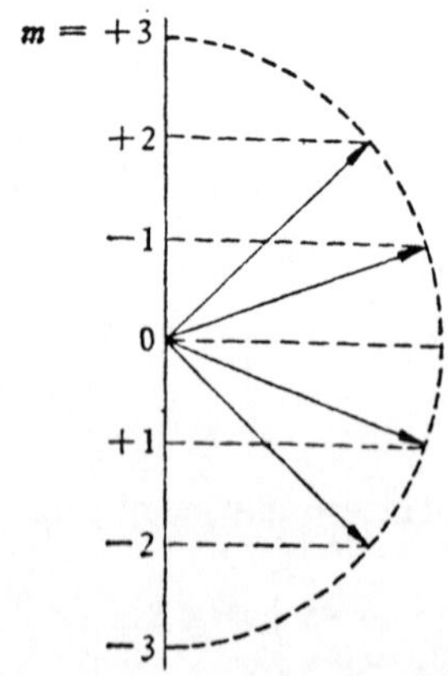

Fig. 2.3. Space quantization.

$$\Delta\nu_B = \frac{eB}{4\pi m_e c}. \tag{2.2}$$

If expressed in terms of wavelength, we have

$$\Delta\lambda_B = \frac{eB\lambda_2}{4\pi m_e c^2}. \tag{2.3}$$

When one adopts the wavenumber as unit, the amount of Zeeman splitting becomes

$$L = \frac{eB}{4\pi m_e c} = 4.67 \times 10^{-5} B \text{ cm}^{-1}. \tag{2.4}$$

This is called a *Lorentz unit*. If the electron charge e, electron mass m_e and light velocity c are all expressed in C.G.S. units, then the unit of the strength of magnetic field B is the gauss (G).

The Zeeman effect can also be interpreted using the quantum theory. For this one has to use the hypothesis of space quantization (Figure 2.3), which asserts that projections of the total angular momentum $J(= L + S)$ in the direction of a magnetic field can only be integers. They are called *magnetic quantum numbers* $m = J, J-1, \ldots, -(J-1), -J$. In a magnetic field both the upper and lower levels are split into sublevels, but any transition between an upper and a lower level should obey the rule: $\Delta m = +1, 0, -1$. Thus the result of transitions is the formation of three components (see Figure 2.4). This is called a *Zeeman triplet*.

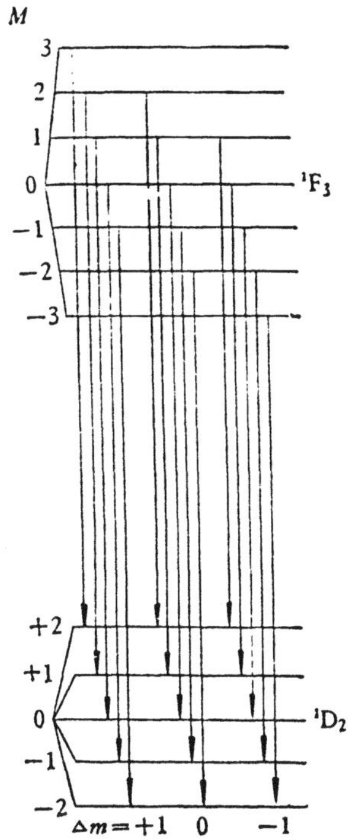

Fig. 2.4. The vector model interpretation of the Zeeman effect.

When the Zeeman effect is interpreted using quantum mechanics, the Landé splitting factor g should be added to the right-hand sides of Eqs. (2.1)–(2.4). Its definition is

$$g = 1 + \frac{j(j+1) + s(s+1) - l(l-1)}{2j(j+1)}. \tag{2.5}$$

Every energy level has its own Landé factor. For a spectral line which exhibits a normal Zeeman effect, the g-factors of its upper and lower levels are equal. The Landé factor is a measure of the sensitivity of a spectral line to the magnetic field. The larger g, the more sensitive the line is. $g = 0$ means that the magnetic field has no influence on the spectral line. Such lines, of course, cannot be used in magnetic measurements. However, as will be stated later, they are most appropriate for the observation of velocity fields.

2.1.2 Patterns of Zeeman Splitting

What has been described above belongs to the normal Zeeman effect. Most of the working lines used in measurements of celestial magnetic fields, such as Fe I $\lambda\lambda$5250, 6302 and 6173, display this effect. In other words, they are Zeeman triplets. However, many more spectral lines exhibit the *anomalous* or *abnormal* Zeeman effect, i.e. they are split by magnetic fields not into two or three lines, but into more components. The H and K lines of singly ionized calcium are such examples. Their patterns of splitting are shown in the central part of Figure 2.5. (As in Figure 2.1, the lengths of the straight lines are proportional to the intensities of the Zeeman components). The splitting patterns of various lines exhibiting abnormal Zeeman effect may be different, but the amount of splitting of every component is a rational multiple of the Lorentz unit L. For instance, the transition $^2D_{5/2} - ^2D_{3/2}$ gives rise to twelve components and their magnitudes of splitting are $\pm\frac{1}{15}L$, $\pm\frac{3}{15}L$ (these values belong to π components); $\pm L$, $\pm\frac{17}{15}L$, $\frac{19}{15}L$, $\pm\frac{21}{15}L$ (σ components). Such a splitting pattern can be expressed as

$$\Delta\nu = \frac{(\pm 1), (\pm 3), \pm 15, \pm 17, \pm 19, \pm 21}{15} L. \tag{2.6}$$

The numbers in parentheses pertain to π components, while those without parentheses relate to σ components.

In the general case the amount of Zeeman splitting is determined by the following formula:

$$\Delta\nu = L(g'M' - g''M''). \tag{2.7}$$

Let $\Delta M = M'' - M'$, then

$$\Delta\nu = L[(g' - g'')M' - g''\Delta M]. \tag{2.8}$$

According to the selection rule, ΔM can only be 0 (π components) and ± 1 (σ components). For lines of a singlet series, $S = 0$, then $L = J$ and $g' = g'' = 1$. In this case, for π components one has $\Delta\nu = 0$ and for σ components $\Delta\overline{\nu} = L$. This is the splitting pattern of the normal Zeeman effect.

2.1.3 Relative Intensities of Zeeman Components

As stated above, the relative intensities of Zeeman components depend on the direction of observation. When the angle γ formed by the line of sight and the magnetic line of force is equal to 90°, i.e. in the transverse

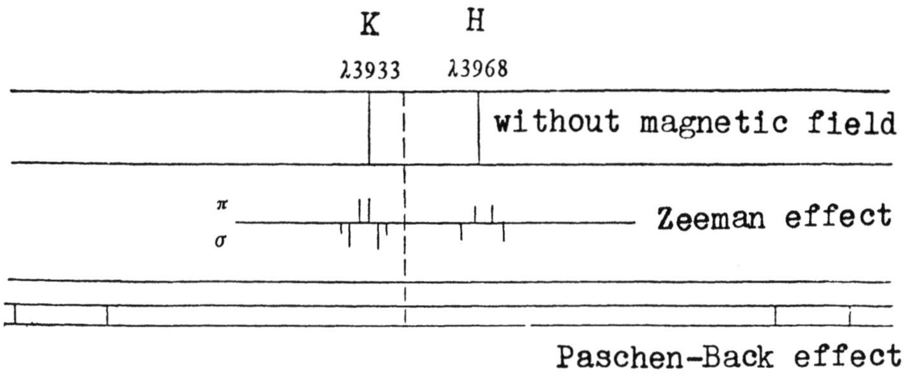

Fig. 2.5. Zeeman and Paschen-Back effects of H, K lines.

case, the π component is twice as intense as each of the σ components. In the longitudinal case ($\gamma = 0$), the intensity of each σ component is doubled in comparison with the transverse case. As for the general case, i.e. for an arbitrary γ, the relative intensities of the Zeeman components are given by the following formula:

$$I_{\sigma_-} : I_\pi : I_{\sigma_+} = \frac{1}{4}(1+\cos^2\gamma) : \frac{1}{2}\sin^2\gamma : \frac{1}{4}(1+\cos^2\gamma). \tag{2.9}$$

It should be noted that this equation can be applied only to Zeeman triplets in emission and it is invalid for absorption lines. This will be discussed further in Section 3.1.4.

2.1.4 Paschen-Back Effect

Strictly speaking, all that has been stated above is valid only in comparatively weak magnetic fields. To be precise, they are much weaker than the fields produced by the spin and orbital motion of electrons, so they have no influence on the internal motions within atoms. However, when external magnetic fields become strong enough, they can interfere with motions in the interior of atoms. In this situation the Paschen-Back effect comes into existence. It may seem improbable, but in such a case the line splitting patterns become even simpler and similar to those of the normal Zeeman effect. For example, the π component given jointly by Ca II H and K is located between these two lines and the σ components are not two single lines, but two pairs (see the lower part of Figure 2.5).

Generally speaking, the Paschen-Back effect rarely appears in astronomical spectra. This is because the magnetic fields of the Sun and ordinary stars are not strong enough. Nevertheless, Maltby (1971) found that the Paschen-Back effect is exhibited by the resonance line Li I λ6708 in the sunspot spectrum. This can be explained by the exceptionally small electron spin and orbital angular momentum of lithium atoms. In this circumstance, even the spot magnetic field, which is not very intense, can disturb their internal motions.

2.2 Inverse Zeeman Effect

2.2.1 Direct and Inverse Zeeman Effects

Strictly speaking, the Zeeman effect described in the preceding section can be applied only to emission lines. This is called the direct Zeeman effect. However, what we observe in astronomical spectra is mostly absorption lines superposed on continuous spectra, i.e. Fraunhofer lines. They display the so-called *inverse Zeeman effect*. It is well known that Fraunhofer lines are formed under a series of specific physical conditions, including the magnetic field, and through a complicated process of radiative transfer. Hence the inverse Zeeman effect is a difficult theoretical topic. Until the present, it has not been investigated sufficiently. What will be related in this section is merely the basic knowledge. Further discussion is reserved until Chapter 11.

As actual experimental results, the pattern of triple splitting and state of polarization of the Zeeman components of an absorption line are shown in Figure 2.6. A quick look might form the impression that they were the same as in the direct Zeeman effect. However, from a careful comparison of Figures 2.1 and 2.6 one can find the following differences. (1) For the longitudinal case the directions of polarization of the corresponding σ components of the direct and inverse effects are opposite to each other. (2) In the transverse magnetic field the directions of polarization of the π components provided by direct and inverse effects are perpendicular to each other. (3) In the transverse field, the two σ components of the inverse effect are only partially polarized, while for the direct effect the polarization is complete. These facts tell us that one should never confuse the inverse Zeeman effect with the direct one. It is sad to say that, before 1956, many astronomers, including G.E. Hale, imposed the direct Zeeman effect upon the Fraunhofer lines of sunspots and some of the results thus obtained are erroneous.

It has to be pointed out that the actual situation is still more com-

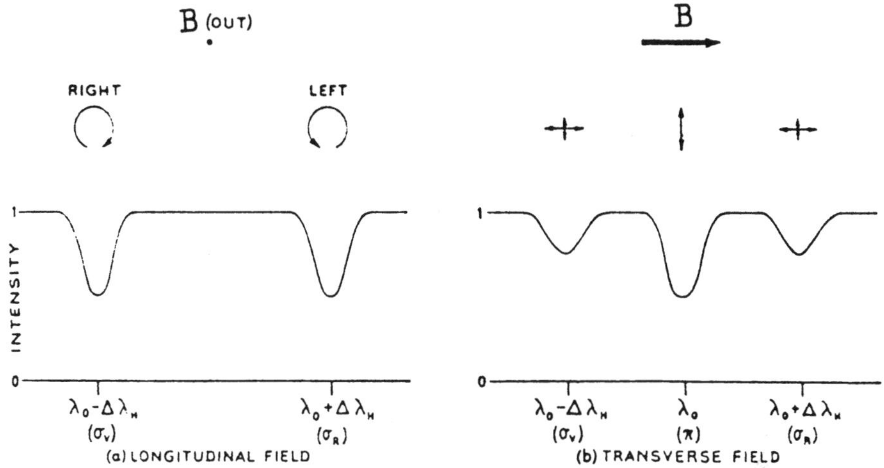

Fig. 2.6. Inverse Zeeman effect.

plex than the case of the pure inverse Zeeman effect. This is due to the following reasons. (1) The matter of celestial bodies (by this we mean mainly stellar atmospheres) can simultaneously absorb, scatter and re-emit radiation. Besides, the scattering may be incoherent. So the process of radiative transfer is quite complicated. (2) In general cases the magnetic field is not sufficiently strong and the Zeeman components are not completely split up. So radiations with different states of polarization may be mixed together. (3) The inhomogeneity of stellar atmospheres and their magnetic fields makes the question still more difficult. Therefore, what we should deal with is in essence the transfer of polarized radiation in a magnetic field. Before establishing such a theory, we need a suitable tool to express the various characteristics of the polarized radiation.

2.2.2 Representation of Polarized Radiation

In general, a beam of polarized radiation can be taken to be a mixture of natural light without polarization and elliptically polarized light. (Both the linear and circular polarizations are special cases of elliptical polarization.) In order to specify the state of polarization, we must simultaneously determine the following four quantities: (1) intensity of radiation (I); (2) degree of polarization (p); (3) azimuthal angle of the plane of polarization (χ); (4) ellipticity of polarized light (β). When the

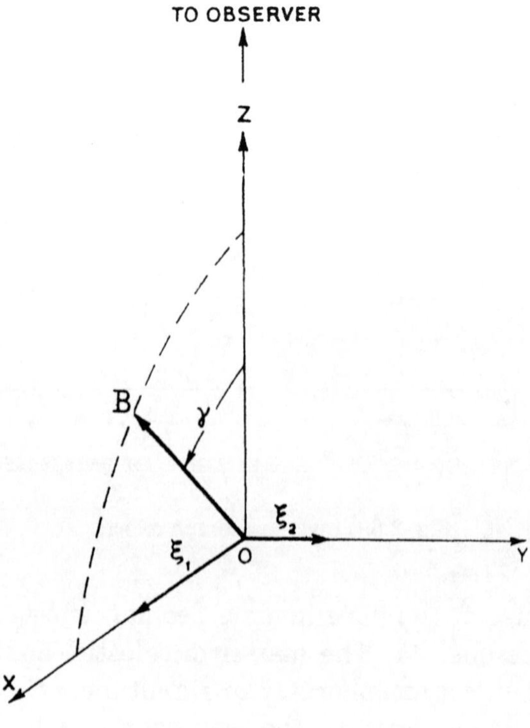

Fig. 2.7. Expression of polarized radiation.

Faraday rotation[2] is ignored, the plane of polarization is fixed and the azimuth does not change. For simplicity we take $\chi = 0$. This implies that the magnetic vector **B** is located in plane XOZ (see Figure 2.7). Besides, we assume OZ to be the direction of the line of sight and γ to be the angle formed by this direction and the vector **B**. Then the motion of an electromagnetic oscillator which produces the polarized light can be decomposed into two mutually perpendicular oscillations along the directions of OX and OY. The expressions of these oscillations are

$$\begin{aligned} \zeta_x &= \zeta_1 \cos(\omega t - \varepsilon_1), \\ \zeta_y &= \zeta_2 \cos(\omega t - \varepsilon_2), \end{aligned} \tag{2.10}$$

where ζ_1 and ζ_2 are amplitudes and ε_1 and ε_2 are phase angles.

To describe polarized radiation, the Stokes parameters (I, Q, U, V) are most widely used. Their definitions are

[2]For an explanation see Section 2.4.

$$
\left.\begin{aligned}
I &= I_1 + I_2, \\
Q &= I_1 - I_2, \\
U &= 2\overline{\zeta_1\zeta_2\cos(\varepsilon_1 - \varepsilon_2)}, \\
V &= 2\overline{\zeta_1\zeta_2\sin(\varepsilon_1 - \varepsilon_2)},
\end{aligned}\right\} \tag{2.11}
$$

where $I_1 = \zeta_1^2$, $I_2 = \zeta_2^2$. The dimensions of all these four parameters are the same, i.e. the intensity of radiation. As stated above, when the Faraday rotation is not taken into consideration, the azimuthal angle χ is constant and may be assumed to be zero. Then by use of the third formula of Eq. (2.14), for this case we can take the third Stokes parameter U to be permanently equal to zero.

An outstanding merit of the Stokes parameters is the following. When several independent beams of polarized light converge together, any one of the Stokes parameters, say Q, of the mixed light beam is the sum of the same parameter belonging to various beams, viz., $Q = Q_1 + Q_2 + \cdots$. This property greatly simplifies the problem of propagation of polarized radiation. (A detailed discussion of the properties of the Stokes parameters can be found in Chandrasekhar (1950).)

The Poincaré sphere is also a good tool for expressing the polarized radiation. It is a sphere with unit length as radius and various points on its surface represent different states of polarization. For instance, the position of point P in Figure 2.8 is determined by the two parameters 2χ and 2ω, where χ is the azimuth of elliptically polarized light and ω is a measure of ellipticity. (Concretely speaking, $\tan\omega = b/a$, a and b being the major and minor axes of the ellipse, respectively). The Cartesian coordinates of P are

$$
\left.\begin{aligned}
x &= \cos 2\omega \cos 2\chi, \\
y &= \cos 2\omega \sin 2\chi, \\
z &= \sin 2\omega.
\end{aligned}\right\} \tag{2.12}
$$

Therefore, point $A(1,0,0)$ on axis OX represents the horizontal linear polarization and $B(-1,0,0)$ is the perpendicular linear polarization. Various points on the great circle in the XOY plane stand for linear polarizations of different directions. The two poles express circular polarization. An arbitrary point on the sphere represents elliptical polarization with a certain ellipticity and azimuth. (For further details, see Schurcliff, 1962).

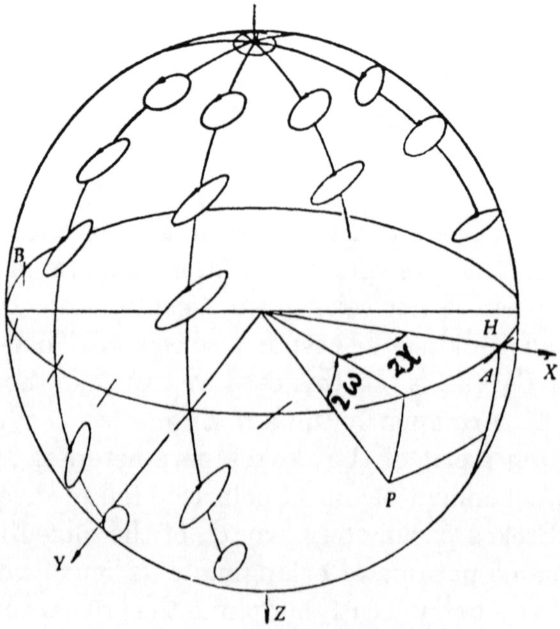

Fig. 2.8. Poincaré sphere.

The main advantage of the Poincaré sphere is the following. When we consider the effect of a polarizing device on radiation, we have only to draw a certain arc on the sphere from one point (P) to another (P'). Then the change of the polarization state of radiation after passing through the polarizing device is $P \to P'$. The observational data obtained with different kinds of magnetographs can be interpreted by means of the Poincaré sphere. (A concrete description may be found in Beckers, 1968c).

It is interesting to point out that the two representations of the state of polarization described above can be mutually transformed from the one to the other. By definition of the Stokes parameters in Eq. (2.11) it is easy see that

$$Q^2 + U^2 + V^2 = I^2. \tag{2.13}$$

This implies that if one constructs a sphere with Q, U and V as coordinates and the origin as center, then the radius is I. So the Stokes parameters can be said to be equivalent to a Poincaré sphere with radius I.

Now let us explain successively the physical meaning of the Stokes parameters. (1) By definition, I is the total intensity of radiation. (2) Q represents linear polarization. This is because if $\zeta_2 = 0$ (i.e., only the linearly polarized light of intensity ζ_1^2, remains), then $U = V = 0$ and Q attains its maximum (i.e., $Q = I$). This is equivalent to point A on axis OX of the Poincaré sphere. (3) For the same reason, U is also a measure of the linear polarization. (4) V stands for the circular polarization. The conditions of this state of polarization are $\zeta_1^2 = \zeta_2^2$ and $\varepsilon_1 - \varepsilon_2 = \pi/2$. The first condition leads to $Q = 0$, and the second to $U = 0$. So $V = I$. This corresponds to the point of intersection of axis OZ with the Poincaré sphere.

Finally, there are fixed relations between the four Stokes parameters and the four physical quantities representing the state of polarization. These are:

$$
\begin{aligned}
I &= I, \\
p &= \frac{\sqrt{Q^2 + U^2 + V^2}}{I}, \\
\tan 2\chi &= \frac{U}{Q}, \\
\sin 2\beta &= \frac{V}{\sqrt{Q^2 + U^2 + V^2}}.
\end{aligned}
\tag{2.14}
$$

Therefore, if all four Stokes parameters are known, the polarization state of radiation is completely determined.

2.2.3 Unno's Theory of Spectral Line Formation in a Magnetic Field

Unno (1956) first proposed the study of the inverse Zeeman effect and the formation of spectral lines in magnetic fields with the Stokes parameters. So the transfer of polarized radiation can be treated with a set of equations of transfer of the Stokes parameters. This has laid a solid foundation for the theory of line formation in the magnetic field. Unno assumed that spectral lines are formed solely by true absorption, and so the source function is the Planck function of black body radiation B [3]. In this situation he established the following set of equations of

[3] All the physical quantities relating to radiation, such as B, I and τ, are functions of either wavelength or frequency. For simplicity, the subscript λ or ν is omitted.

transfer of the Stokes parameters[4]:

$$\begin{cases} \cos\theta \dfrac{\mathrm{d}I}{\mathrm{d}\tau} = (I + \eta_I)I + \eta_Q Q + \eta_V V - (1 + \eta_I)B, \\ \cos\theta \dfrac{\mathrm{d}Q}{\mathrm{d}\tau} = \eta_Q I + (1 + \eta_I)Q - \eta_Q B, \\ \cos\theta \dfrac{\mathrm{d}V}{\mathrm{d}\tau} = \eta_V I + (1 + \eta_V)V - \eta_V B. \end{cases} \tag{2.15}$$

Here θ is the angle formed by the direction of radiation and the normal to the solar surface, τ is the optical depth of the continuum, and η_I, η_Q, η_V are defined by the following expressions:

$$\begin{cases} \eta_I = \frac{1}{2}\eta_p \sin^2\gamma + \frac{1}{4}(\eta_l + \eta_r)(1 + \cos^2\gamma), \\ \eta_Q = \left[\frac{1}{2}\eta_p - \frac{1}{4}(\eta_l + \eta_r)\right] \sin^2\gamma, \\ \eta_V = \frac{1}{2}(-\eta_l + \eta_r)\cos\gamma. \end{cases} \tag{2.16}$$

The quantity η_λ is equal to K_λ/K, i.e. the ratio of the selective absorption coefficient within the spectral line and the continuous absorption coefficient. For a Zeeman triplet, the linearly polarized light (denoted with the symbol p) as well as left-handed (l) and right-handed (r) circularly polarized light have the following relations:

$$K_p(v) = K_l(v \mp v_B) = K_r(v \pm v_B) = K_\lambda, \tag{2.17}$$

where $v \equiv \Delta\lambda/\Delta\lambda_D$, $v_B \equiv \Delta\lambda_B/\Delta\lambda_D$ and $\Delta\lambda_D$ is the Doppler half-width of the spectral line under consideration. $\Delta\lambda_B$ is given by Eq. (2.1). It can be seen from Eq. (2.17) that K_p, K_l and K_r are the same as K_λ, i.e. the selective absorption coefficient in the absence of the magnetic field. The unique difference lies in the wavelength displacement of K_l and K_r. When light rays form an angle γ with the magnetic field, the coefficients of selective absorption become $K_p \sin^2\gamma$, $K_l(1 + \cos^2\gamma)/2$ and $K_r(1 + \cos^2\gamma)/2$, respectively.

In order to obtain algebraic solutions of Eq. (2.15), Unno adopted the following three additional assumptions.

(1) The magnetic field is homogeneous, viz., $B =$ constant;

(2) η does not vary with depth. In other words, the Milne-Eddington atmospheric model is used.

[4]Due to the reason stated above, U is taken to be always equal to zero. So there are only three parameters and three equations.

(3) $B = B_0(1 + \beta_0\tau)$, i.e. the source function is a linear function of optical depth.

Under these conditions, he obtained the following solutions of Eq. (2.15):

$$\begin{aligned}
\eta_I(\theta) &= \frac{I(0,\theta)}{I_c(0,\theta)} = \frac{1}{1+\beta_0\cos\theta}\left\{1+\beta_0\cos\theta\frac{1+\eta_I}{(1+\eta_I)^2-\eta_Q^2-\eta_V^2}\right\},\\
\eta_Q(\theta) &= \frac{Q(0,\theta)}{I_c(0,\theta)} = -\frac{\beta_0\cos\theta}{1+\beta_0\cos\theta}\frac{\eta_Q}{(1+\eta_I)^2-\eta_Q^2-\eta_V^2}, \qquad (2.18)\\
\eta_V(\theta) &= \frac{V(0,\theta)}{I_c(0,\theta)} = -\frac{\beta_0\cos\theta}{1+\beta_0\cos\theta}\frac{\eta_V}{(1+\eta_I)^2-\eta_Q^2-\eta_V^2},
\end{aligned}$$

where $I_c(0,\theta)$ is the intensity of emergent continuous radiation in the direction θ. Therefore, this set of solutions provides us with the emergent values of the three Stokes parameters in units of $I_c(0,\theta)$.

Unno's algebraic solutions are important. We can use them to explain the principles of measurements of sunspot magnetic fields with quarter-wave plane and half-wave plate analyzers (see Section 3.1), to interpret the observational data of magnetographs (see Section 3.2 and 3.3) and to calculate the degrees of polarization and relative intensities of various Zeeman components (Section 5.2.3 in Bray and Loughhead (1964).)

However, it should be emphasized that Unno's algebraic solutions are not sufficiently accurate. The author and his coworkers (Ye Shi-hui *et al.*, 1978a) abandoned some of Unno's simplifying hypotheses, solved his system of Equations (2.15) by numerical integration, and calculated the theoretical I-, Q- and V-profiles of the magneto-sensitive line Fe I λ6302.499. By comparison of these profiles with the corresponding ones given by Unno's solutions, it can be said that our numerical solution is closer to reality. Figure 2.9 provides two examples. On the right side, in the case of purely transverse Zeeman effect, the I-profile should have three components. This agrees with the pattern of splitting given by our numerical solution. Nevertheless, Unno's algebraic solution leads to a pattern of only two components. On the left, our profile has a central residual intensity of 0.42, which is close to the observed value. The corresponding value given by Unno's solution is only 0.01. In fact, there is no solar spectral line which is so deep at its center.

Many astronomers have tried to improve Unno's theory, so that it might also be applied to the cases of inhomogeneous magnetic fields, non-LTE, arbitrary atmospheric models, anomalous Zeeman effect, magneto-optical effects, etc. These problems will be discussed in Section 11.1.

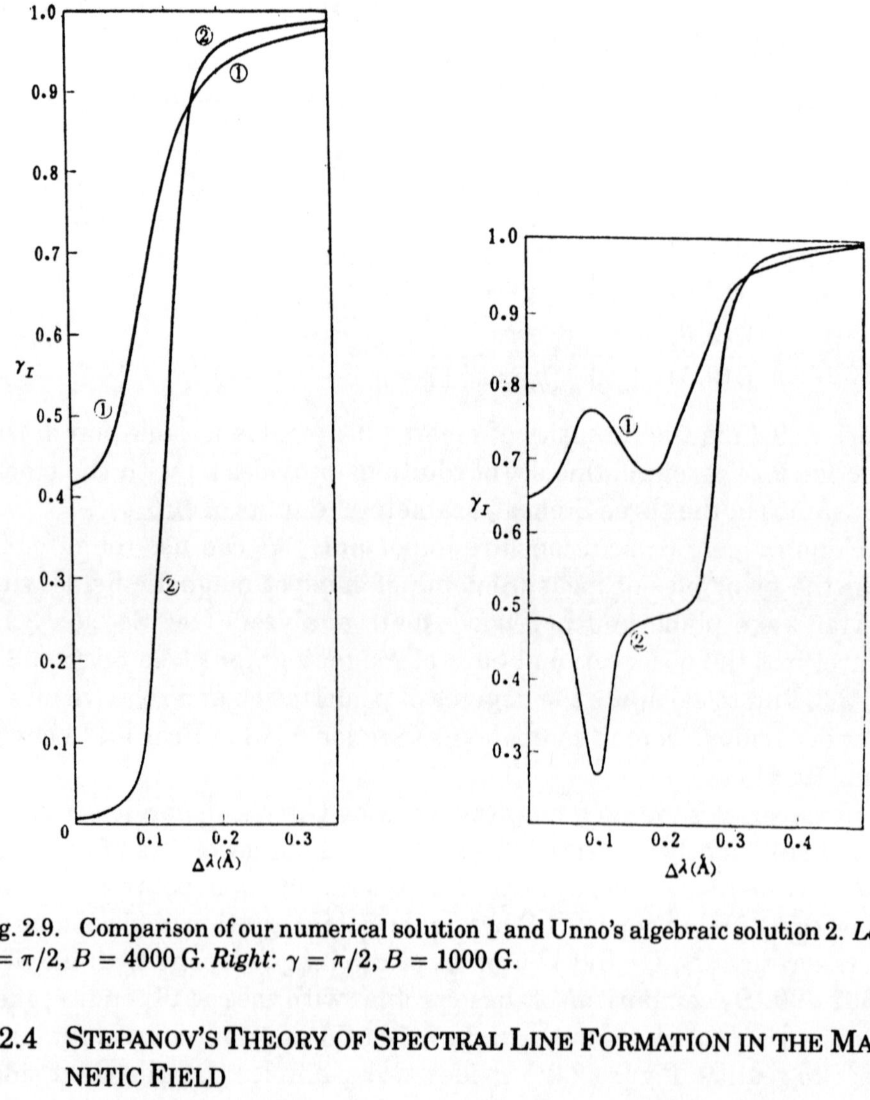

Fig. 2.9. Comparison of our numerical solution 1 and Unno's algebraic solution 2. *Left*: $\gamma = \pi/2$, $B = 4000$ G. *Right*: $\gamma = \pi/2$, $B = 1000$ G.

2.2.4 Stepanov's Theory of Spectral Line Formation in the Magnetic Field

Almost at the same time as Unno, Stepanov (1958a, b) used another method to attack the problem of inverse Zeeman effect and spectral line formation in magnetic fields. He applied the classical magneto-optical theory to solve the following equation of motion of electrons in a magnetic field:

$$\ddot{\mathbf{r}} + \gamma\dot{\mathbf{r}} + \frac{e}{m_e c}\mathbf{B} \times \dot{\mathbf{r}} + \omega_0^2 \mathbf{r} = \frac{e}{m_e}\mathbf{E}, \tag{2.19}$$

and obtained an expression for the absorption coefficient in the magnetic field, i.e.

$$s_{\pm} = \tfrac{1}{2}\Big[(s_1 + s_2) + \tfrac{1}{2}(2s_0 - s_1 - s_2)\sin^2\gamma$$

$$\pm\sqrt{\tfrac{1}{4}(2s_2 - s_1 - s_2)^2 \sin^2\gamma + (s_1 - s_2)^2 \cos^2\gamma}\Bigg] . \tag{2.20}$$

Here s_0, s_1 and s_2 are the absorption coefficients of the three components in the normal Zeeman effect, while s_+ and s_- are the absorption coefficients of the two polarized radiations, which are orthogonal to each other.

When a source of continuous spectrum is placed in a magnetic field **B**, which is parallel to the direction of propagation of radiation, and an absorbing device is located in another magnetic field **B′**, which has the same intensity as **B** but a direction opposite to it, then the gas in the absorbing device is completely transparent to the radiation. However, if **B** and **B′** have the same direction, the radiation is absorbed during its passage through the absorbing device and in this case the longitudinal Zeeman effect appears. According to these experimental facts, Stepanov suggested that in stellar atmospheres there are two radiation fields, I_+ and I_-, which possess mutually perpendicular directions of oscillation, and also two kinds of atoms, which have the absorption coefficients s_+ and s_- defined above. Then the two equations of transfer for these two radiation fields and two kinds of atoms may be written to be

$$\begin{aligned}
\cos\theta \frac{dI_{\nu+}}{d\tau_{\nu+}} &= (1 + \eta_{\nu+})I_{\nu+}(\theta) - \frac{j_{\nu+}}{K_\nu} - \frac{1}{2}B_\nu , \\
\cos\theta \frac{dI_{\nu-}}{d\tau_{\nu-}} &= (1 + \eta_{\nu-})I_{\nu-}(\theta) - \frac{j_{\nu-}}{K_\nu} - \frac{1}{2}B_\nu ,
\end{aligned} \tag{2.21}$$

Their unique difference from an ordinary transfer equation is that the source function is not B_ν, but $\frac{1}{2}B_\nu$. Therefore, various existing methods of solution of transfer equations can be utilized. Combining their solutions according to the formula $I_\nu = \frac{1}{2}(I_{\nu+} + I_{\nu-})$, one may study the influence of magnetic field strength B and its direction γ on line profiles, equivalent widths, curves of growth, etc.

It is interesting to note the following finding of Rachkovskii (1961a, b). Although the two systems of equations of Unno and Stepanov are quite different, they are in essence the same and can be transformed from one to the other. By comparison, the former possesses more generality, while the latter is easier to treat. In fact, the overwhelming majority of astronomers prefer Unno's equations, which have been generalized to the cases of nonuniform magnetic fields, arbitrary atmospheric models, noncoherent scattering, etc. These topics will be described in Chapter 11.

2.3 Hanle Effect

In recent years in the magnetic observations of the Sun, in particular solar prominences and corona, the Hanle effect has begun to play an important role. Although its theory seems to be insufficiently mature, its latent capacity is quite evident. This effect represents the influence of the magnetic field on the state of polarization, which is caused by *coherent scattering*. When a spectral line is formed by this physical process, the radiation often exhibits a linear polarization. If there is a magnetic field in the region of line formation and if the magnetic sublevels are not well separated by the Zeeman splitting, then the quantum interference among sublevels of the upper level of transition may give rise to some changes of the state of polarization. To be precise, they are

(1) *depolarization*, i.e., decrease of the degree of polarization p,

(2) *rotation of the plane of polarization.*

Hence if these phenomena can be detected and accurately measured, one may get some information on the magnetic field.

According to the theoretical study of Stenflo (1991a), the depolarization factor $p/p_{\max}$ and the angle of rotation β of the polarization plane in the two cases $p_{\max} = 1$ and $p_{\max} \to 0$ are shown in Figure 2.10. These two observable quantities are functions of the parameter Ω and the magnetic field azimuth χ. Ω is proportional to the magnetic field strength B and is defined as follows

$$\Omega = \frac{2g_b\omega_L}{\Gamma}, \tag{2.22}$$

where the Larmor frequency $\omega_L = 8.803 \times 10^6\, B\ \mathrm{s}^{-1}$, g_b is the Landé factor of the upper level and Γ is the natural width of transition.

In principle the Hanle effect is quite different from the Zeeman effect. First, the latter gives rise to a state of polarization which did not exist, while the former can only alter such a state. Second, the degree of polarization is not increased, as in the case of the Zeeman effect, but, on the contrary, is decreased by the Hanle effect. This can be seen from the following equation:

$$p = \frac{p_0}{1 + (g_b eB/\gamma m_e c)^2}, \tag{2.23}$$

where p_0 is the degree of polarization in the absence of the magnetic field and γ is the damping constant.

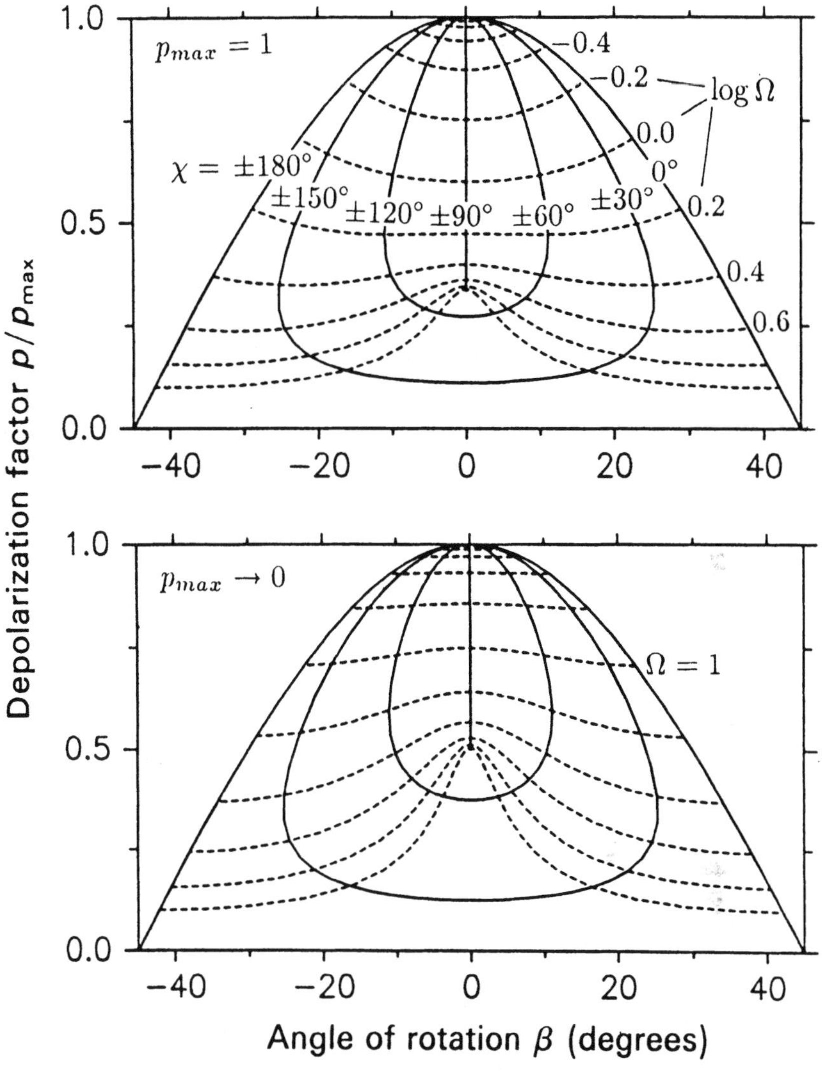

Fig. 2.10. Hanle effect.

The Hanle effect has some attractive merits, and the most important one is the following. The action of the Zeeman effect changes with the ratio of the amount of Zeeman splitting to the natural width of the spectral line under consideration: the lower the ratio, the weaker the Zeeman effect. However, the Hanle effect is proportional to the ratio of Zeeman splitting to the damping width of the line. Because the damping width is much smaller than the natural width, the sensitivity of the Hanle effect to the magnetic field is much higher than that of the Zeeman effect. Secondly, turbulence may cause cancellation of magnetic streams with different polarities, and then the almost complete disappearance of the Zeeman effect. But in this situation the Hanle effect still exists. Besides, the Zeeman splitting is proportional to the square of wavelength. Thus the Zeeman effect can hardly be detected and measured in the ultraviolet and X-ray lines, which are very important in the study of the chromosphere and corona. It is pleasant to say that this is not so for the Hanle effect.

On the other hand, however, the Hanle effect has some disadvantages. One of them lies in the difficulty of measuring very weak linear polarization, which is often less than one percent and so very likely submerged by instrumental polarization. Another shortcoming is connected with the complexity of the theory of radiative transfer including the Hanle effect (see, e.g., Stenflo, 1978a, b, 1988; Landi Degl'Innocenti, 1983). Until now, this effect has mainly been used in the magnetic field observation of solar prominences, because for them there is no background radiation from the solar disk and so the problem of radiative transfer is highly simplified.

In fact, the radiation of prominences arises chiefly from the coherent scattering of solar radiation. In the measurements of prominence magnetic fields with the Hanle effect some interesting results have been achieved. They will be described in Section 5.4.

Finally, it should be noted that by means of the Hanle effect only two quantities — viz. the amount of depolarization and the angle of polarization plane — are used to derive the magnetic field information. Therefore, at least at the present stage, this effect can hardly be adopted to determine unambiguously the three components of the magnetic vector. This difficulty may be overcome by simultaneous observations in two spectral lines, which have different sensitivities to the Hanle effect. Keeping this in mind, we can say that the Hanle effect is very promising in the observation of cosmic magnetic fields.

2.4 Faraday Rotation

In 1845 the distinguished British physicist Michael Faraday discovered that when a beam of linearly polarized light passes through a medium with a magnetic field, the plane of polarization continuously rotates. This is called the *Faraday rotation* or the *Macaluso–Corbino effect*. In recent decades it has often been applied to the magnetic measurements of interstellar and intergalactic matter as well as the solar corona. Their magnetic fields are too weak to be observed at certain discrete places by means of the Zeeman or some other effect. But due to their extremely large scales the amount of Faraday rotation accumulated

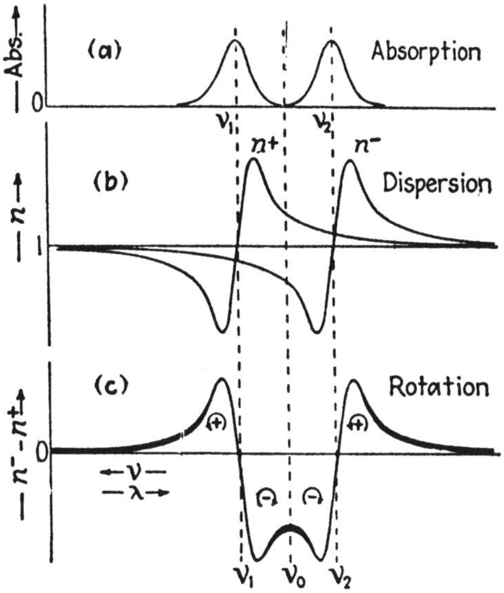

Fig. 2.11. Physical interpretation of Faraday rotation.

during the passage of radiation of a distant object, say an extragalactic radio source, through the intergalactic and interstellar space onto our instrument can be reliably measured. The longitudinal magnetic field strength can be readily calculated from the amount of Faraday rotation.

The physical conditions of appearance of the Faraday rotation are as follows. As stated in Section 2.1, under the action of the longitudinal Zeeman effect a spectral line is split into two components with frequencies ν_1 and ν_2. Both of them are circularly polarized, but the directions of rotation are opposite to each other. (See the upper part of Figure 2.11.) Now it has to be added that, due to the *anomalous dispersion*, the index of refraction is reversed across a spectral line. This is the same for both Zeeman components (middle part of Figure 2.11). A beam of linearly polarized light may be decomposed into two beams of circularly polarized light with opposite directions of rotation. So, as can readily be seen in the lower part of the accompanying figure, in the frequency regions $\nu < \nu_1$ and $\nu > \nu_2$, we have $n^+ < n^-$. In this case the speed of light with index of refraction n^+ is larger than that with n^-. Then the plane of polarization rotates in one direction. In the region between ν_1 and ν_2 the plane of polarization rotates in the opposite direction. Thus the Faraday rotation comes into existence.

The angle of rotation of the polarization plane ϕ is given by the following formula (for its mathematical derivation see, e.g., Zeldovich *et al.*, 1983):

$$\mathrm{d}\phi = \frac{\nu_B}{2c}\left(\frac{\nu_p}{\nu}\right)^2 \mathrm{d}x, \tag{2.24}$$

where the Larmor frequency $\nu_B = eB/4\pi m_e c$, the plasma frequency $\nu_p = 4\pi n_e e^2/m_e$, and x is the path length of radiation. Therefore, the amount of Faraday rotation is proportional to the magnetic field strength, the electron density and the distance from the source of radiation and inversely proportional to the square of frequency.

It is more convenient to express the rotation angle in terms of the squared wavelength, viz.,

$$\phi = (RM)\lambda^2 + \phi_0, \tag{2.25}$$

where RM means the 'rotation measure' and ϕ_0 is the initial angle of inclination of the polarization plane. In the astrophysical literature RM is usually expressed as

$$RM = \frac{|e|^3}{\pi m_e^2 c^3}\int n_e B_{\parallel}\,\mathrm{d}x = 0.81\int n_e B_{\parallel}\,\mathrm{d}x. \tag{2.26}$$

In this equation the unit of radiation path length is the parsec, that of electron density is cm^{-3}, and that of magnetic field strength is μG. In order to determine unambiguously both RM and ϕ_0 in Eq. (2.25), one has to take measurements on at least two wavelengths. If the distance to the source of radiation and the electron density are known, the longitudinal magnetic field strength $B_{\parallel}$ can be computed with the observed RM by means of Eq. (2.26). Application of this method has secured much interesting information about the coronal, interstellar and intergalactic magnetic fields. This will be described later in this book.

2.5 Radio Observations

Besides the visible light, radio waves can also be used for magnetic measurements and some methods have been established for this. They can be briefly outlined in the following.

2.5.1 Non-Burst Solar Radio Emission

The solar radio radiation consists of two parts, i.e. the slowly varying component, which is called the non-burst emission, and bursts of various types. The former is generated mainly by two physical processes:

thermal bremsstrahlung and *gyromagnetic emission*. In most cases the former is the dominating mechanism. Gelfreich (1972) has shown that for this radiation the degree of circular polarization can be expressed as

$$p = k\frac{\nu_B}{\nu}\,|\cos\gamma|, \tag{2.27}$$

where

$$k = \frac{\partial(\log T_B)}{\partial(\log \nu)}\,. \tag{2.28}$$

The gyrofrequency $\nu_B = 280B$ MHz, and T_B is the brightness temperature. Eq. (2.27) allows the following expression to be deduced:

$$B_{\parallel} \propto V \bigg/ \frac{\partial I}{\partial \nu}\,. \tag{2.29}$$

Here V is the Stokes parameter which represents the circular polarization and $\partial I/\partial\nu$ is the intensity gradient of the spectrum. If both these quantities are determined by observation, the longitudinal field strength $B_{\parallel}$ can be readily computed.

2.5.2 Solar Radio Bursts

There are several types of solar radio bursts. Although their physical mechanisms are quite complex and still not fully understood, they are all connected with the magnetic field. Therefore, in principle it is feasible to extract from their observations some information about magnetic fields in the regions of their generation. A few examples follow.

The variations of the intensity of a radio burst with both time and frequency are exhibited by its *dynamical spectrum*. When the temporal and spectral resolutions are sufficiently high, some interesting fine structures can be revealed. An instance is the so-called "zebra pattern", which consists of parallel drifting bands changing with a period of approximately one second. As suggested by Rosenberg (1972), this period is connected with the Larmor or electron gyrofrequency, ν_B, defined above. When ν_B is determined by the observed zebra pattern, the field strength B may be easily calculated.

Another kind of fine structure is the *fiber bursts*, which appear on the background of type IV continua. A fiber burst is composed of a lot of neighboring emission and absorption ridges of short duration. Either the instantaneous bandwidth or the drift rate of fibers may be used to derive the magnetic field strength (Stenflo, 1978b).

It is also feasible to utilize the solar radio bursts of type II, which are generally believed to be produced by MHD shock waves. In the dynamic spectra of these bursts there are parallel drifting bands at the fundamental and second harmonic of the plasma frequency. These bands are sometimes split up and each band becomes two parallel ones. The amount of splitting and the disturbance speed given by the frequency drift rate can also be used to determine the magnetic field intensity (Smerd *et al.*, 1975).

2.5.3 Occultation of External Radio Sources

When the radio waves from a spacecraft or a distant source, e.g. the Crab Nebula, pass through the solar corona, they undergo scattering in the coronal magnetic field. This can lead to an increase of the apparent size of the radio source as well as a decrease of its intensity. By analysis of these effects it is possible to study the general configuration of the coronal magnetic field out to a distance of several tens of solar radii. (For details, see Högbom, 1960.)

The Faraday rotation introduced in the preceding section can also be used to understand the magnetic structure of the solar corona. For instance, Stelzried *et al.* (1970) observed this effect between $4R_\odot$ and $12R_\odot$ at 14.4 cm during the passage of the spacecraft Pioneer 6 behind the Sun. Their observations agree well with theoretical calculations. Another example is the Faraday rotation of the polarized radiation of the Crab Nebula. By its measurement Golnev *et al.* (1969) obtained a magnetic field strength equal to 10^{-2} G for the corona at distance $5R_\odot$.

Finally, it is interesting to note that the Zeeman effect of radio waves can also be used for magnetic measurements. Except, the historic observation of interstellar magnetic field with the Zeeman effect of the 21 cm hydrogen line (Verschuur, 1968, 1971), Kazes and Crutcher (1986) used the 1665 and 1667 MHz lines of OH to measure the magnetic field of molecular clouds. Even for low signal-to-noise ratios, such observations can still be carried out effectively (Sault *et al.*, 1990).

2.6 Synchrotron Radiation

Under the action of the Lorentz force, electrons move in spirals around magnetic field lines and emit radiation continuously. For non-relativistic electrons this is called *cyclotron radiation*, and for relativistic electrons (i.e. electrons with velocities close to that of light) — *sychrotron radiation*. The latter is so called, because it was first detected and verified in synchrotron accelerators. (Sometimes, especially in the Russian literature, it is also called *magnetobremsstrahlung*.) Now it is important to point out that such a radiation is emitted by many celestial bodies. One of them is the Crab Nebula, which is the remnant of the famous supernova of July 4, 1054. (In Chinese historical annals this is the "guest star" which appeared in May of the first year of the Zhihe Reign of the Emperor Renzong of the Sung Dynasty.) Besides, the solar radio bursts of type IV may also be due to the synchrotron radiation.

Let us assume that the energy distribution of relativistic electrons obeys the law

$$n_{re}\,\mathrm{d}E = KE^{-\gamma}\,\mathrm{d}E. \tag{2.30}$$

Here n_{re} is the number density of electrons in the interval $(E, E + \mathrm{d}E)$, the index γ generally lies between 2 and 3, and the coefficient K depends on the total number of relativistic electrons per cubic centimeter and the energy range under consideration. Then the emission coefficient of the synchrotron radiation is

$$\sigma_I \propto K\nu^{-(\gamma-1)/2} B_{\perp}^{(\gamma+1)/2}\,, \tag{2.31}$$

where ν is the frequency of radiation and $B_{\perp}$ is the strength of the transverse component of the magnetic field. The coefficient of proportionality, omitted in the above expression, depends solely on the value of γ.

The study of the magnetic field using synchrotron radiation is confronted with the difficulty of uncertainty in the determination of n_{re}. Not long ago we possessed only some indirect information, given by the observation of cosmic rays near the Earth. Only recently, due to the space measurement of cosmic diffuse gamma radiation, it has become possible to determine n_{re} in the Galaxy at great distances from the Sun. (A detailed description can be found in Ruzmaikin *et al.*, 1988.)

The plasma in the region of generation of the synchrotron radiation may cause an abrupt drop of its energy spectrum below the critical frequency, ν_c. This is called the *Razin effect*, and the relation between the critical frequency and magnetic field strength is given by $\nu_c = 20n_e/B$. (For a detailed description, see Ginzburg and Syrovatskii (1965).) This effect was observed, for instance, by Bioschot and Clavelier (1967) in the corona at $1R_{\odot}$ from the solar limb. Their measurements yielded $\nu_c = 400$ GHz. With n_e known to be 10^7 cm^{-3}, it was readily found that $B = 0.5$ G.

2.7 Morphological Studies

In the universe there are many places where the criterion (1.3) is satisfied. In other words, the physical state and motion of plasmas in these regions are governed by the magnetic field. So the magnetic configuration is embodied in the intensity pattern or velocity field, which can be observed much more easily. For example, spots and faculae are such places in the solar photosphere. In the entire solar chromosphere and corona, and also in interstellar space, the situation is similar. This is the physical prerequisite for the following morphological methods of obtaining magnetic field information.

(1) *Comparison of solar monochromatic images and magnetic maps.* As shown by a large amount of observational data, solar monochromatic images, observed mainly in the cores of strong Fraunhofer lines, agree quite well in configuration with magnetic maps. This is especially pronounced in the calcium monochromatic images scanned in H and K Ca II. As is well known, radiation of the central parts of these lines comes from the chromosphere. Even weak magnetic fields of about 2.5 G may affect the details of H-K monochromatic images. According to a statistical study by Leighton (1965), the correlation coefficient of calcium monochromatic images and magnetic maps is as high as 99.9%! A similar effect is exhibited by the monochromatic images scanned at the head of cyanogen molecules (CN) (Chapman and Sheeley, 1968). Its wavelength is 3883 Å. Therefore, for morphological work the solar monochromatic images can be used as magnetic maps. Veeder and Zirin (1970) compared Hα monochromatic images and the magnetograms of Mt. Wilson Observatory and found some interesting regularities. However, the question may be not so simple. As pointed out by Chung-Chieh Cheng *et al.* (1973), magnetograms are in general recorded in weak Fraunhofer lines which are formed in the photosphere, while monochromatic images are mainly observed with the radiation of the central parts of intense lines, which comes from the chromosphere. Thus their similarity should require that the magnetic fields in these two atmospheric layers are basically the same, but this may not always coincide with reality. As noted by Wang Jing-xiu and Shi Zhong-xian (1992), in a region of flare activity the configuration of the chromospheric magnetic field may differ significantly from that of the photospheric field.

(2) *Propagation of fast disturbances.* In solar active features we often see some phenomena which move with velocities larger than that of sound. An example is the *penumbral waves*. These may be interpreted as MHD disturbances, which propagate with the Alfvén velocity v_A(=

$B/\sqrt{4\pi\rho}$). If the density of the active feature ρ is known and v_A is determined from monochromatic observation, then it is easy to obtain the value of B. This method has been applied, e.g. by Jensen (1983), to the magnetic field observation of quiescent prominences.

(3) *Oscillation of solar prominences.* Sometimes, in particular after the occurrence of flares, prominences may exhibit periodic oscillations. This is manifested by systematic changes of the Doppler displacement of the Hα line. According to Hyder (1966) and Kleczek and Kuperun (1969), this phenomenon is connected with the magnetic field. In each of these two works an expression of the relation between the frequency of oscillation f and magnetic field strength B was derived. To be precise, Hyder obtained the following equation:

$$B_r^2 = \pi \rho r_0^2 (4\pi^2 f^2 + \alpha^2), \tag{2.32}$$

where B_r is the radial component of the magnetic field strength, ρ and r_0 are the density and scale height of the prominence, respectively, and α is the characteristic decay constant ($\simeq 10^{-3}$–10^{-4} s^{-1}). The formula given by Kleczek and Kuperun is

$$B = 4\pi h f \sqrt{\pi\rho}\,, \tag{2.33}$$

where h is the height in the prominence. If, besides f, the quantities ρ and h are also determined from observation, B can be computed. By this method Hyder obtained $2 \leq B_r \leq 30$ G, and Kleczek and Kuperun $B = 9$ G. In order of magnitude these values agree with the results of measurements by Zeeman splittings (see Section 5.6).

(4) *Magnetic fields of diffuse nebulae and spiral galaxies.* Shajn (1955) pointed out that the elongated form and orientation of diffuse nebulae in our Galaxy can be explained by the action of the magnetic field. Although this idea is not always correct, it played an important role in the study of the galactic magnetism. Not long ago, Sofue (1987) noted that the dark filaments (or dust lanes) in the central parts of some spiral galaxies are connected with their poloidal magnetic fields. Being twisted by galactic rotation, these fields may yield outflows of disk matter into galactic halos.

2.8 Theoretical Calculations and Other Methods

Besides measurements by various effects described above, it is also possible to get some information on celestial magnetic fields by means of theoretical computations. Two such instances are as follows.

2.8.1 CALCULATION OF THE CORONAL MAGNETIC FIELD

Due to the extremely high temperature ($\sim 10^6$ K) and very weak magnetic field ($\sim 10^{-1}$–10^{-2} G), the Doppler widths of coronal spectral lines are much larger than the amounts of their Zeeman splittings. So it is practically impossible to measure the coronal magnetic field by the traditional method based on the Zeeman effect. Although some information may be obtained from radio observations, the accuracy is not sufficiently high. In this situation many astronomers prefer the theoretical extrapolation of the photospheric magnetic field, which can be measured more reliably. As mentioned before, owing to its very low density, the coronal plasma is frozen in the magnetic field. Then both the trajectory of mass motion and the electric current should be parallel to magnetic lines of force. Bearing this in mind, we may assume that the coronal magnetic field is *force-free*. The necessary and sufficient condition for the existence of such a field is

$$\nabla \times \mathbf{B} = \alpha(\mathbf{r}, t)\mathbf{B}, \tag{2.34}$$

where α is a scalar factor. In the general case it is a function of the spatial position $\mathbf{r}$ and time t. When α is a constant — viz. when it is not a function of t — the force-free field is said to be stationary. Furthermore, if $\alpha = 0$, we have a current-free field, which is also called a *potential field*. For this special case of a force-free field, the magnetic energy drops to its minimum and there is no twisting of magnetic lines of force. For the outer part of the corona with distance from the solar surface larger than $2.5R_\odot$, the solar wind may give rise to magnetic disturbances and the coronal field cannot be taken to be a potential one. But for the inner corona with distance smaller than $2.5R_\odot$, the potential field seems to be a good approximation.

The common approach of theoretical computation is to solve Eq. (2.34) simultaneously with the following equation

$$\nabla \cdot \mathbf{B} = 0, \tag{2.35}$$

which implies that the magnetic field does not possess sources. The observed photospheric magnetic field[5] is used as the boundary condition. This method has been applied by many authors to the investigation of the structure of the coronal magnetic field. For example, Altschuler and Newkirk (1969) established the potential field model of the corona

[5]Recently, even vector magnetograms have been utilized and an algorithm using numerical techniques has been developed to solve 3-D MHD equations (Mikić *et al.*, 1989).

in spherical coordinates. A recent work is the study of the twisting and braiding of coronal field lines caused by the random shuffling of magnetic footpoints in the photosphere (van Ballegooijen, 1990). A few other interesting examples will be presented in Section 6.3.

2.8.2 Equipartition of Magnetic and Kinetic Energies

Under the action of convective motions, the density of magnetic energy and that of kinetic energy of a plasma should have the same order of magnitude, viz.,

$$\frac{B^2}{8\pi} \simeq \frac{1}{2}\rho v^2. \tag{2.36}$$

If both the matter density ρ and the velocity of convective cells v are known from observation, the above equation can be used to calculate the approximate value of B. This method has been applied to estimate the magnetic field strength of supergranules (Parker, 1963; Clark and Johnson, 1967) and the results seem to be rational. By the same method, the magnetic field strengths of the solar interior and the convective zone are found to be 10^9 G and 10^5 G, respectively.

After obtaining the horizontal component of velocity by geometric analysis of the observed trajectory of prominence knots, Ballester and Kleczek (1984) and Ballester (1984) applied the principle of equipartition of energy to calculate the strength of the longitudinal magnetic field of prominences. Their results are also consistent with the direct measurements of Zeeman splittings.

Another kind of energy equipartition, which is widely used in the study of galactic magnetic fields, will be described in Section 10.3.

Apart from all the methods introduced above, there are some other ways of observing celestial magnetic fields. Due to the limitation of space, we can only briefly mention their main ideas.

(1) *Polarization of starlight by interstellar dust.* The scattering of light by nonspherical dust particles may give rise to linear polarization whose degree is expressed by

$$p = p_0 d \sin^2\theta, \tag{2.37}$$

where p_0 is a constant determined by the properties of dust particles, θ is the angle between the line of sight and the orientation axis of particles which is taken to be parallel to the magnetic field direction, and d is the distance to the star. According to the study of nearly 7000 stars by Mathewson and Ford (1970), the dust particles in our Galaxy exhibit a regular large-scale orientation, which is like the pattern of iron fillings in the vicinity of a magnet. This allows us to understand the configuration and structure of the Galactic magnetic field.

(2) *Maser radiation.* "Maser" is the acronym for microwave amplification by stimulated emission of radiation. It is produced by an atomic, molecular or ionic system with inverted populations of the upper and lower levels, when the system is excited by a microwave radiation field. For the radiation emitted by astronomical masers (e.g. the circumstellar SiO masers), magnetic fields may cause high linear polarization. So from the observed degree of polarization it is possible to determine the magnetic field strength in the region of formation of the maser radiation. Western and Watson (1984) concluded that fractional linear polarizations larger than one third can be produced by magnetic fields alone. Moreover, if the degree of linear polarization of the maser radiation is greater than one third, the magnetic field may be much stronger than previously recognized. For instance, a fractional linear polarization of 50% requires a magnetic field strength larger than 10 G.

(3) *Polarization of molecular lines at radio frequencies.* When the optical properties of the source are anisotropic, the radiation of radio lines of some molecules, such as CO, may become polarized. This also happens under the action of magnetic fields. Therefore, this kind of polarization can be used to measure the magnetic field of the emitting region. For this purpose, both emission lines (Goldreich and Kylafis, 1981, 1982) and absorption lines (Kylafis, 1983; Kylafis and Shapiro, 1983) may be adopted. Nevertheless, as far as we know, this effect has not yet been discovered observationally.

(4) *Circular polarization of the continuous spectrum.* For some stars with exceedingly high density, such as white dwarfs, their radiation is basically the continuous spectrum, with very few spectral lines. So the common method of magnetic observations by the Zeeman effect can hardly be used. In this case, one may apply Kemp's (1970) theory of circular polarization of the continuum to derive the magnetic field strength. As revealed by observation, some white dwarfs do possess continuum circular polarization of about 1–3% (see, e.g., Angel and Landstreet, 1971; West, 1989) and their magnetic fields may be as intense as 10^5–10^7 G. In the presence of such superstrong magnetism, many interesting phenomena come into existence. A detailed account of this will be given in Section 9.5.3.

(5) *In situ measurements.* Since very ancient times, astronomers could study the celestial objects solely through their radiation received

on the Earth. However, the situation has drastically changed in recent decades. Due to the advent of the space era, it has become possible to carry out magnetic measurements on or near the Moon and some planets. It has also been possible to make *in situ* observations of the magnetic field of the solar wind with magnetometers abroad the Pioneer, Voyager and other spacecraft. The results of these interesting experiments will be described in Chapter 8.

CHAPTER 3

Techniques for the Measurement of Magnetic Fields of Celestial Bodies

Magnetic fields (and their inevitable offspring fast particles) are found everywhere in the universe where we have the means to look for them.
E.N. Parker, *Cosmical Magnetic Fields* (1979).

In the preceding chapter we presented a comprehensive account of the principles underlying the observation of cosmic magnetism. On this basis, in this chapter we proceed to discuss some technical questions. We are aware of the fact that there are many kinds of instruments specially designed and constructed for magnetic measurements of heavenly objects (in particular, the Sun) and that most of them are quite sophisticated in their optical layout, mechanical structure and electronic circuits. In this chapter we want only to give a general description of some selected samples of these facilities. For the numerous technical details, the reader is referred to the specific literature. Furthermore, the magnetic observations of stars are related to some special questions, which will be treated in Section 9.1.

3.1 Apparatus for Measuring Sunspot Magnetic Fields

3.1.1 BASIC PRINCIPLES

Sunspots possess intense magnetic fields with strengths up to several thousand gauss. Hence the method and apparatus for their measurements are comparatively simple. Early in the present century, Hale and his associates (Hale, 1908; Hale and Nicholson, 1938) established the

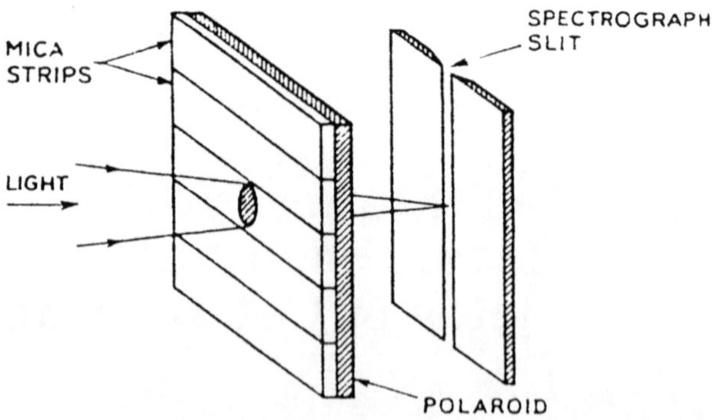

Fig. 3.1. Optical scheme of a polarization analyzer.

classical method of observation, which is still used up to the present. Its essential points are as follows.

A high resolution spectrograph (or spectroscope) is used to observe a certain magneto-sensitive spectral line, which exhibits the normal Zeeman effect. In front of the spectrograph slit is placed a *polarization analyzer*, which consists of a quarter-wave ($\lambda/4$) plate and a polaroid (Figure 3.1). The former is utilized to transform circularly polarized light into linearly polarized light. Moreover, it contains a grid of narrow mica strips, which are cut so that the adjoining ones have mutually perpendicular optical axes. The latter allows light beams with different directions of polarization to pass through in order. In consequence, on the focal plane of the spectrograph the image of the observed spectral line exhibits a saw-toothed pattern, as shown in Figure 3.2 (left part).

Strictly speaking, the polarization analyzer described above can be adopted solely for the observation of spots in the central region of the solar disk. This is because for such spots we observe mainly the *longitudinal* magnetic field and so the two σ components are circularly polarized. However, for spots, which are located near the solar limb, what we measure is chiefly the *transverse* field and in this case all the π and σ components are linearly polarized. Then we have to use a mosaic of half-wave ($\lambda/2$) plates, which allows the π and σ components to pass through alternately (Figure 3.2 (right)). Nevertheless, in either of these two cases, or in an intermediate situation, it is easy to measure the amount of Zeeman splitting $\Delta\lambda_B$ and to compute the total strength of the sunspot magnetic field B with Eq. (2.1).

The spectral lines suitable for magnetic observations in the spectral

TABLE 3.1
Working lines of magnetic field measurements

Wavelength	Element	Rowland intensity		Excitation potential (eV)		Transition	Pattern of	$g\lambda^2 \times 10^{10}$
x101ł (Å)		Sunspot	Solar disc	Lower level	Upper level		Zeeman splitting	(cm^2)
5131.478	Fe	3	2	2.213	4.618	a^5P_1–y^5P_1	$\frac{(0)5}{2}$	66
5247.576	Cr	5	3	0.957	3.309	a^5D_0–z^5P_1	$\frac{(0)5}{2}$	69
5250.218	Fe	3	2	0.121	2.471	a^5D_0–z^7D_1	$\frac{(0)3}{1}$	83
6173.348	Fe	5	55	2.213	4.212	a^5P_1–y^5D_0	$\frac{(0)5}{2}$	95
6258.578	V	2	−1	0.261	2.233	$a^6D_{\frac{1}{2}}$–$z^6D_{\frac{1}{2}}$	$\frac{(0)10}{3}$	132
6302.508	Fe	4	5	3.671	5.629	z^5P_1–e^5D_0	$\frac{(0)5}{2}$	100
6733.162	Fe	0	1	4.618	6.451	y^5P_1–f^5D_0	$\frac{(0)5}{2}$	114
8468.417	Fe	4	2	2.213	3.67	a^5P_1–$a^5P_1^0$	$\frac{(0)5}{2}$	180

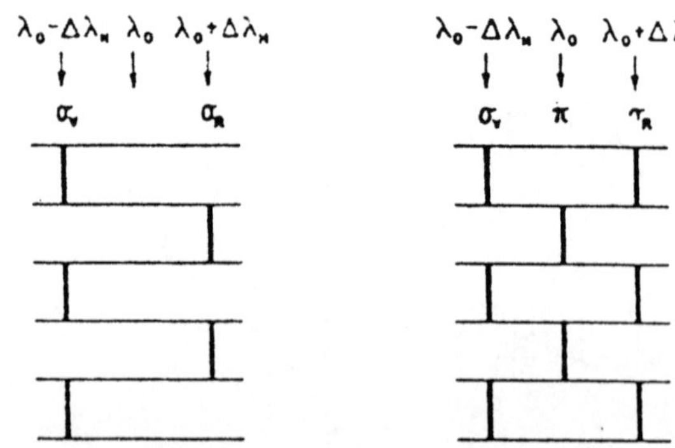

Fig. 3.2. Patterns of splitting in the longitudinal case (*left*) and transverse case (*right*).

range 5000–9000 Å are listed in Table 3.1. In order to increase the accuracy of measurements, it is desirable to use a spectral line with $g\lambda^2$ as large as possible. The line with the longest wavelength ever used in spot magnetic observation, as far as we know, is Fe I λ15649 (Rabin and Graves, 1989). It has $g = 3.0$ and $g\lambda^2 = 735 \times 10^{-10}$ cm^2. However, it must be noted that a working line should be selected not solely by its value of $g\lambda^2$. One has still to consider its temperature sensitivity, blending with neighboring lines and some other factors. These questions will be discussed in the last section of this chapter. For many years, the spectral lines most frequently used in the magnetic observations of sunspots are the three iron lines with wavelengths 5250, 6173 and 6302 Å.

3.1.2 Sources of Error

Since the method set up by Hale *et al.* is both simple and effective, it has been used in routine measurements by many astronomical institutions after the Mt. Wilson Observatory. Although all observers adopt the same method, the results of their measurements may differ significantly. Besides accidental errors, there are evident systematic discrepancies. For instance, for sunspots with magnetic fields stronger than 1000 G, the values measured at Potsdam are often 1.5–2 times larger than those of Mt. Wilson. But for fields more intense than 3000 G, the former, on the contrary, are 1.3–1.5 times smaller than the latter. Severny and Stepanov (1957) performed a detailed analysis of these circumstances. As pointed out by them, systematic errors have two main sources. First, due to the scattering of light in the atmosphere and the instrument as well as the oscillation of the solar image, the solar radiation outside the spot and without appreciable Zeeman splitting may diffuse into the spot image. Then the π component becomes more intense and the σ components appear to be weaker. If the magnetic field is not strong enough and so the Zeeman components are not completely split off, the measured $\Delta\lambda_B$ becomes smaller than the intrinsic value. This is the same for the calculated B. In order to avoid this trouble, one may use lines which are weak in the photospheric spectrum but enhanced in the umbral one. An example of such "purely" umbral lines is Ti I λ6064.6 Å with $g = 2.00$ (Zwaan and Brants, 1981). Second, the magnetic field inside a spot may be different at various places. Generally speaking, by the magnetic field strength of a spot we mean its maximum strength. However, the atmospheric agitation and scattering cause the light rays from different parts of a spot to be mixed together, and so the measured **B** is less than the maximum value. At each observing site the atmospheric conditions, the size of solar image and the accuracy of the guiding system may all differ from those of the others. So these two causes may give rise to different systematic errors.

Apart from the sources of error described above, the line displacements due to the *Evershed effect* (see Section 4.5) can blend with $\Delta\lambda_B$ and this makes the measurements even more difficult. Besides, when a spot is situated at different positions on the solar disk, the radiation we receive comes from various atmospheric layers. Therefore, when one and the same spot is observed consecutively over several days, and even if its magnetic field did not change at all, the depth gradient of its field strength may lead to unequal values of B. To sum up, for the accurate measurements of sunspot magnetic fields, many questions still have to be studied further. Their detailed discussion can be found in von Klüber (1948) and Stepanov and Severny (1962).

3.1.3 THEORETICAL INTERPRETATION

Unno's theory, introduced in Section 2.2.3, can be used to interpret the action of the polarization analyzer not only in the passage and suppression of the Zeeman components, but also in the variations of their profiles (see Figure 3.3).

The X and Y axes in Figure 3.3 are the same as in Figure 2.7. f and s represent the *fast* and *slow* axes of the $\lambda/4$ plate, respectively. o and e are the *ordinary* and *extraordinary* axes of the polaroid or Nicol prism. The angle between o and f as well as that between f and e are all equal to 45°. Let the intensities of the emergent ordinary and

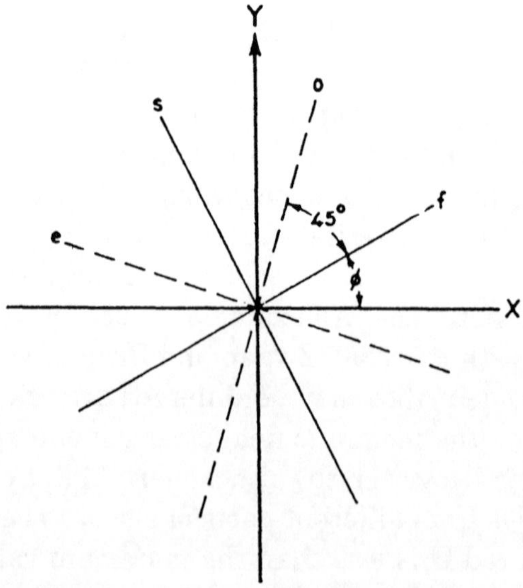

Fig. 3.3. Principles of polarization analyzer.

extraordinary radiation be denoted by $I_{o,\lambda/4}$ and $I_{e,\lambda/4}$, respectively. It is more convenient to use the depths of line profiles, i.e.

$$d_{o,\lambda/4} = \frac{\frac{1}{2}I_c - I_{o,\lambda/4}}{\frac{1}{2}I_c}, \qquad d_{e,\lambda/4} = \frac{\frac{1}{2}I_c - I_{e,\lambda/4}}{\frac{1}{2}I_c}, \tag{3.1}$$

where I_c is the intensity of the continuum. According to Unno's algebraic solution of the equations of transfer of Stokes parameters, we have

$$\begin{aligned} d_{o,\lambda/4} &= \frac{\beta_0 \cos\theta}{1+\beta_0\cos\theta}\left\{1 - \frac{1+\eta_I-\eta_V}{(1+\eta_I)^2-\eta_Q^2-\eta_V^2}\right\}, \\ d_{e,\lambda/4} &= \frac{\beta_0 \cos\theta}{1+\beta_0\cos\theta}\left\{1 - \frac{1+\eta_I+\eta_V}{(1+\eta_I)^2-\eta_Q^2-\eta_V^2}\right\}. \end{aligned} \tag{3.2}$$

For a purely longitudinal magnetic field, viz., for the case $\gamma = 0$, the above expressions become

$$\begin{aligned} d_{o,\lambda/4} &= \frac{\beta_0 \cos\theta}{1+\beta_0\cos\theta}\frac{\eta_r}{1+\eta_r}, \\ d_{e,\lambda/4} &= \frac{\beta_0 \cos\theta}{1+\beta_0\cos\theta}\frac{\eta_l}{1+\eta_l}. \end{aligned} \tag{3.3}$$

It then follows that in ordinary light there is only the σ_R component and in extraordinary light, σ_V. If the $\lambda/4$ plate is rotated through 90°, i.e., the o and e axes in Figure 3.3 are interchanged, then the situation is exactly reversed. This is the theoretical explanation for the action of the $\lambda/4$ analyzer in the alternate passage and suppression of the two σ components.

When the magnetic field is so strong that all the Zeeman components are completely split up, one may use Eqs. (2.16), (2.17) and (3.2) to derive the following expressions for the depths of the Zeeman components:

$$\begin{aligned} d_{\sigma_V,\lambda/4} &= \frac{\beta_0\cos\theta}{1+\beta_0\cos\theta}\,\frac{\frac{1}{4}\eta_0(1\mp\cos\gamma)^2}{1+\frac{1}{2}\eta_0(1+\cos^2\gamma)}\,, \\ d_{\pi,\lambda/4} &= \frac{\beta_0\cos\theta}{1+\beta_0\cos\theta}\,\frac{\frac{1}{2}\eta_0\sin^2\gamma}{1+\eta_0\sin^2\gamma}\,, \\ d_{\sigma_R,\lambda/4} &= \frac{\beta_0\cos\theta}{1+\beta_0\cos\theta}\,\frac{\frac{1}{4}\eta_0(1\pm\cos\gamma)^2}{1+\frac{1}{2}\eta_0(1+\cos^2\gamma)}\,. \end{aligned} \tag{3.4}$$

Here η_0 is the value of η_λ at the line center in the case of absence of a magnetic field. This system of formulae tells us that the relative intensities of Zeeman components is connected with the angle γ. Therefore, it is in principle feasible to determine the direction of the magnetic field with the relative intensities of Zeeman components. This question will be discussed further in the following subsection.

3.1.4 SEARES' FORMULA

Before the advent of the theory of inverse Zeeman effect, Hale and Nicholson (1938) carried out routine observations of the direction of sunspot magnetic fields with the following Seares' formula:

$$d_{\sigma_V,\lambda/4} : d_{\pi,\lambda/4} : d_{\sigma_R,\lambda/4} = \tfrac{1}{4}(1\mp\cos\gamma)^2 : \tfrac{1}{2}\sin^2\gamma : \tfrac{1}{4}(1\pm\cos\gamma)^2\,. \tag{3.5}$$

In other words, they calculated γ with the observed relative intensities of π and σ components. However, it should be stressed that this formula was established for the *direct* Zeeman effect of emission lines (Seares, 1913). Hence it is doubtful whether they can be applied to the *inverse* effect displayed by absorption lines. For some time it was believed that, for the following two reasons, Seares' formula could still be adopted in the case of the inverse effect. First, for very weak lines with $\eta_0 \ll 1$, the expressions (3.4) can be reduced to (3.5). Second, Stepanov (1958b) concluded from his theoretical calculations for Fe I λ6173 and with $B = 2500$ G that the error caused by the misuse of Seares' formula to absorption lines is quite small and does not exceed 4°. Nevertheless, we (Ye Shi-hui *et al.*, 1978b) pointed out that both these arguments are questionable. At first, Unno's algebraic solution is rather crude and, moreover, spectral lines with $\eta_0 \ll 1$ do not exist in practice. Second, Stepanov's computations were performed only for one line and one value of B, so his conclusion may be incorrect in the general case. We derived the expressions of the residual intensities of "o" and "e" light as

$$\begin{aligned} r_o &= r_I + r_V \, , \\ r_e &= r_I - r_V \, . \end{aligned} \tag{3.6}$$

They enabled us to compute the ratios of the central depths of Zeeman components as functions of γ. The results of our calculations are represented by the solid lines in Figure 3.4. The broken curves are given by Seares' formula and the dotted ones represent Stepanov's calculations. The following conclusions can be drawn from this figure:

(1) The results of our computations differ greatly from those given by Seares' formula.

(2) According to our theoretical calculations the ratios of the relative intensities of Zeeman components should depend not only on γ, but also on B. However, Seares' formula gives no relation with the magnetic field strength.

(3) Stepanov's calculations yield a relation between $\log(\mathrm{d}\sigma_1/\mathrm{d}\sigma_2)$ and γ close to that given by Seares' formulae, but the case for $\log(\mathrm{d}\sigma/\mathrm{d}\pi) \sim \gamma$ is quite different. This implies that the former is possibly an accidental coincidence.

In conclusion we may say that the method of measuring the direction of sunspot magnetic fields with Seares' formula is not a good one. More accurate methods will be introduced in Section 4.2.

3.2 Photoelectric Magnetographs

The method for measuring sunspot magnetic fields described in the last section can hardly be applied to local fields outside spots and, much more, to the general field of the Sun. This is because the Zeeman splittings caused by their comparatively weak fields with strengths of 1–10^2 G are only about 10^{-5}–10^{-3} Å, i.e. much smaller than the widths of spectral lines used in magnetic observations. (For instance, Fe I $\lambda 6302$ has $\Delta\lambda = 0.13$ Å). Beginning with Hale, many astronomers tried to set up some other methods to measure solar magnetic fields outside

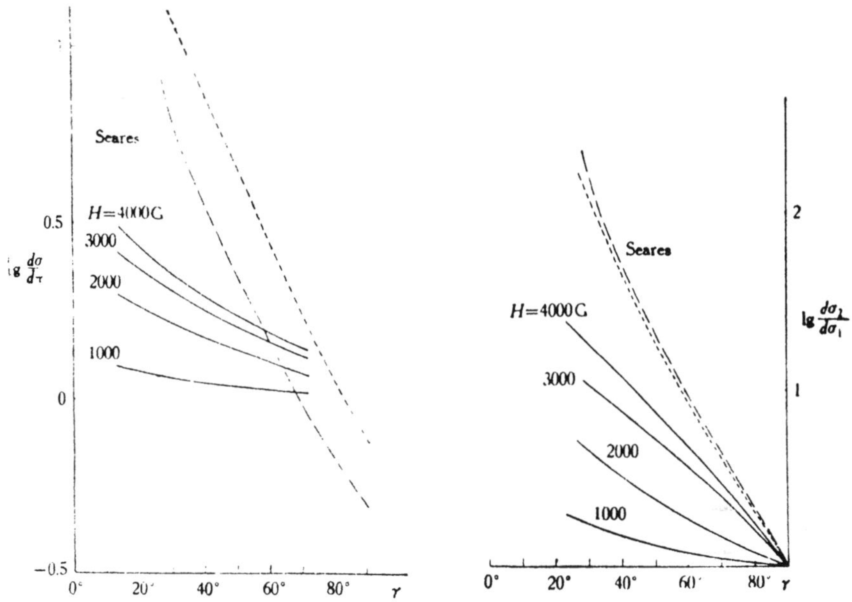

Fig. 3.4. Relation between γ and relative intensities of Zeeman components.

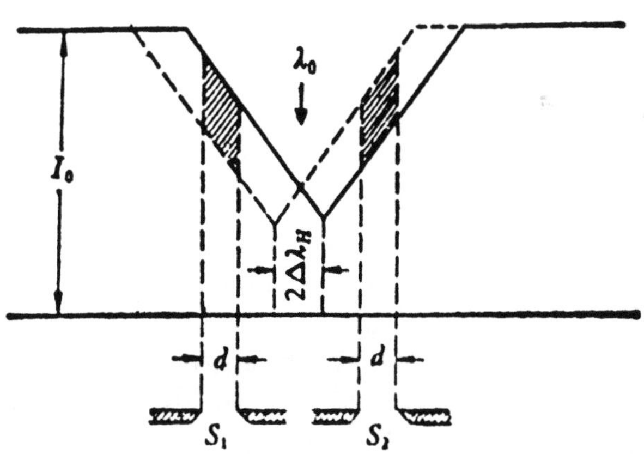

Fig. 3.5. Optical scheme of photoelectric magnetograph.

spots. This was achieved in 1952 by H.D. and H.W. Babcocks' invention of the *photoelectric magnetograph* (Babcock, 1953). This may be said to have revolutionized the observation of celestial magnetic fields, and this has led to a series of important discoveries.

3.2.1 FUNDAMENTAL IDEA

Although the electronic circuits and mechanical structure of magnetographs may be quite complex, their basic principles are rather simple. Under the action of a weak magnetic field, the profiles of the two σ components are close to each other. They are shown in Figure 3.5 by solid and dotted lines, respectively. On the focal plane of a spectrograph two identical exit slits S_1 and S_2 are placed at symmetric positions in the wings of a magneto-sensitive spectral line. Before the entrance slit of the spectrograph there is an automatic modulator made with an ADP crystal.[1] This device can turn the incident light to be left- or right-handed circularly polarized, alternately, with a frequency of more than one hundred times per second. Thus the solid and dotted profiles in Figure 3.5 appear in turn and the light flux passing through S_1 or S_2 is constantly changing. It may be readily shown that this change is given by the following expression:

$$\delta F = 2dI_c \frac{\partial r_\lambda}{\partial \lambda} \Delta\lambda_B, \tag{3.7}$$

where d is the slit width, I_c is the intensity of the neighboring continuum and r_λ is the residual intensity of the spectral line. Substituting Eq. (2.1) into the above equation, we have

$$\delta F = 9.34 \times 10^{-5} I_c d \frac{\partial r_\lambda}{\partial \lambda} g\lambda^2 B. \tag{3.8}$$

This tells us that δF is proportional to B. If two identical photomultipliers are placed behind S_1 and S_2 and their signals are sent to a difference amplifier, then after narrow-band amplification it is possible to record the value of δF and also that of B. Eq. (3.8) shows that δF is also proportional to I_c. Therefore, the limb darkening of the Sun and the variation in the terrestrial atmosphere can also influence the results of measurements. It should be pointed out that B here represents the component of the magnetic field strength along the line of sight. Hence, Eq.(3.8) has, strictly speaking, to be rewritten as

$$\delta F = 9.34 \times 10^{-5} I_c d \frac{\partial r_\lambda}{\partial \lambda} g\lambda^2 B \cos\gamma. \tag{3.9}$$

In passing, it is worthwhile noting that the quantity δF is the difference between two light fluxes and it is not influenced by instrumental

[1] Its chemical composition is ammonium dihydrogen phosphate (NH_4PO_4). When a voltage of a certain amount is supplied to its terminals, it can let only one kind of

polarization. Hence, in the measurement of the longitudinal magnetic field, it is not necessary to correct for this source of error. However, for the observation of the transverse field the situation is different. This will be discussed later in this chapter.

3.2.2 CONSTRUCTION

Although the solar photoelectric magnetographs belonging to various institutions are different in many respects, there is still much in common. In the following we take as a sample the instrument built in 1958 at the Crimean Astrophysical Observatory and describle its structure in some details.

The block diagram of this magnetograph (Nikulin *et al.*, 1958; Nikulin, 1960) is given in Figure 3.6. After reflection by the plane mirror M_3, the image of a spectral line is sent into the magnetograph through a compound system consisting of three slits (Figure 3.7). Behind the slits there are photomultipliers C_1, C_2 and C_3. The signals from C_1 and C_2 are sent to the difference amplifier A for registration of the magnetic field strength. C_3 is used to record the intensity at line center and so to trace the brightness field or the monochromatic image.

Compared with the Babcocks' instrument, the Crimean magnetograph introduced a series of innovations:

(1) It has a *radial velocity compensator* which can completely eliminate the spectral line displacements produced by all causes, with the exception only of the magnetic field, such as the solar rotation, local motions of the Sun, mechanical vibration of the instrument, etc. This ensures that the magnetographic recordings belong solely to the magnetic field. The principle of the compensator is as follows. The constant signals of C_1 and C_2 are sent to another difference amplifier B. Once the spectral line is displaced, the amplifier emits a signal. Under the action of this signal a mechanical device rotates a glass plate (line-shifter) in front of the magnetograph, so that the spectral line returns to its original position.

(2) The line-shifter provides a measure of the radial velocity, and the signal of C_3, the intensity distribution. Thus by means of the Crimean magnetograph one may study the magnetic, velocity and brightness fields of one and the same solar region.

---

polarized light pass through. With periodic variation of voltage, light beams in various states of polarization are alternately transmitted. In recent years another kind of crystal, called KDP (potassium dihydrogen phosphate), has been adopted.

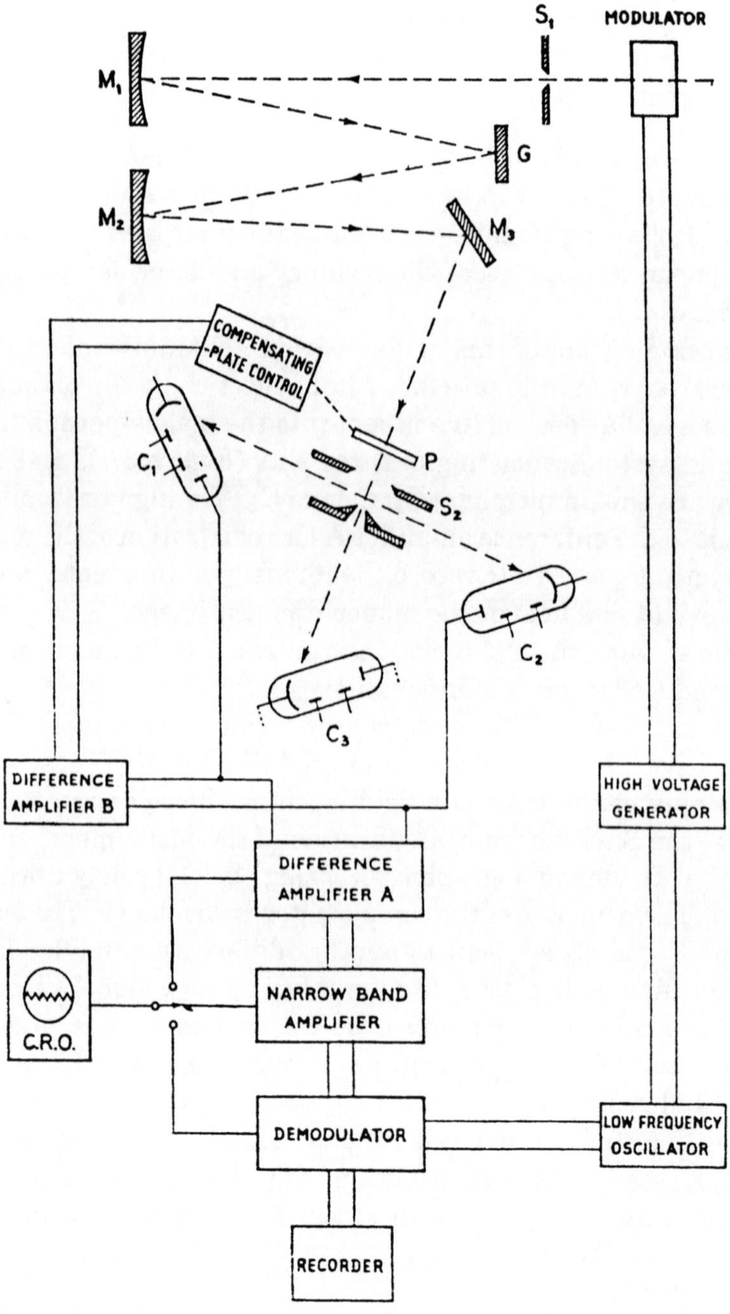

Fig. 3.6. Optical and electronic systems of the Crimean photoelectric magnetograph.

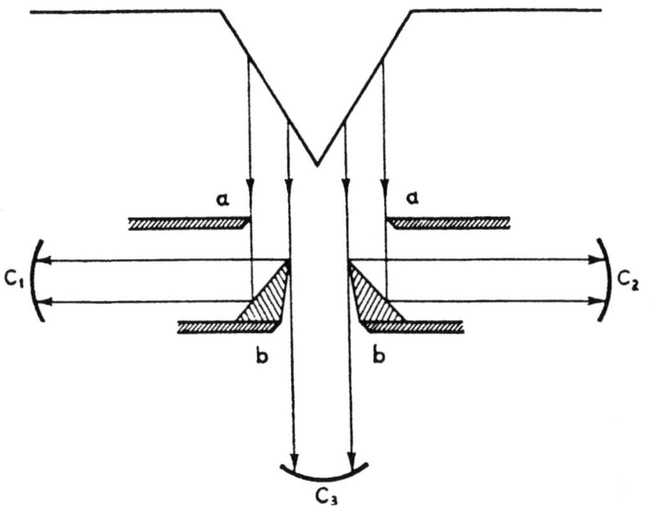

Fig. 3.7. Compound exit slit system.

(3) The Doppler displacements caused by the solar rotation are used for calibration. According to Eq. (3.7), δF is proportional to $\Delta\lambda$. The $\Delta\lambda_B$ given by the solar general magnetic field is negligible, so the quantity $\Delta\lambda$ is determined mainly by the plasma motions on the solar surface. By means of a fine adjustment device the solar image can be moved across the spectrograph slit along the solar equator. A recording of the type of Figure 3.8 is then obtained. Although the random motions in local regions may give rise to irregular fluctuations of the spectral line displacement, the Doppler shift caused by the solar rotation and the corresponding magnetographic records show evident regularities. Because the velocity of solar rotation and the Doppler displacement produced by it are all known quantities, one may reduce magnetographic recordings to absolute values.

3.2.3 Interpretation of Magnetographic Recordings

Strictly speaking, Eqs. (3.7–3.9) are very crude. More accurate formulae can be set up on the basis of the theory of spectral line formation in magnetic fields. In the last section we obtained expressions of the line depths in ordinary and extraordinary lights, $d_{o,\lambda/4}$ and $d_{e,\lambda/4}$, given by a quarter-wave analyzer. In the following, instead of line depth d we shall use line intensity $I = 1 - d$ as well as the residual intensity $r = I/\frac{1}{2}I_c$. Besides, due to the action of the photoelectric modulator, we

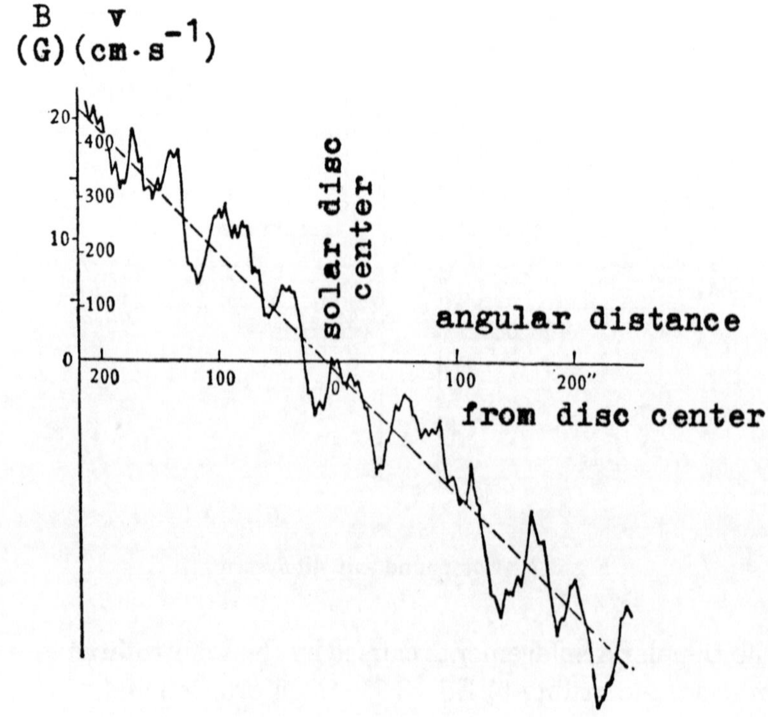

Fig. 3.8. Calibration by solar rotation.

have alternately the profiles of ordinary and extraordinary light (see Figure 3.9). Then at the exit slits S_1 and S_2 the changeable part of the residual intensity is $\Delta r = r_{o,\lambda/4} - r_{e,\lambda/4}$. The signal of the magnetograph is proportional to Δr. So we have[2]

$$\delta i = cI_c \Delta r, \tag{3.10}$$

where c is a constant. By use of Eq. (3.2) it is easy to obtain

$$\delta i = 2cI_c \frac{\beta_0 \cos\theta}{1 + \beta_0 \cos\theta} \frac{\eta_V}{(1+\eta_I)^2 - \eta_Q^2 - \eta_V^2} \,. \tag{3.11}$$

In order to confer greater accuracy on this formula, we may use Eq. (2.17) and let $\eta_p(v) = \eta(v)$, $\eta_l = \eta(v - v_B)$, $\eta_r = \eta(v + v_B)$. As the first approximation, in the case of very weak magnetic fields, we can take

[2]The slit is assumed to be infinitely narrow. If its width d is taken into consideration, this equality has to be modified to be $\delta i = cI_c \int_0^d \Delta r \, d\lambda$.

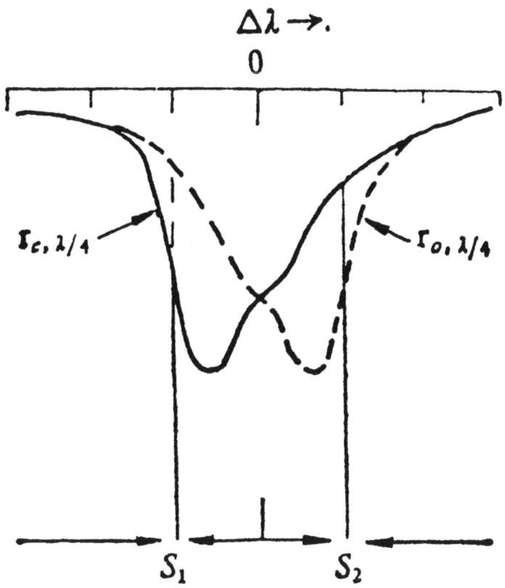

Fig. 3.9. "o" and "e" profiles at magnetograph slits.

$$\eta_l = \eta(v) - v_B \tfrac{\mathrm{d}\eta}{\mathrm{d}v}, \qquad \eta_r = \eta(v) + v_B \tfrac{\mathrm{d}\eta}{\mathrm{d}v}. \tag{3.12}$$

By substituting these expressions into (2.16), it is easy to get

$$\eta_I = \eta(v), \quad \eta_Q = 0, \quad \eta_V = v_B \frac{\mathrm{d}\eta}{\mathrm{d}v} \cos\gamma. \tag{3.13}$$

Then from (3.13) and (3.11) we obtain

$$\delta i \simeq 2cI_c \frac{\beta_0 \cos\theta}{1 + \beta_0 \cos\theta} v_B \cos\gamma \frac{\mathrm{d}\eta/\mathrm{d}v}{[1 + \eta(v)]^2} . \tag{3.14}$$

This shows that only for very weak magnetic fields we have $\delta i \propto B \cos\gamma$.

Although Eq. (3.14) is more reliable than (3.7–3.9), it is still a rough approximation. This is not only because in its derivation we adopted the first order expansion (3.12), but also due to the fact that Unno's theory contains a series of simplifying assumptions and so is not strictly correct. The second reason was explained in some detail in Section 2.2.3. It was also stated there that for the improvement of the theory of spectral line formation in the magnetic field a lot of problems, such as the nonuniformity of magnetic fields, have to be attacked. So the correct interpretation of magnetographic observations is quite a difficult project.

3.2.4 NEW INSTRUMENTS

The progress of science and technology is unlimited. Since the construction of the Crimean photoelectric magnetograph in 1958 there have appeared many other, still better instruments. In this subsection only two kinds among them will be briefly introduced.

3.2.4.1 Multichannel magnetographs

The main merit of the photoelectric magnetograph is its high sensitivity. To be precise, it can measure longitudinal magnetic fields as weak as 1 G and radial velocities as small as 1–10 $m\,s^{-1}$. However, it can be used to observe only one single point on the solar surface at each moment. Hence its temporal and spatial resolutions are rather low. In order to improve this situation, the former Kitt Peak National Observatory in the late sixties made a 40-channel magnetograph (Livingston and Harvey, 1971), which was later reconstructed to have 512 channels (Trumbo *et al.*, 1974; Smithson, 1975). By use of a double array of diodes along the spectrograph slit and of fiber optics, there are 40 or 512 points on the exit slit and simultaneous observations can be performed at all these places. So, in this sense, a multi-channel magnetograph may be said to be equivalent to the joint operation of many photoelectric magnetographs. It scans the solar surface with the whole slit, and both the spatial and temporal resolutions are significantly improved. Concretely speaking, the 40-channel magnetograph can yield a magnetic map of the whole solar disk with 21 scans, which altogether last 40 minutes. In contrast, about one hour is needed for a common magnetograph to scan an active region of moderate size.

Figures 3.10 and 3.11 are two Kitt Peak 512-channel magnetograms of the full solar disk. The former was obtained in an epoch of very low level of solar activity, and the latter in a period with a high level of activity. Each magnetogram is the combination of four scans. The black and white regions represent magnetic fields of different polarities. It is interesting to note:

(1) the quiet Sun possesses very fine background magnetic fields,

(2) the active Sun has a number of large bipolar magnetic regions and the arrangements of their polarities in the two hemispheres are opposite to each other.

These phenomena will be discussed in detail later in this book.

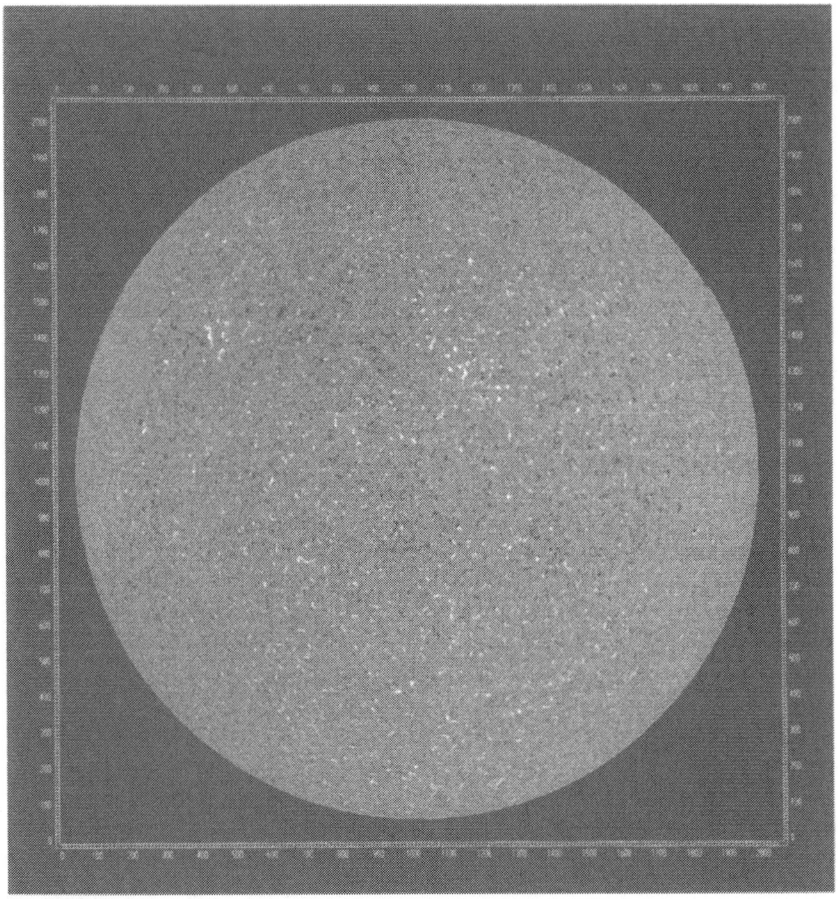

Fig. 3.10. Full-disk magnetogram of the quiet Sun, observed on Dec. 27, 1985. *(Courtesy of W.C. Livingston).*

3.2.4.2 Double-Line Magnetographs

In various layers of the solar atmosphere the physical conditions, including the magnetic field strength and orientation, may be significantly different. For the study of the origin and mechanism of solar activity it is necessary to observe simultaneously the magnetic fields in different layers, e.g. in both the photosphere and the chromosphere. With this aim a double-line magnetograph was constructed in the mid-sixties at the Crimean Astrophysical Observatory (Severny, 1966). In basic principles it is almost the same as common photoelectric magnetographs. The unique difference is that the double-line magnetograph

Fig. 3.11. Full-disk magnetogram of the active Sun, observed on February 12, 1989 *(Courtesy of W.C. Livingston).*

has two gratings placed in the parallel beams after the collimator of the spectrograph, and so magnetic observations can be carried out at the same time on any two spectral lines, such as Fe I $\lambda 5250$ and Hα. The former is formed mainly in the photosphere and the core of the latter in the chromosphere.

Observations with the double-line magnetograph have revealed that in most (about 70%) cases the basic configurations of the photospheric and chromospheric magnetic fields are similar. But the structure of the latter is not so fine as that of the former. In some (approximately 30%) cases, the chromospheric fields are more intense and possess polarities which are opposite to those of the photospheric fields. (As mentioned in Section 2.7, this is so in particular in regions of flare activity.) It is of still more interest to discover that sometimes in the intermediate layer between the photosphere and the chromosphere (observed in, e.g., Ca I λ6103) there are isolated monopolar magnetic fields, but no other fields can be observed both above and below them. This also shows the complexity of the structure of solar magnetic fields.

The Potsdam magnetograph (Staude *et al.*, 1991) is also a double-line magnetograph. Moreover, it is a vector magnetograph of the spectrographic type (Section 3.4.4). Its error of measurement of the longitudinal field is 6 G and that of the transverse field is much larger, 120 G.

3.3 Videomagnetographs

Thanks to the very rapid progress of technology, a new kind of solar magnetographs has come into existence. Their outstanding merit lies in their very high temporal resolution. In a single moment they can display magnetic maps of the solar surface on TV screens or allow them to be recorded on photographic films or magnetic tapes. This is very favorable for the magnetic observation of solar active features, especially flares and eruptive prominences.

Such instruments are called videomagnetographs (VMGs). Their structure differs greatly from that of photoelectric magnetographs. The main characteristics are as follows:

(1) Instead of spectrographs they use birefringent filters to isolate radiation within one or two magneto-sensitive lines. So it is not necessary to scan with a slit, and the entire solar disk or a part of it can be observed simultaneously.

(2) The subtraction of solar images formed in right- and left-handed polarized lights can be performed automatically by means of electronic devices.

(3) With electronic computers the observational data may be immediately reduced to magnetic field information.

The first VMG was built at the Culgoora Observatory of Australia (Ramsey *et al.*, 1971). Its monochromatic filter is composed of three automatically controlled Fabry-Perot interferometers and has a passband width of 0.05–0.1 Å. The passband can be adjusted to any wavelength in the range 4000–6000 Å. This magnetograph is mounted on a refracting telescope with aperture of 20 cm and focal length of 170 cm and can observe a part of the solar disk (about 1/4 of the solar radius). After semi-automatic subtraction by photographic or television methods, a magnetogram may be obtained with an exposure time of 0.3 s and spatial resolution of about 2 arcseconds.

In the VMG of the Big Bear Solar Observatory (BBSO) (Mosher, 1976; Zirin, 1986) a Lyot filter is used to form monochromatic images and a KDP modulator is placed before the filter. The image is recorded by a CCD operating at video rates, digitized on-line and accumulated in a memory device. When the KDP voltage is reversed, a new exposure is made with the reversed signal. This process continues at the 30 frame/sec video rate. The BBSO's VMG is very sensitive, because the images can be added up almost indefinitely and so the signal-to-noise ratio is very high.

The VMG of the Marshall Space Flight Center (MSFC) and that of the Huairou Station of the Beijing Astronomical Observatory (called the Magnetic Field Telescope) are all vector magnetographs, which will be treated in the next section.

3.4 Vector Magnetographs

The early photoelectric magnetographs could measure merely the *longitudinal* magnetic field ($B_{\|}$), i.e. the component parallel to the line of sight. This, of course, cannot satisfy fully the needs of scientific research, because the magnetic field itself is three-dimensional. For many years astronomers have been engaged in the construction of vector magnetographs. The first one was made in the beginning of the sixties at the Crimean Astrophysical Observatory (Stepanov and Severny, 1962). It could measure the *transverse* field ($B_{\perp}$), viz. the component perpendicular to the line of sight, but not simultaneously with $B_{\|}$.

In order to determine both the angle γ formed by the line of sight and magnetic field lines and the azimuthal angle χ of the transverse component (see Figure 3.12), the observation of one and the same point on the Sun should be made twice with different positions of the $\lambda/2$ modulator. Hence this instrument was inconvenient to operate and ceased to be used not long after its invention. In recent years many new vector magnetographs have been constructed, and they can measure the three quantities B, γ and χ at practically the same time. Before introducing these instruments, we would like to explain to the reader why the transverse component is especially important in the observation of

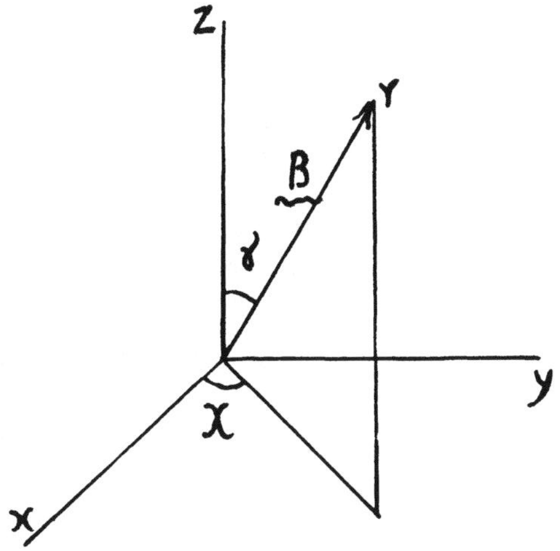

Fig. 3.12. Magnetic vector.

solar magnetic fields.

3.4.1 SIGNIFICANCE OF OBSERVATION OF THE TRANSVERSE COMPONENT

As we have just said, the transverse component should be measured, because the magnetic field itself has a 3-D structure. This statement is naturally correct, but not the complete story by far. The importance of the measurement of $B_\perp$ may be illustrated by its close relation with solar flares and some other topics. It is generally believed that the tremendous amount of energy of flares comes from the surrounding magnetic field. Therefore, it is natural to think that after the outburst of a flare the structure of the local magnetic field should change significantly. As shown by the systematic study of Severny (1958a, b; 1960b), such variations really took place for a lot of flares, but notable exceptions were found by some other authors. (A detailed account of this contradictory situation will be given in Section 5.5.3.) It is interesting to say that in recent years we have gradually become aware of the essence of the matter. Flares are produced mainly by the rapid changes of the transverse magnetic field, while the longitudinal field remains practically unchanged. This is because the release of energy caused by the *shear*, *compression*, *reconnection* and maybe other forms of variation of the local magnetic field is connected chiefly with $B_\perp$. On this the observation of Hagyard (1990) may be taken to be convincing evidence. With the MSFC Vector Magnetograph she studied four flares, which all occurred near the *neutral line* of local fields. Moreover, the locations of flares are characterized by the following requirements:

(1) $B_\perp$ attains its maximum;

(2) the angular shear $\Delta\phi$ is almost a right angle.

The quantity $\Delta\phi$ is defined as the difference between the calculated azimuth of the photospheric potential field ϕ_p and the observed azimuth ϕ_0 at the same place (see Figure 3.13) In Hagyard's expression, $\Delta\phi$ represents the "nonpotentiality" of the transverse field. It implies the degree of departure of the magnetic field from the potential field, which contains minimum magnetic energy. The larger $\Delta\phi$, the more energy is stored in the magnetic field, and the greater is the possibility of flare occurrence. The two criteria stated above are shown in Figure 3.14. The abscissa is the length along the neutral line and the symbol $\odot$ indicates the location of flare eruption. It is readily seen that flares occur at places where both B_T and $\Delta\phi$ are largest. Hagyard emphasized that both of these quantities can be determined only with measurements of the transverse field. In recent years the study of *neutral current sheets* and *magnetic reconnection* (e.g., Tandberg-Hanssen and Emslie, 1988) has achieved much progress. Here we want only to stress that the vertical current sheet closely concerned with the occurrence of flares may be calculated merely with the transverse field strength, according to the formula $j_z = (c/4\pi)[\partial B_y/\partial x - \partial B_x/\partial y]$. This also tells us that the transverse component is of particular importance and significance in the observation of solar magnetic fields.

3.4.2 FUNDAMENTAL DIFFICULTY

The observation of the solar vector magnetic field is a very hard work. Now we want to stress that the measurement of $B_\perp$ is still more difficult than that of $B_\parallel$. This is because the error is much larger, the influence of instrumental polarization should be taken into account, the azimuthal angle χ has to be determined and, in doing so, the Faraday rotation must be considered. The so-called 180°-*ambiguity* should also be treated in the observation of $B_\perp$, while this has no concern with $B_\parallel$. However, in spite of all these troubles, Stenflo (1985a, b) pointed out that the fundamental difficulty lies neither in the observational techniques, nor in the theoretical interpretation, but is embodied in the following matter. The solar magnetic field consists of very small elements or *flux tubes*, which cannot be directly discerned by any existing telescopes, and, moreover,

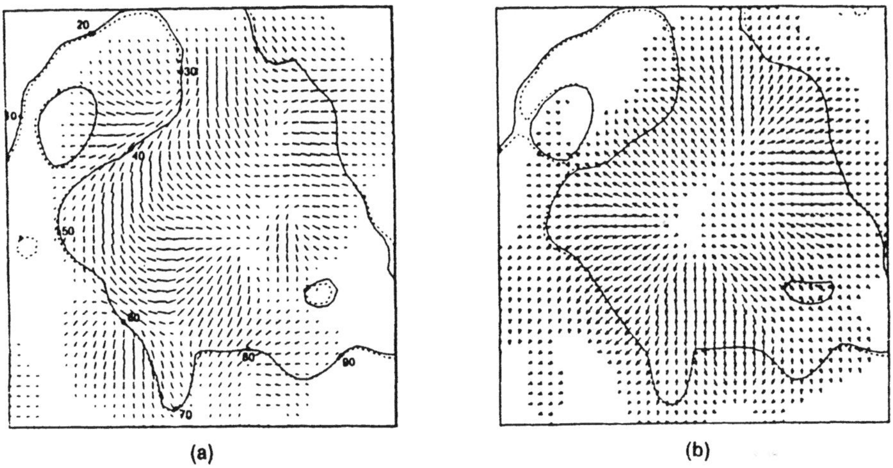

Fig. 3.13. Transverse magnetic field of AR 4711 on February 3, 1986, at 1642 UT. (a) Observed field. (b) Potential field.

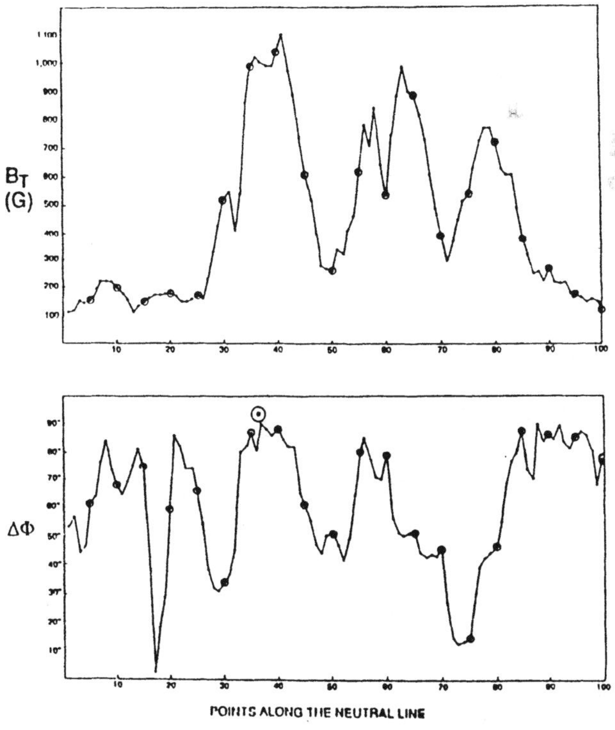

Fig. 3.14. Angular shear $\Delta\phi$ and transverse field strength B_T along neutral line.

the influences of such fine structures on the measurements of $B_{\parallel}$ and $B_{\perp}$ are quite different. Now let us explain this concept in detail.

For a Zeeman triplet, the intensities of the circularly polarized components σ_1 and σ_2 caused by $B_{\parallel}$ can be expressed as

$$I_{\sigma_{1,2}} = \tfrac{1}{2}(I \pm V). \tag{3.15}$$

If the magnetic field strength B does not vary with depth, the profiles of these components are

$$I_{\sigma_{1,2}} = \tfrac{1}{2}I(\lambda \pm \Delta\lambda_B). \tag{3.16}$$

From these two equalities one may readily obtain

$$V = \tfrac{1}{2}[I(\lambda + \Delta\lambda_B) - I(\lambda - \Delta\lambda_B)]. \tag{3.17}$$

Taking Taylor expansions for $I(\lambda + \Delta\lambda_B)$ and $I(\lambda - \Delta\lambda_B)$ and adopting the *weak-field approximation* (WFA), i.e. $\Delta\lambda_B$ is less than the half-width of the spectral line $\Delta\lambda_D$, we have

$$V = \Delta\lambda_B \left[\frac{\partial I}{\partial\lambda} + \frac{1}{6}(\Delta\lambda_B)^2 \frac{\partial^3 I}{\partial\lambda^3} + \cdots\right] \simeq \Delta_B \frac{\partial I}{\partial\lambda}. \tag{3.18}$$

In a transverse magnetic field the spectral line is split into three linearly polarized components π, σ_1 and σ_2. Their intensities are

$$I_{\pi,\sigma} = \tfrac{1}{2}(I \pm Q). \tag{3.19}$$

The plus sign belongs to the π component, and the minus sign to the σ components. Their profiles have the following expressions:

$$\begin{aligned} I_{\pi,\sigma} &= \tfrac{1}{2}I(\Delta\lambda), \\ I_\sigma &= \tfrac{1}{4}[I(\lambda + \Delta\lambda_B) + I(\lambda - \Delta\lambda_B)]. \end{aligned} \tag{3.20}$$

From Eqs. (3.19) and (3.20) we obtain

$$Q = \tfrac{1}{2}\left\{I(\lambda) - \tfrac{1}{2}[I(\lambda + \Delta\lambda_B) + I(\lambda - \Delta\lambda_B)]\right\}. \tag{3.21}$$

By taking Taylor expansions we have

$$\begin{aligned} Q &= -\tfrac{1}{4}(\Delta\lambda_B)^2\left[\frac{\partial^2 I}{\partial\lambda^2} + \tfrac{1}{12}(\Delta\lambda_B)^2\frac{\partial^4 I}{\partial\lambda^4} + \cdot\right] \\ &\simeq -\tfrac{1}{4}(\Delta\lambda_B)^2\frac{\partial^2 I}{\partial\lambda^2}. \end{aligned} \tag{3.22}$$

A similar expression can be obtained for the Stokes parameter U. The formulae (3.18) and (3.22) tell us that the parameter V, which represents circular polarization, is proportional to $\Delta\lambda_B$, and thus to B, while Q and U, which stand for linear polarization, are proportional to $(\Delta\lambda_B)^2$. In other words, V increases with the magnetic flux, and Q or U with the magnetic energy density. In recent decades a large amount of observational material has shown that the solar magnetic field possesses very delicate and complicated structures. It is natural to think that such structures affect the magnetic flux $(\alpha\Delta\lambda_B)$ and magnetic energy density $[\alpha(\Delta\lambda_B)^2]$ in different ways. Therefore, the influence of the fine structure of magnetic fields on V and Q (or U) is also quite dissimilar.

Furthermore, in the case of WFA one may derive the strength of the longitudinal component $B_{\|}(\equiv B\cos\gamma)$ with the intensity of V and obtain the values of $B_{\perp}(\equiv B\sin\gamma)$ and χ with Q and U. The concrete expressions are as follows.

$$\begin{aligned} Q &= C_Q B^2 \sin^2\gamma \cos 2\chi \frac{\partial^2 I(\lambda)}{\partial\lambda^2}, \\ U &= C_U B^2 \sin^2\gamma \sin 2\chi \frac{\partial^2 I(\lambda)}{\partial\lambda^2}, \\ V &= C_V B \cos\gamma \frac{\partial I(\lambda)}{\partial\lambda}, \end{aligned} \tag{3.23}$$

where C_Q, C_U and C_V are constants, which remain to be determined by calibration. From the above it follows that the effects of the fine structure of magnetic fields on the measurements of $B_{\|}$ and $B_{\perp}$ may be quite different.

Finally, it is worth noting that the error of observation of $B_{\perp}$ is much larger than that of $B_{\|}$. This is because Q (or U) $\propto (\Delta\lambda_B)^2$ and $V \propto (\Delta\lambda_B)$. This justifies the first difficulty of measurement of $B_{\perp}$ mentioned at the beginning of this subsection. Combining the contents of these two subsections, we find that the 'more important' is just the 'more difficult'. This is the challenge that nature offers us!

3.4.3 Basic Principles

It follows from Eq. (3.23) that the three characteristic quantities of the magnetic vector, B, γ and χ, can be determined if the three Stokes parameters Q, U and V are known from observation. Hence magnetic field observations are in essence *polarimetric measurements*. All vector magnetographs, no matter how different they are in construction, may be regarded as *polarimeters*, and their basic principles are practically the same. This can be briefly described as follows.

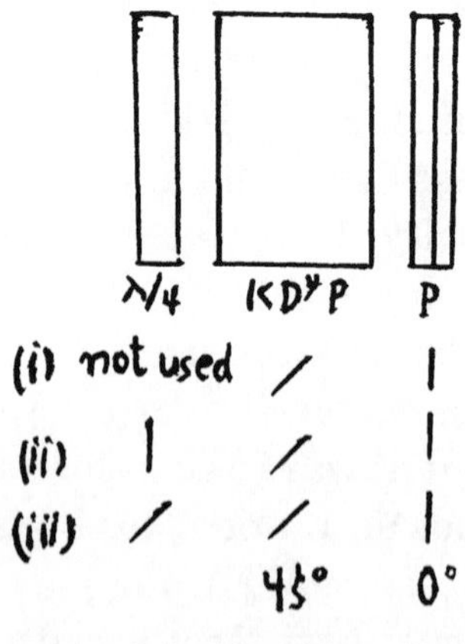

Fig. 3.15. Polarization analyzer of Huairou magnetograph.

After forming the solar image with a telescope, the sunlight is sent into a polarization analyzer, which in general consists of a $\lambda/4$ plate, a modulator and a polaroid. By adjusting the azimuth of the optical axis of the $\lambda/4$ plate (θ) and the phase retardation of the modulator (δ), one may get various combinations of the Stokes parameters, such as $I \pm Q$, $I \pm U$ and $I \pm V$. From these combinations the individual parameters can be determined at practically the same time, and then the quantities B, γ and χ may be calculated. Vector magnetograms are obtained by means of electronic computers. Since these principles can be realized in both photoelectric magnetographs and VMGs, there are two kinds of vector magnetographs; some with birefringent filters and the others with spectrographs. In the following the Magnetic Field Telescope of the Huairou Station of the Beijing Astronomical Observatory (Ai Guo-xiang and Hu Yue-feng, 1986) and the vector magnetograph of the Okayama Astrophysical Observatory (Makita, 1986a) are taken to be examples of these two types of instruments and will be introduced in some detail.

(1) The Huairou Magnetic Field Telescope. The image-forming system is a vacuum refractor with an aperture of 36 cm and a focal length of 280 cm. The birefringent filter has two pass-bands centered at Fe I λ5324.191 and Hβ with half-widths of 0.15 and 0.12 Å, respectively. The polarization analyzer is shown in Figure 3.15. The effects of its three constituents on the state of polarization of the incident radiation can be expressed with Müller matrices as follows.

$$
S = \begin{bmatrix} \frac{1}{2} & \frac{1}{2} & 0 & 0 \\ \frac{1}{2} & \frac{1}{2} & 0 & 0 \\ 0 & 0 & 0 & 0 \\ 0 & 0 & 0 & 0 \end{bmatrix} \times \begin{bmatrix} 1 & 0 & 0 & 0 \\ 0 & \cos\delta & 0 & -\sin\delta \\ 0 & 0 & 1 & 0 \\ 0 & \sin\delta & 0 & \cos\delta \end{bmatrix} \times
$$

$$
\begin{bmatrix} 1 & 0 & 0 & 0 \\ 0 & \cos^2 2\theta & \frac{1}{4}\sin^2 4\theta & -\sin^2 2\theta \\ 0 & \frac{1}{4}\sin^2 4\theta & \sin^2 2\theta & \cos^2 2\theta \\ 0 & \sin^2 2\theta & -\cos^2 2\theta & 0 \end{bmatrix} \times \begin{bmatrix} I \\ Q \\ U \\ V \end{bmatrix}
$$

$$
= 0.5I + 0.5Q(\cos\delta\cos^2 2\theta - \sin\delta\sin 2\theta) +
$$

$$
+0.5U(\tfrac{1}{4}\cos\delta\sin^2 4\theta + \sin\delta\cos^2 2\theta) - 0.5V\sin^2 2\theta\cos\delta. \quad (3.24)
$$

The polarization analyzer may be operated in the following three cases.

(i) The $\lambda/4$ plate is not used, and then the above expression becomes

$$
S = 0.5I + 0.5Q\cos\delta - 0.5V\sin\delta. \quad (3.25)
$$

The KD*P crystal provides a modulation of the form of rectangular waves with δ alternately equal to $\pi/2$ and $-\pi/2$. Then we have at one moment the signal S_+, and at another moment S_-, respectively, given by

$$
S_\pm = 0.5I \mp 0.5V. \quad (3.26)
$$

This leads to

$$
V = S_- - S_+ . \quad (3.27)
$$

This formula allows us to measure the longitudinal field strength $B_\parallel$ with the magnetographic signals S_+ and S_-.

(ii) The $\lambda/4$ plate is used and the azimuth of its axis is taken to be equal to zero. The same modulation of δ is adopted. Then we have

$$
S_\pm = 0.5I \mp 0.5U, \quad (3.28)
$$

and

$$
U = S_+ - S_- . \quad (3.29)
$$

(iii) When θ is 45°, we get

$$S_{\pm} = 0.5I \mp 0.5Q, \tag{3.30}$$

and

$$Q = S_{-} - S_{+}\,. \tag{3.31}$$

After evaluating the parameters Q and U, one may proceed to determine the strength and azimuth of the transverse field with the formulae:

$$B_{\perp} = A\sqrt{Q^2 + U^2}\,, \tag{3.32}$$

$$\chi = \frac{1}{2}\tan^{-1}\frac{Q}{U}\,, \tag{3.33}$$

where A is a constant which can be determined by calibration.

In recent years the Huairou Magnetic Field Telescope has accumulated a large number of solar magnetograms of fine quality. Two samples are shown in Figures 3.16 and 3.17. Besides, this instrument is also a VMG. When a 2-D CCD camera is attached to it (Ming Changrong *et al.*, 1989), real-time videomagnetograms can be secured. Two such examples are shown in Figure 3.18.

(2) *The Okayama Vector Magnetograph*. This is attached to a 65 cm solar coudé telescope and a 10 m Littrow spectrograph (Figure 3.19). Before entering the magnetograph, the solar light is reflected at large angles on two mirrors, and so the amount of instrumental polarization may be large (several percent). In order to eliminate it, two compensators are used. The polarization analyzer consists of a $\lambda/4$ plate and a Wollaston prism. The former rotates with a frequency of 40 Hz. So instead of photoelectric modulation, mechanical modulation is adopted in this magnetograph. The latter separates two spectral bands around Fe I λ5250. Six photomultipliers are used to record the intensities of radiation in the red and violet wings of this line as well as in the neighboring continua of the two spectra (Figure 3.20). The intensity of each magnetographic signal is

$$I_{\text{obs}} = \frac{1}{2}\left[I + \frac{1+\cos\delta}{2}Q + \right.$$
$$+ \frac{1-\cos\delta}{2}(Q\cos 4\Omega t + U\sin 4\Omega t) +$$
$$\left. + V\sin\delta\sin 2\Omega t + R\cos 2\Omega t\right], \tag{3.34}$$

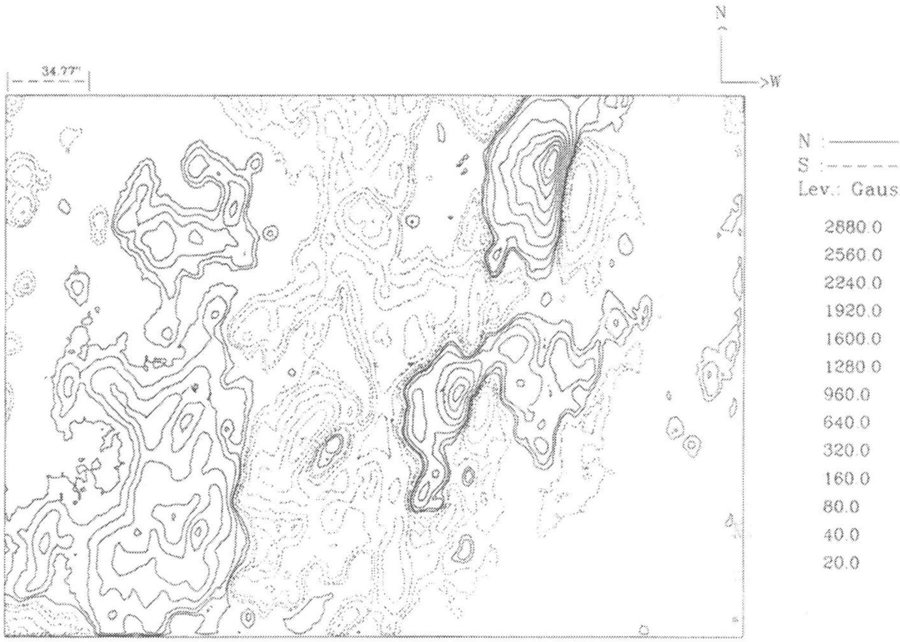

Fig. 3.16. A longitudinal magnetogram obtained on August 28, 1990 in Fe I λ5324. Solid and dotted curves are isogauss contours with different polarities. The scale of magnitude of $B_{\parallel}$ is on the right side (Courtesy of Ai Guoxiang).

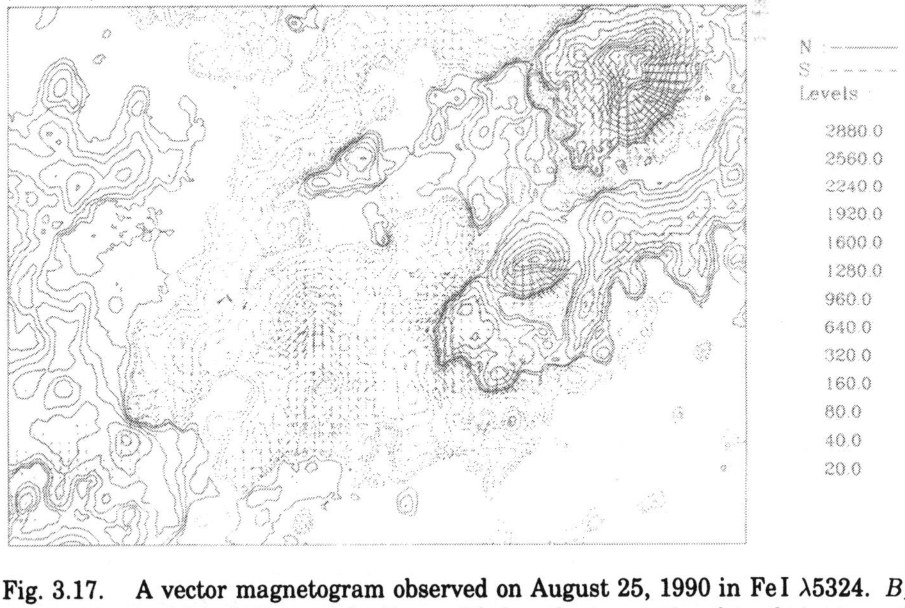

Fig. 3.17. A vector magnetogram observed on August 25, 1990 in Fe I λ5324. $B_{\perp}$s are represented by short straight lines with lengths proportional to their strengths (Courtesy of Ai Guoxiang).

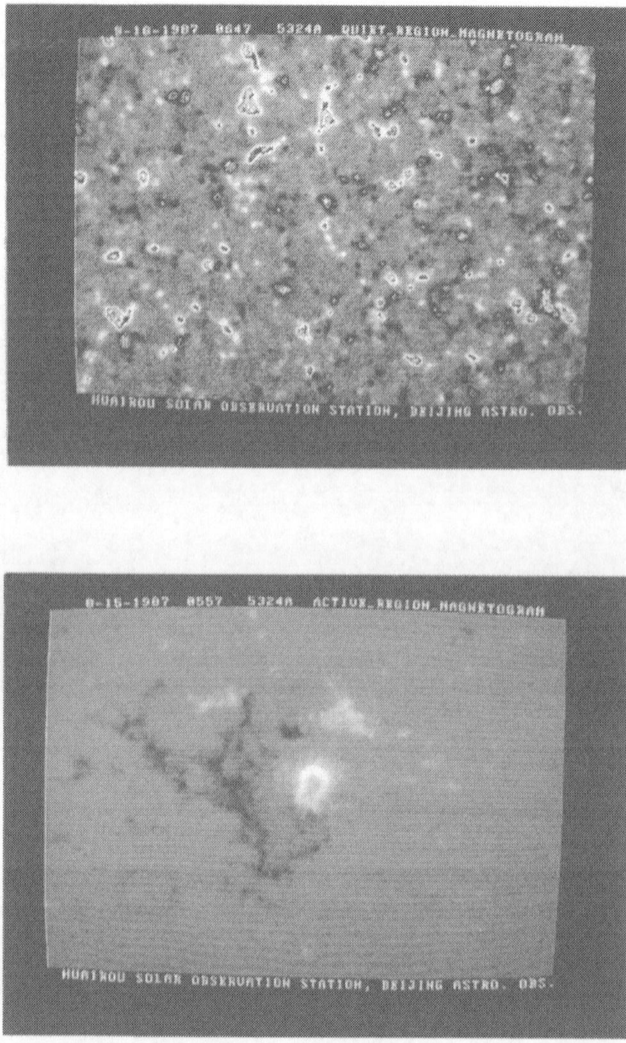

Fig. 3.18. Videomagnetograms of a quiet region (upper panel) and an active region (lower panel) (Courtesy of Ai Guoxiang).

where δ, Q and R are, respectively, the phase retardation, angular frequency and reflection coefficient of the rotating wave plate. In front of the spectral image there is a line-shifter, which can make the signals of the two wings equal and thus provides recordings of the velocity field. Because only one point on the Sun can be observed at each moment, the scanning of a region with 50×45 points requires 80 minutes. After end of the observation, the vector magnetogram and dopplergram may be automatically constructed in 15 minutes. Under the assumption of

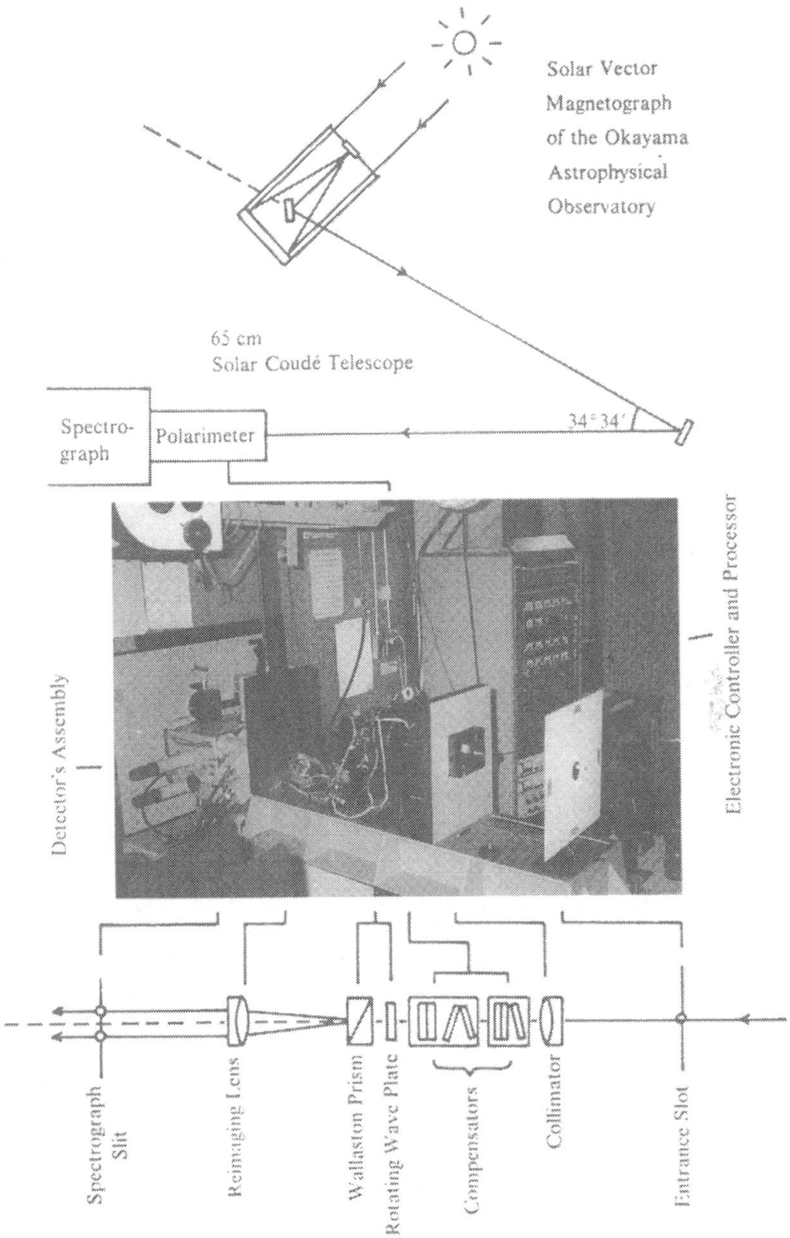

Fig. 3.19. Optical scheme and polarization analyzer of Okayama Vector Magnetograph.

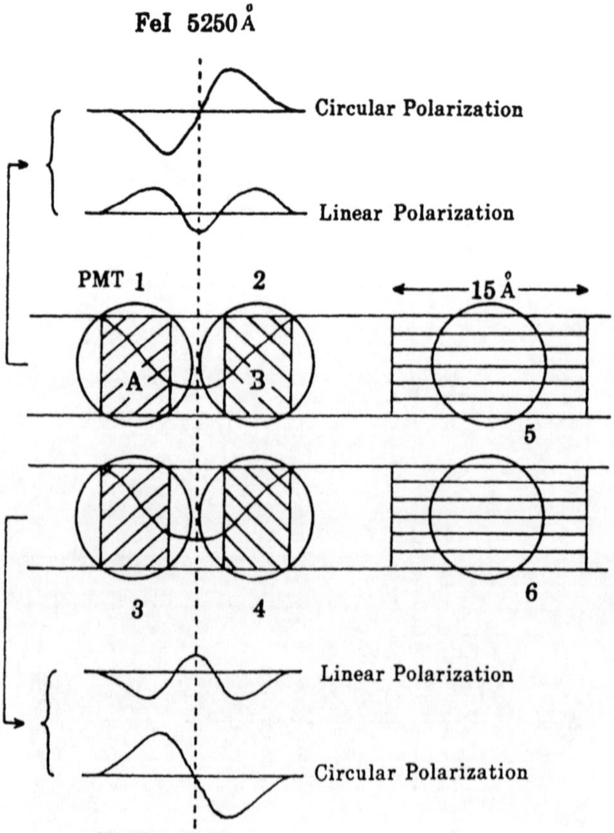

Fig. 3.20. Arrangement of photomultipliers of Okayama Magnetograph.

the potential field, the distribution of magnetic lines of force can also be traced out. The accuracies of polarimetric measurements within the spectral line and in the continuum are, respectively, 10^{-3} and 10^{-4}.

3.4.4 Comparison of Vector Magnetographs of Two Types

The two instruments introduced above belong, respectively, to the vector magnetographs with birefringent filters and those with spectrographs. With regard to the first category, we can still mention the vector magnetographs of the BBSO, MSFC and the Applied Physics Laboratory of the Johns Hopkins University (JHU/APL). As for the second kind, there are vector magnetographs at the Mees Solar Observatory (MSO) of the University of Hawaii, the Solar Observatory "Einsteinturm" at Potsdam and the Sayan Observatory near Irkutsk, as well as the reconstructed instrument in the Crimea. Now we want to emphasize that facilities of these two types are similar in principle. To be precise, in all of them telescopes are used to form solar images, polarization analyzers to determine the Stokes parameters, and computers to calculate the characteristic quantities of the magnetic vector and to trace out magnetograms. The unique difference lies in the way of isolating the radiation within one or two magneto-sensitive spectral lines. It is clear that for this either filters or spectrographs may be adopted. Due to this difference, instruments of the first kind can obtain magnetograms at once for a large region or even the entire solar disk, while those of the second type have to perform point-to-point scanning. Therefore, the temporal and spatial resolutions of vector magnetographs of the first category are by far higher than those of the second kind. So for the morphological study of flares and some other rapidly changing active features, the vector magnetographs with monochromatic filters are undoubtedly better than those with spectrographs.

However, the instruments of the first type have their shortcomings. First, due to technical limitations, the passband widths of filters are not small enough. For instance, the half-width of the filter of the JHU/APL magnetograph (Rust and O'Byrne, 1991) is 0.167 Å. If this number is taken to be the half-width of the instrumental profile and observation is made in the region of 6000 Å, the spectral resolution is readily found to be 3.7×10^4. This is much less than the corresponding values of common solar spectrographs. In this situation the profiles of the Stokes parameters inevitably suffer serious distortion by the filter. As examples, the two sets of Stokes profiles shown in Figure 3.21 are misshapen into the corresponding ones in Figure 3.22. Secondly, the transparency of birefringent filters is in general less than 10%. So after passing through a filter the intensities of all the Stokes parameters are diminished by about one order of magnitude. Therefore, for precise quantitative studies the vector magnetographs with filters are inferior to those with spectrographs.

In order to get a clear impression, Ronan *et al.* (1992) compared the results of observations with the MSFC and MSO vector magnetographs. As stated above, they belong to instruments of the first and second categories. Measurements were made of solar active regions and on the same days. A point-to-point comparison of the magnetograms leads to the following conclusion. The results of observations agree well enough, although the two instruments are different in construction, work in various spectral lines and are located at highly separated sites. More concrete information can be obtained by inspection of Figures 3.23 and 3.24.

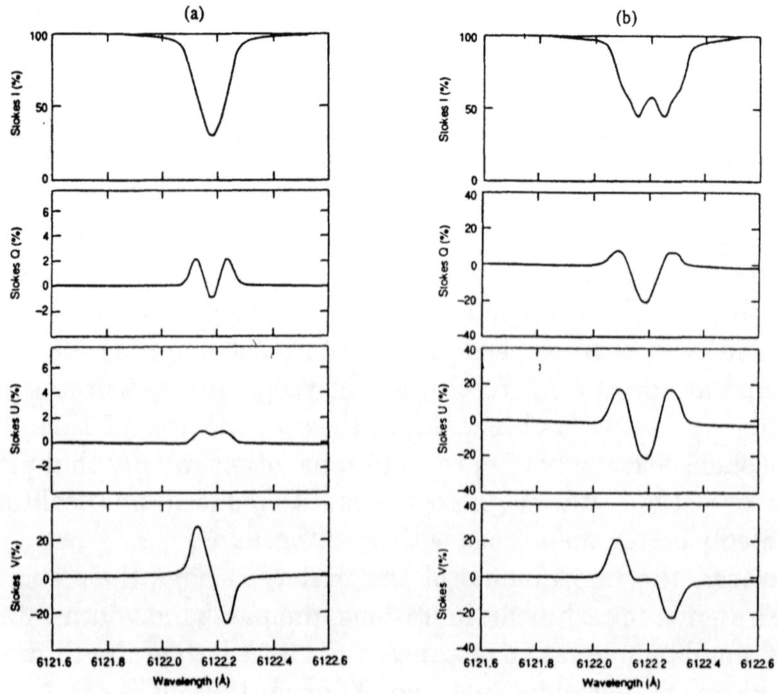

Fig. 3.21. Intrinsic profiles of Stokes parameters. *(a)* $B = 1000$ G, $\gamma = 30°$, $\chi = 0°$. *(b)* $B = 3000$ G, $\gamma = 67°$, $\chi = 30°$.

3.4.5 Methods of Calibration and Problem of 180° Ambiguity

In order to get the absolute values of the three components of the magnetic vector, B, γ, χ, from the observed Stokes parameters Q, U and V with some formulae, such as Eq. (3.23), it is necessary to determine the proportional constants C_Q, C_U and C_V. This is called the process of *calibration*. The classical method based upon the solar rotation was described in Section 3.2.2. In recent years many astronomers have prefered the theoretical calibration. The essence of this method is to calculate the *theoretical relations* between the longitudinal and transverse field strengths and the intensities of Stokes parameters with some suitable atmospheric model. For instance, under the condition of WFA Rust and O'Byrne (1991) found the following formulae:

$$V = -\mu_\lambda B \cos\gamma \frac{\mathrm{d}I}{\mathrm{d}\lambda}\,,$$

$$Q = \left(\frac{\mu_\lambda B \sin\gamma}{2}\right)^2 \frac{\mathrm{d}I}{\mathrm{d}\lambda}\frac{3}{\Delta\lambda_{\mathrm{off}}}\,, \qquad (3.35)$$

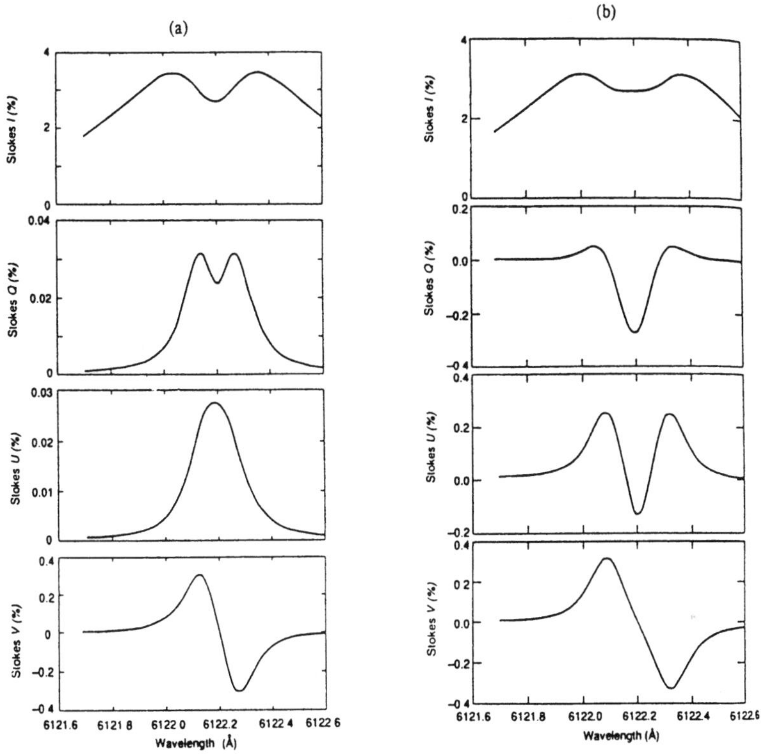

Fig. 3.22. Stokes profiles corresponding to those in Fig. 3.21, but distorted by the filter of the JHU/APL magnetograph.

where $\mathrm{d}I/\mathrm{d}\lambda$ is the slope of the line profile, $\Delta\lambda_{\mathrm{off}}$ is the wavelength distance between the filter passband and line center and μ_λ is the magnetic moment of the atomic transition. Similar expressions, which appear in the following, have been given by Jefferies, Lites and Skumanich (1989) and Jefferies and Mickey (1991).

$$
\begin{aligned}
|\nu_b| \cos\gamma &= -V \left(\frac{\mathrm{d}I}{\mathrm{d}\nu}\right)^{-1}, \\
\left(\frac{\nu_b \sin\gamma}{2}\right)^2 &= \left|\frac{H(a,\nu)'}{H(a,\nu)''}\sqrt{Q^2+U^2}\left(\frac{\mathrm{d}I}{\mathrm{d}\nu}\right)^{-1}\right|,
\end{aligned}
\qquad (3.36)
$$

where $\nu_b = (eBg)/(4\pi mc^2)(\lambda^2/\Delta\lambda_D)$ and H' and H'' are the first and second derivatives, with respect to ν, of the Voigt function.

For further information of theoretical calibration see, e.g., Staude (1980) and Staude and Hofmann (1988).

Now let us discuss a question of another category. When both the strength and azimuth of the transverse magnetic field are known, it is still necessary to determine the sense of its arrowhead. Because no

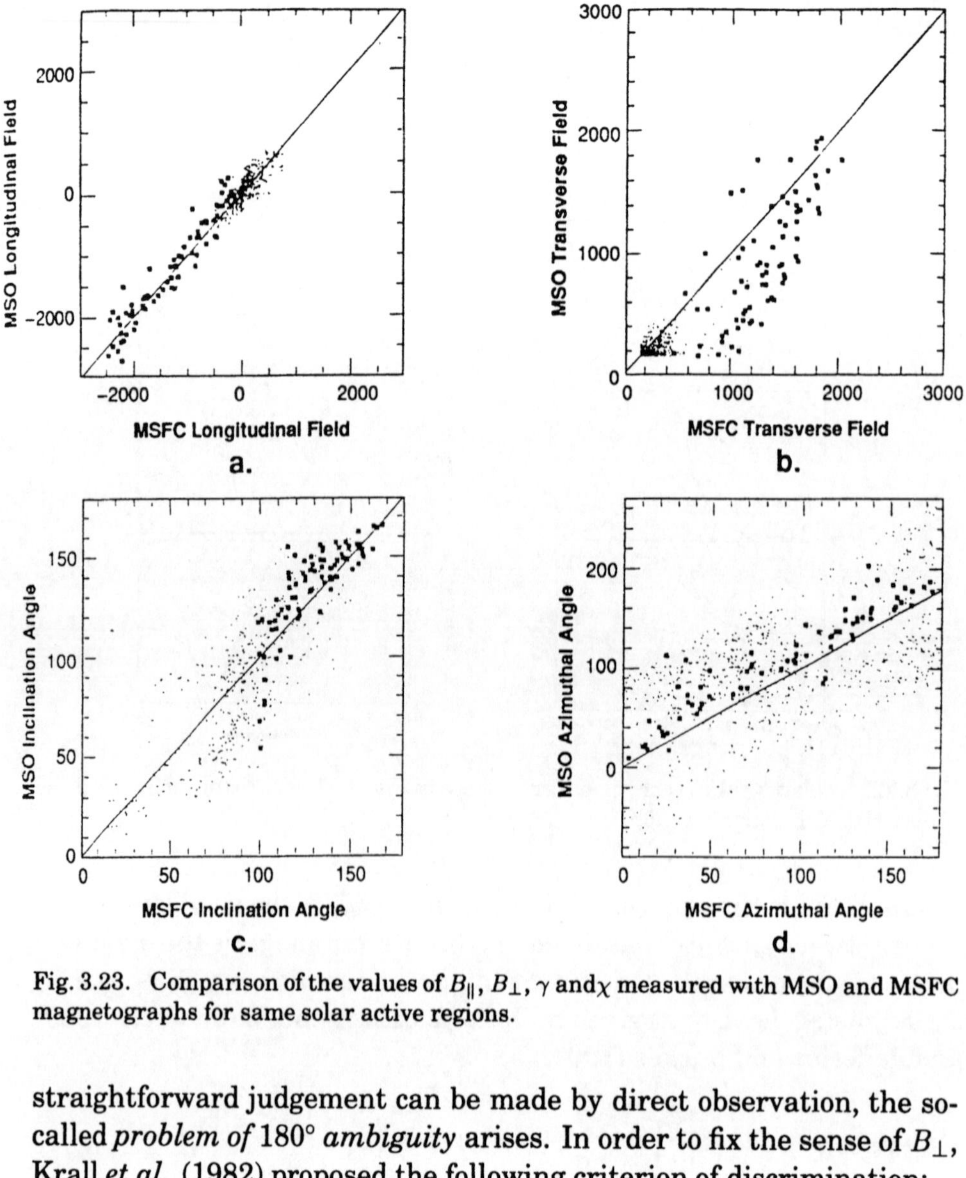

Fig. 3.23. Comparison of the values of $B_{\|}$, $B_{\perp}$, γ and χ measured with MSO and MSFC magnetographs for same solar active regions.

straightforward judgement can be made by direct observation, the so-called *problem of* 180° *ambiguity* arises. In order to fix the sense of $B_{\perp}$, Krall *et al.* (1982) proposed the following criterion of discrimination:

$$B_{\perp} \cdot \nabla B_z < 0. \tag{3.37}$$

This implies that the direction of increase or decrease of $B_{\perp}$ is opposite to that of ∇B_z. However, Wu Linxiang and Ai Guoxiang (1990) pointed out that this condition is not always true and that there is a lot of counterevidence. They then established another criterion, i.e.

$$\pm B_z[\Delta(B_{\perp}\cos\theta) + \Delta(B_{\perp}\sin\theta)] \leq 0, \tag{3.38}$$

where θ is the angle of inclination of $B_{\perp}$ with respect to the x-axis.

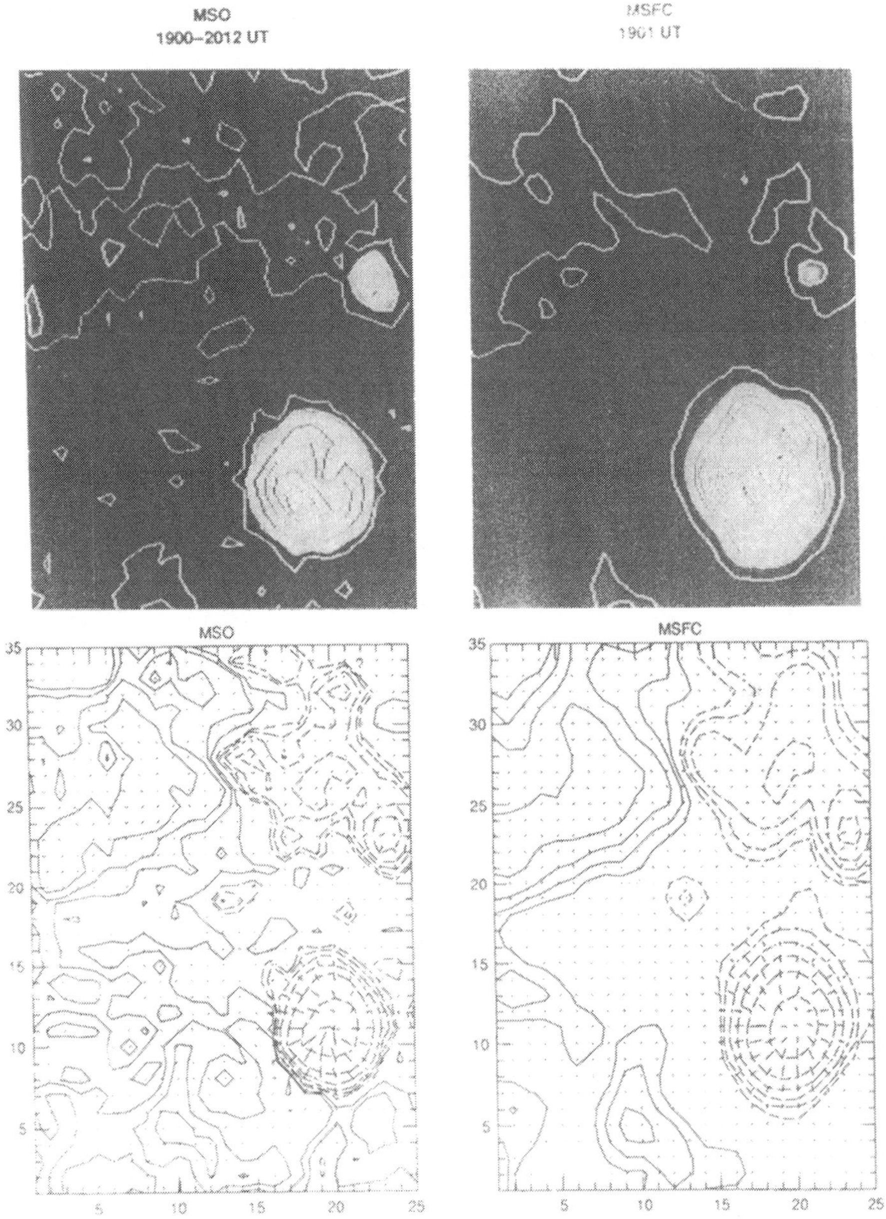

Fig. 3.24. Comparison of transverse (upper panel) and vector magnetograms (lower panel) observed for same solar active regions on June 8, 1985 with MSO and MSFC magnetographs.

3.5 Stokesmeters

Stokesmeters are a new kind of instrument, sometimes called 'scanning Stokes polarimeters'. Instead of the integrated intensities of Stokes parameters, their profiles are measured and then the three components of the magnetic vector as well as some additional information can be secured.

3.5.1 Basic Principles

The structure of stokesmeters has much in common with that of vector magnetographs. Both types of instruments use telescopes to form solar images, polarization analyzers to acquire the Stokes parameters, as well as electronic computers to calculate the components of of the magnetic vector. The unique major difference lies in the following. By rotation of a diffraction grating the Stokes profiles, i.e., distribution of intensities of the four Stokes parameters within a magneto-sensitive line, can be traced out. A set of such profiles is provided by Figure 3.25. They are yielded by a stokesmeter, called Stokes-I, which was built by the High Altitude Observatory and installed on the 40 cm coronograph of the Sacramento Peak Observatory. (For details see Baur *et al.*, 1981.) After this instrument the Stokes-II (Baur *et al.*, 1981) and the Advanced Stokes Polarimeter (Lites and Skumanich, 1989) were built in succession. A 12μ Stokes polarimeter is now under construction at NASA Goddard Space Flight Center and is planned to be installed at the McMath Telescope on Kitt Peak in 1993 (Deming, 1992).

3.5.2 Derivation of Information on the Magnetic Vector from Stokes Profiles

It is in principle feasible to infer from Stokes profiles the full information about the vector magnetic field, as well as much knowledge on the velocity and radiation fields. However, this is in practice a very difficult project, because the Stokes profiles are formed through an extremely complicated process of radiative transfer in a magnetized atmosphere. In the study of this process one has to consider many intricate physical factors, such as the inhomogeneities of the atmosphere and magnetic field, the deviations from LTE and magneto-optical effects. For many years astronomers of various countries have attacked this problem and a series of methods have been established. According to Landi Degl'Innocenti (1985) there are altogether six ways leading to the solu-

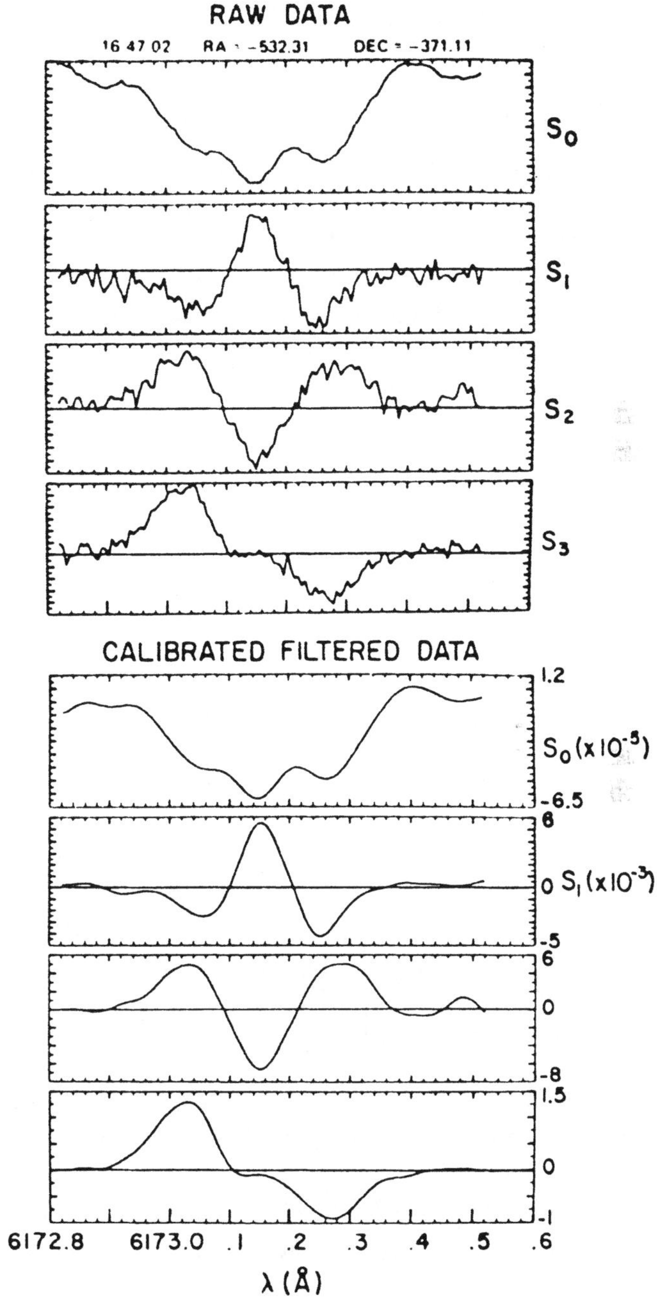

Fig. 3.25. Stokes profiles of a sunspot around Fe I 6173 Å. (S_0, S_1, S_2 and S_3 stand for the commonly used symbols I, Q, U and V.)

tion of the problem. Now, due to shortage of space, only two of them are introduced in the following.

(1) *Inversion routine*. By means of least squares fitting of the observed Stokes profiles to the theoretical ones, calculated with Unno's (1956) algebraic solution or a more elaborate solution of the transfer equations of Stokes parameters, it is possible to select the most probable values of B, γ and χ. The first attempt is the AHH process, established by Auer, Heasley and House (1977). Because Unno's solution is rather crude, Landolfi, Landi Degl'Innocenti and Arena (1984) improved this technique by taking into consideration magneto-optical effects, damping, inhomogeneity of the magnetic field and some other factors, and set up the ALL process. This method has been further developed by Skumanich and Lites (1987). As an application of this technique, we may mention the work of Arena, Landi Degl'Innocenti and Noci (1990). They performed a nonlinear least squares fit of the Stokes profiles of a sunspot observed by Stokes-II with the analytical solution of Unno and Rachkovsky. In this process the following eight parameters were determined:

(i) η_0,

(ii) $\Delta\lambda_B$,

(iii) $\Delta\lambda_D$,

(iv) γ,

(v) χ,

(vi) damping constant,

(vii) source function,

(viii) wavelength of the center of the line.

It is impressive that, besides the magnetic field, the main characteristics of the radiation and velocity fields can also be deduced.

(2) *Characteristic feature technique*. We (Ye Shi-hui and Jin Jie-hai, 1983a) established another method, which allows one to determine the magnetic vector with some features of the Stokes profiles. To be precise, our technique consists of the following three steps:

(i) Determination of B with the wavelength distance of $V_{\max}$ from the line center. Using numerical solutions of the Unno–Beckers' equations[3]

[3]See Section 4.2.2.

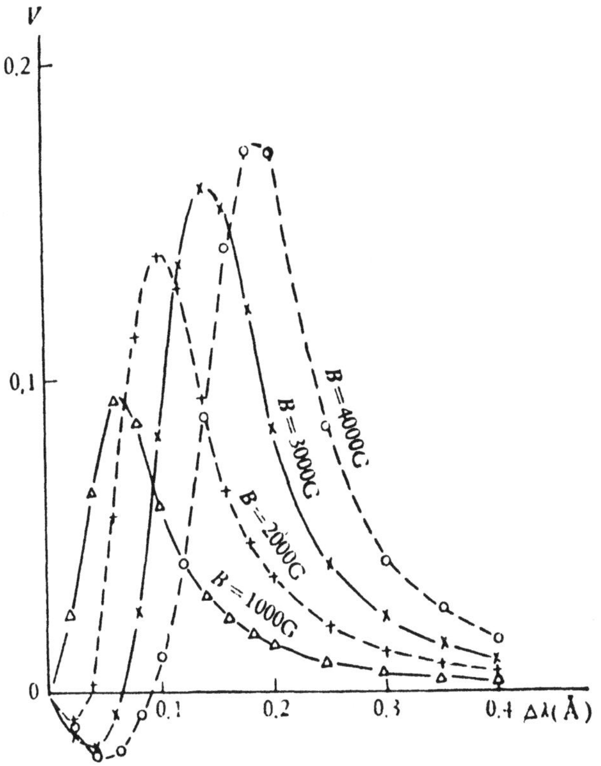

Fig. 3.26. Change of V-profile with B, calculated with Kneer's model and $\theta = 0°$, $\gamma = \pi/3$, $\chi = 0$ for Fe I λ6302.

we calculated a large number of Stokes profiles with which some regularities can be found. As shown in Figures 3.26 and 3.27, the maximum of the V-profile increases with B but depends very little on the variation of γ. Then for a certain value of γ (e.g. $\gamma = 0$), we can get the relation between δ (the wavelength difference between the maximum of V and the line center) and B (see Figure 3.28). When the value of δ is measured with an observed V-profile, then the magnetic field strength B may be readily found.

(ii) Inference of γ from V_{max}. Figure 3.29 shows the relations between V_{max} and γ for various Bs. So when B is known already, one may easily get the value of γ with the V_{max} measured with the observed V-profile.

(iii) Derivation of χ from Q_0. It is evident in Figure 3.30 that Q_0 is very sensitive to the change of χ. So the former can be taken to be the characteristic quantity of the latter. When both B and γ are known from the above two steps, one may use an appropriate curve of the type shown

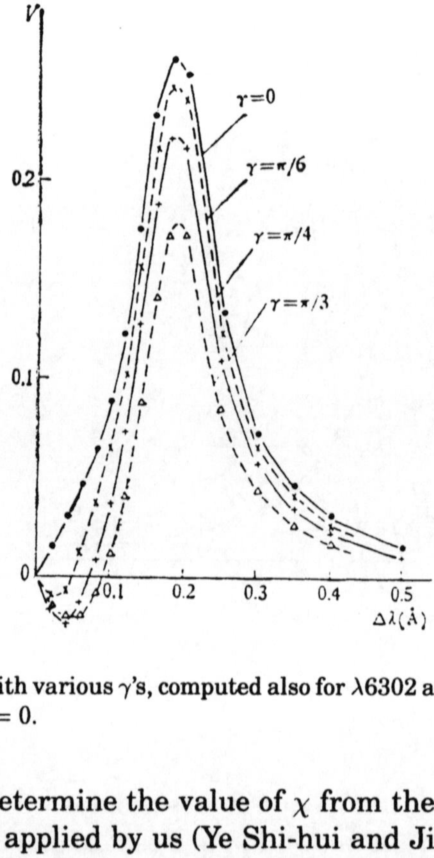

Fig. 3.27. V-profiles with various γ's, computed also for $\lambda 6302$ and with Kneer's model. $\theta = 0°$, $B = 4000$ G, $\chi = 0$.

in Figure 3.31 to determine the value of χ from the observed Q_0. The above method was applied by us (Ye Shi-hui and Jin Jie-hai, 1985) to the Stokes profiles of two large unipolar spots observed at the Okayama Astrophysical Observatory. These observational data had been treated earlier by Kawakami (1983) with the eigenvalue technique invented by Makita (1979). It is interesting to note that our results differ greatly from those of Kawakami. In particular, our Bs are about twice as large as his. Checking with direct measurements by Zeeman splittings and the empirical relation between Bs and spot areas, we have found that our values of B are closer to reality.

3.5.3 Stokes Polarimetry

This is a new area in the study of cosmic (in particular, solar) magnetism. Its goal is to infer from Stokes profiles the structure, filling

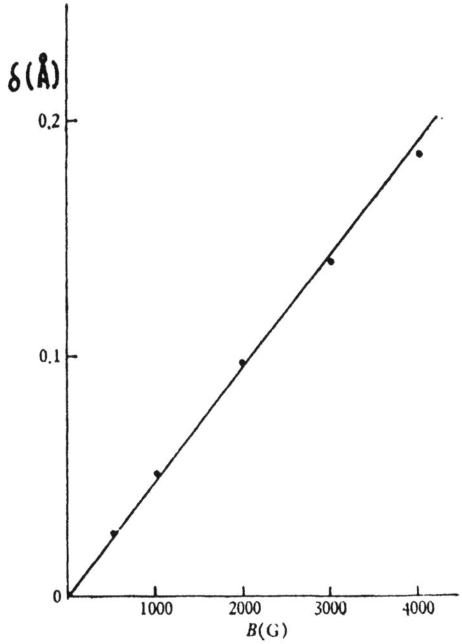

Fig. 3.28. Relation between δ and B for $\gamma = 0$.

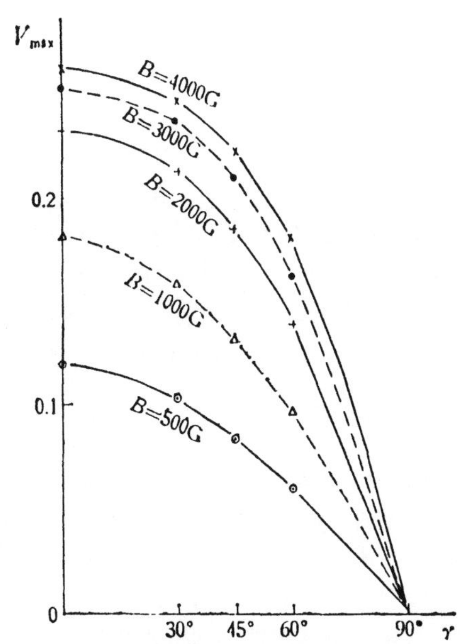

Fig. 3.29. Relations between V_{max} and γ for different B's.

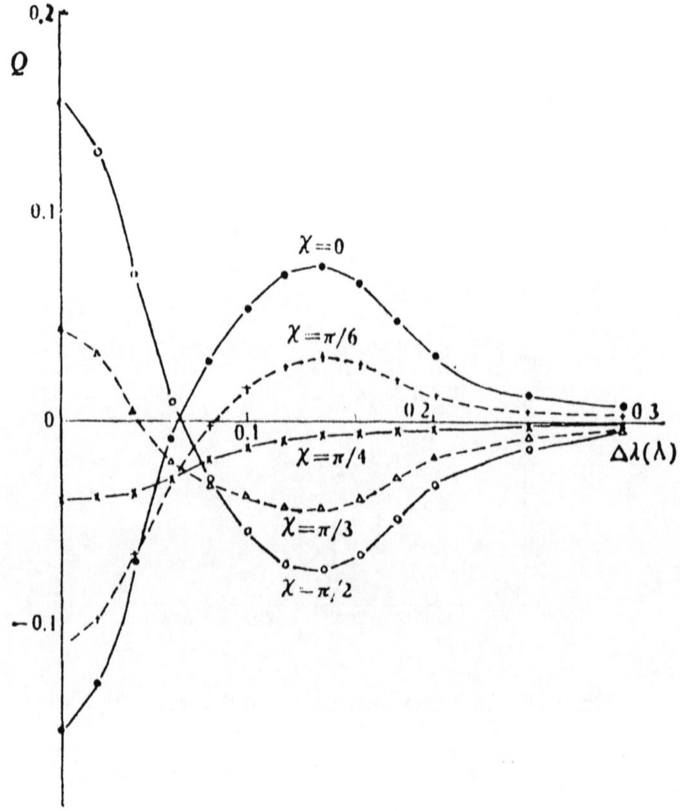

Fig. 3.30. Change of the Q-profile with χ, calculated with $\theta = 0$, $B = 3000$ G and $\gamma = \pi/4$.

factor and physical parameters of the magnetic elements or fluxtubes, which cannot be discerned with existing telescopes. Detailed accounts of this branch of science can be found in, e.g. Stenflo (1985a, b; 1986a), and what follows is merely a very brief sketch.

As stated in Section 3.4.2, the fundamental difficulty of solar vector magnetometry lies in the existence of unresolved fine magnetic structures. The essence of the problem, as we have stressed, is that the influences of fine structures on the measurements of $B_{\|}$ and $B_{\perp}$ are quite different. Now we want to emphasize that by use of the Stokes polarimetry this difficulty may be expected to be overcome. So we can say that this new branch of science has great potentialities and bright prospects in the investigation of celestial magnetic fields.

As an example of Stokes polarimetric studies, we would like to describe the *line-ratio technique*. A pair of spectral lines, which are formed

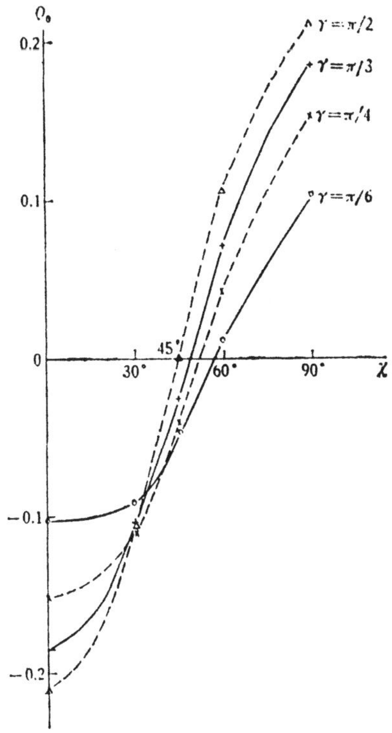

Fig. 3.31. Relation between Q_0 and χ for different γ's.

under practically identical physical conditions, may be chosen for a comparative study. For instance, both Fe I 5247.0585 Å and 5250.2171 Å belong to the first multiplet of neutral iron, and their intensities and excitation potentials are nearly the same. The unique difference is that their Landé splitting factors are unequal, i.e. $g = 2.0$ and 3.0, respectively. Then by use of Eqs. (3.18) and (2.1) there should be $V_{5250} : V_{5247} = 3 : 2$. Nevertheless, the observed V-profiles (see Figure 3.32) of a bright plage show that this ratio is close to unity. This implies that more accurate calculations are needed. Stenflo took more terms in the Taylor expansions of (3.18) and for a series of field strengths computed the ratio $V_{5250}/1.5V_{5247}$ and its change with wavelength. As can be seen in Figure 3.33, for $B = 500$ G the theoretical curve is approximately close to the horizontal straight line, which represents the case $V_{5250}/1.5V_{5247} = 1$. So for comparatively weak fields this equation is roughly true. But with the increase of B, the deviation becomes quite large and the observed curves are near to the theoretical one of $B = 1500$ G. This is much larger than the commonly adopted values

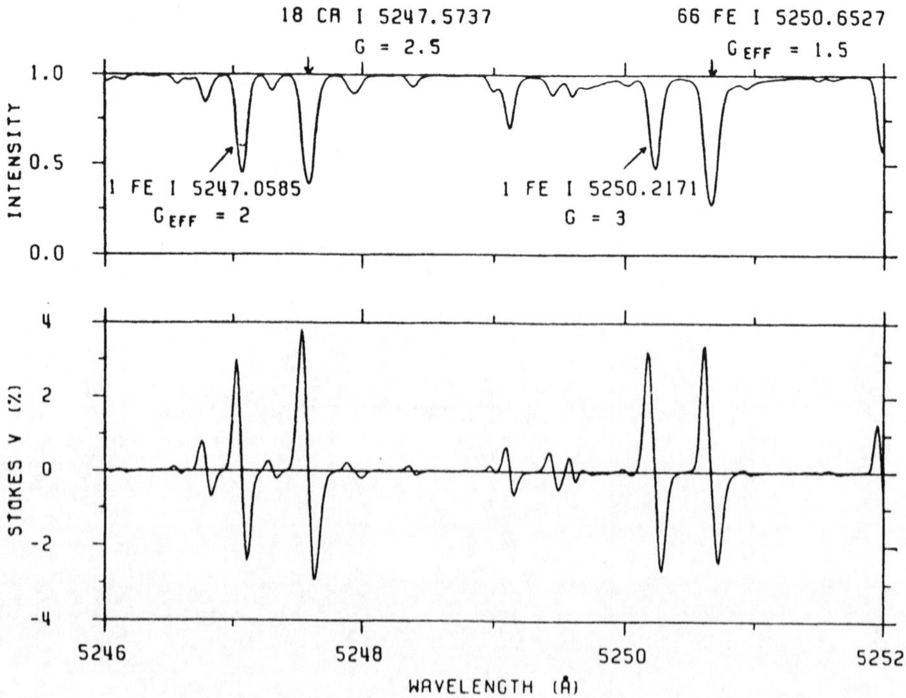

Fig. 3.32. *I*- and *V*-profiles (upper and lower panels, respectively) of Fe I λλ5247 and 5250 observed in a bright plage.

for plages. Therefore, Stokes polarimetry can provide us with more accurate results.

Apart from the line-ratio technique, we may mention the derivation of the change of *B* with height in fluxtubes from the variation of distance between the two peaks of *V*-profile on the solar disk, and also the inference of motions in fluxtubes from the change of *V*-profile asymmetry across the disk of the Sun.

The merits of the Stokes polarimetry may be briefly explained as follows. First, very much information can be extracted from Stokes profiles. Their broadening, asymmetry, distance between peaks, variation on the solar disk, etc. can all be used for analyses. So we may infer from them not only the components of the magnetic vector, but also some thermodynamic parameters. Second, the *Q*-, *U*- and *V*-profiles can be thought to be formed with radiation coming almost solely from flux tubes or magnetic elements, while the intermediate areas make practically no contribution. (By contrast, the *I*-profiles, with which the ordinary magnetometry is performed, are concerned with every corner of the observed region). Third, by separate analyses of *Q*-, *U*- and

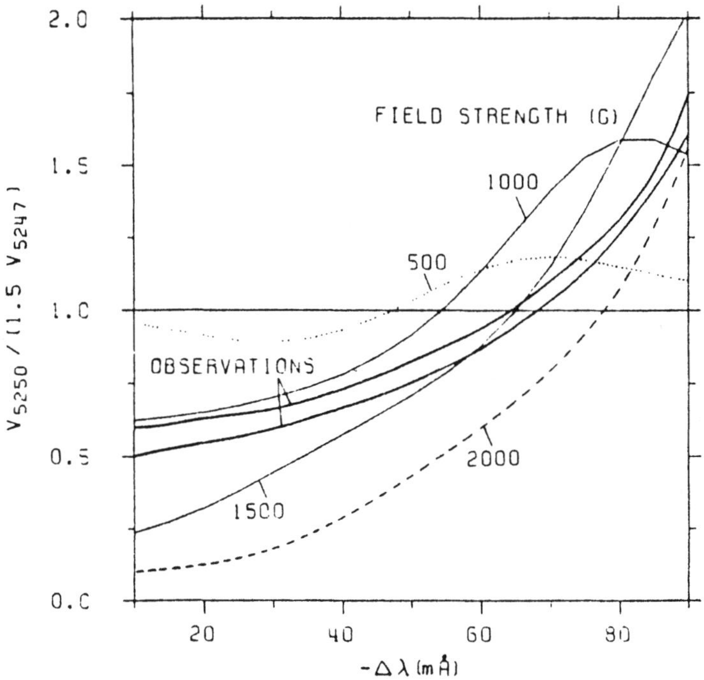

Fig. 3.33. Observed and theoretical fractions $V_{5250}/1.5V_{5247}$ as functions of wavelength.

V-profiles the longitudinal and transverse fields can be studied independently and so the difficulty due to different effects of magnetic fine structures on them may to a great extent be avoided.

In conclusion, we would like to say that in the near future the measurements with stokesmeters and analyses of observational data by Stokes polarimetry may become the mainstream in solar and even cosmic magnetometry.

3.6 Influence of Stray Light and Instrumental Polarization

The measurements of celestial magnetic fields are indeed difficult. In addition to the difficulties in instrumentation and theory, there are some external sources of error which should be excluded in the process of observation. Two such important factors are the *stray light* and *instrumental polarization*.

There are many kinds of stray light and the causes of their formation are rather complex. Generally speaking, due to the influence of the aureole, atmospheric agitation and scattering within the imaging system of the instrument, the light at a certain point of the solar image formed before the magnetograph entrance slit does not exclusively belong to the corresponding place on the Sun, but also comes from some other parts of the solar disk. For instance, a fraction of radiation of the sunspot umbra is diffused to the outside, while some light from the penumbra and the neighboring photosphere can be scattered into the umbra. If the proper umbral magnetic field is purely longitudinal, the scattered light may add a transverse component to the umbral field. So a false π-component appears in the observed pattern of Zeeman splitting. The photospheric natural light may also be scattered into the umbra and the unshifted line can be confused with the π-component. All these phenomena make the magnetometry of sunspots quite complicated. The correction for the stray light can be realized by the following method, due to Wanders (1934). The relation between the *intrinsic intensity* J and the *observed intensity* I is given by

$$I(x,y) = \int_{-\infty}^{\infty}\int_{-\infty}^{\infty} J(\xi,\eta)U(\xi - x, \eta - y)\,\mathrm{d}\xi\mathrm{d}\eta, \tag{3.39}$$

where (x, y) are the coordinates of the point under consideration and (ξ, η) are those of an arbitrary point. The *spread function* U may be taken to be an exponential function, i.e.

$$U(\xi - x, \eta - y) = \frac{a}{\pi} \exp a\left[-(\xi - x)^2 - (\eta - y)^2\right], \tag{3.40}$$

where a is the *coefficient of blurring*. Its value depends on the atmospheric conditions of the observing site and also on the instrument. It can be determined with the intensity distribution of the solar radiation and aureole across the solar limb. To be precise, if there were neither atmospheric agitation and scattering nor instrumental scattering, the intensity at the very limb should change abruptly, as shown by the line 1 in Figure 3.34. In reality, however, due to agitation and scattering, the intensity distribution is represented by one of the curves 2, 3, 4, ..., and each of them corresponds to a definite value of the coefficient a. Therefore, by actual measurements of the intensity distribution near the solar limb one may determine the value of a.

In fact, the forms of sunspots and their magnetic regions may be quite irregular, and magnetic fields are very probably inhomogeneous. Hence the exact correction for the stray light is quite difficult. Nevertheless, thanks to the technical progress in recent years, it is now possible to realize *automatic image restoration* by means of the *discrete*

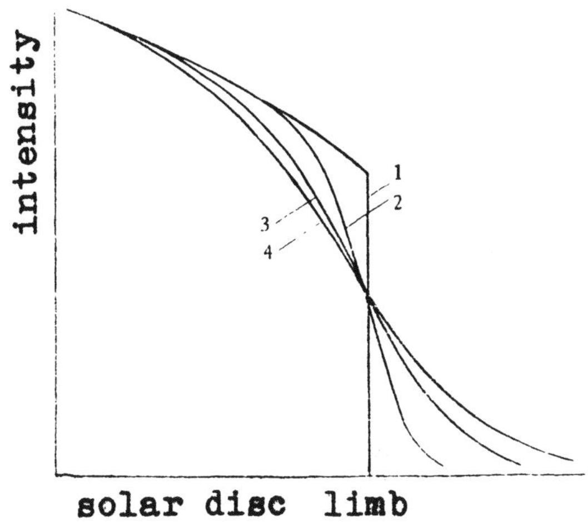

Fig. 3.34. Determination of the coefficient of blurring.

Fourier transform. With a point spread function known from observation, this process can be performed with a computer. After restoration magnetograms have much higher spatial resolution and Figure 3.35 presents an example. (For a detailed description of the principles and technology, see, e.g., Yang Wei-min, 1990.)

It is still more difficult to correct for the influence of instrumental polarization. In most cases the magnetic fields of celestial bodies are observed with the polarization of their radiation. In the process of polarimetric measurements, one has to consider the additional, parasitic polarization due to the oblique reflection of light on mirrors, their metallic coatings, the defects of lenses, etc. When this kind of polarization is mixed with that of the radiation of heavenly bodies, large errors may arise in magnetometry. (More detailed descriptions have been given by, e.g., Dombrovskij (1973), Gehrels (1974) and Stenflo (1976).) Besides the number of mirrors, the amount of instrumental polarization increases with the angle of reflection. The coelostat of the Ondrejov Observatory may be taken as an instance. Near midday its instrumental polarization occupies only about 1% of the total observed polarization, but at large hour angles the percentage may be as high as 6% (Bumba and Topolova-Ruzickova, 1962). Therefore, it is advisable to make observations at noon, to diminish the number of mirrors and to use the refracting system. However, the instrumental polarization can be compensated for with a tilted plane-parallel glass plate before the polarization analyzer. It is also feasible to calculate the amount of instrumental polarization at the time of observation. The method, adopted at the Sayan Observatory (Grigoriev *et al.*, 1980; Grigoriev, 1991), can be briefly described as follows.

The polarization due to the telescope may be expressed with a matrix of the fourth rank $\mathbf{L} = \{\alpha_{ij}\}$. After passing through the telescope, the incident radiation $\mathbb{I}_0 = \{I_0, Q_0, U_0, V_0\}$ becomes

$$\mathbb{I} = \begin{vmatrix} I \\ Q \\ U \\ V \end{vmatrix} = \mathbf{L}\mathbb{I}_0 = \begin{vmatrix} \alpha_{11}I_0 + \alpha_{12}Q_0 + \alpha_{13}U_0 + \alpha_{14}V_0 \\ \alpha_{21}I_0 + \alpha_{22}Q_0 + \alpha_{23}U_0 + \alpha_{24}V_0 \\ \alpha_{31}I_0 + \alpha_{32}Q_0 + \alpha_{33}U_0 + \alpha_{34}V_0 \\ \alpha_{41}I_0 + \alpha_{42}Q_0 + \alpha_{43}U_0 + \alpha_{44}V_0 \end{vmatrix}. \tag{3.41}$$

In order to determine the coefficients α_{ij}'s, the polarized light beams

$$\mathbb{I}_1 = \begin{vmatrix} I \\ 0 \\ 0 \\ 0 \end{vmatrix}, \quad \mathbb{I}_2 = \begin{vmatrix} I \\ I \\ 0 \\ 0 \end{vmatrix}, \quad \mathbb{I}_3 = \begin{vmatrix} I \\ 0 \\ I \\ 0 \end{vmatrix}, \quad \mathbb{I}_4 = \begin{vmatrix} I \\ 0 \\ 0 \\ I \end{vmatrix} \tag{3.42}$$

are successively sent into the telescope and the signals from four channels of the magnetograph $\{I, Q, U, V\}$ are recorded. Then by use of Eq. (3.41) one has a system of 16 equations, which can be solved with the 16 coefficients α_{ij}'s. With the known matrix **L** as well as the measured $\mathbb{I}$, one may calculate $\mathbb{I}_0$ with Eq. (3.41).

A similar method has been applied by Makita *et al.* (1991) to the Domeless Solar Telescope of the Hida Observatory.

3.7 Selection of Magneto-Sensitive Spectral Lines

The contemporary observations of cosmical magnetic fields are mainly based on the Zeeman effect of spectral lines. It is well known that there are many lines in astronomical spectra. For example, in the visible region of the solar spectrum the total number of lines exceeds 24,000. So it is necessary to consider which lines are most suitable for celestial magnetic measurements.

Of course, we prefer to the spectral lines which display conspicuous Zeeman effect, i.e. magneto-sensitive lines. Moreover, we are interested chiefly in those lines that exhibit the normal Zeeman effect. This is because at present it is only for them that we have a well developed theory of formation in magnetic fields. Although a few lines displaying an abnormal Zeeman effect are sometimes used (for instance, in the

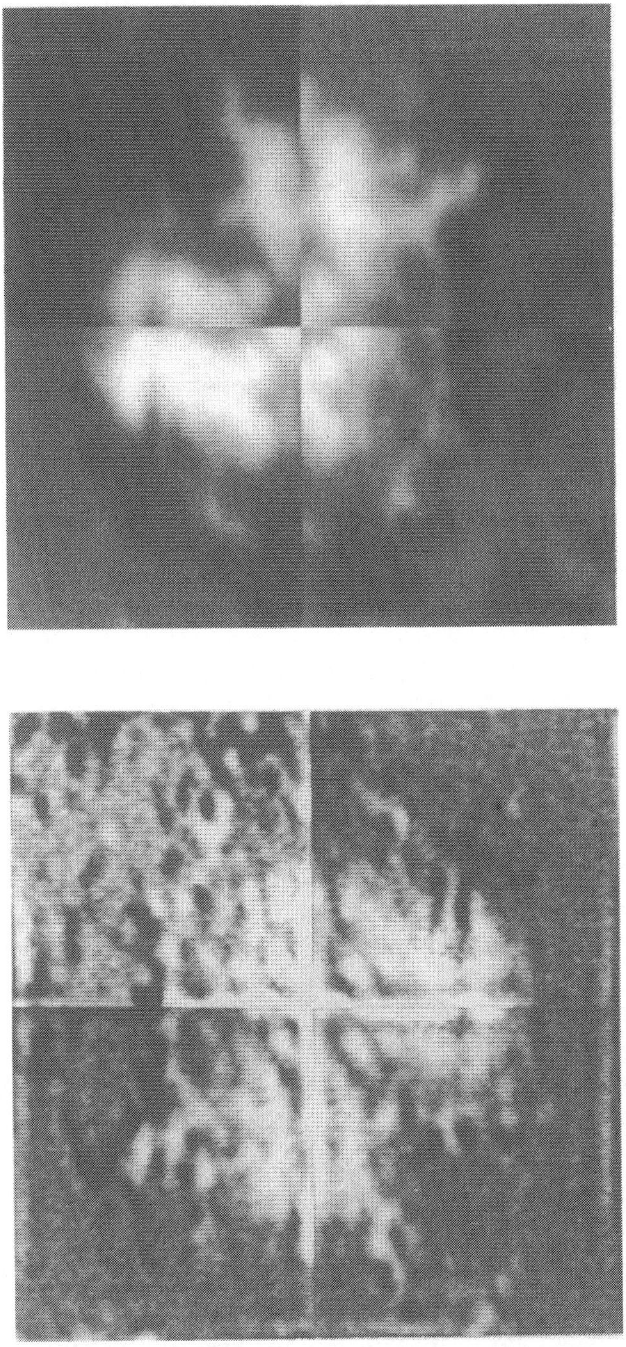

Fig. 3.35. An observed magnetogram of a sunspot (upper panel) and the corresponding restored magnetogram (lower panel). Each picture has been compiled from four separate pictures.

magnetometry of solar prominences the strongest emission line Hα is often adopted), it is hard to give a quantitative interpretation of the observational data. The spectral lines exhibiting the normal Zeeman effect, i.e. Zeeman triplets, are still too many. In order to select the most appropriate lines, one has to make a further choice according to the following principles:

(1) As shown by formula (2.1), the amount of Zeeman splitting $\Delta\lambda_B$ is proportional to the Landé factor g which represents the magnetic sensitivity of a spectral line. So it is profitable to use lines with as large a g as possible. Generally speaking, the lines with $g \geq 2.5$ are sensitive enough to magnetic fields.

(2) Also by (2.1), $\Delta\lambda_B$ is proportional to λ^2. Hence, it is advantageous to utilize lines in the red and infrared regions. In recent years the near infrared range around 1.6μ has aroused much interest among astronomers (Stenflo, 1992a). This is because the opacity minimum is located in this spectral region. This implies that we can see deeper into the atmosphere (and we may then observe more intense magnetic fields) than in any other parts of the solar spectrum. Besides, this region is readily accessible to fairly standard polarimetric techniques.

(3) It is desirable that line profiles are both sharp and deep, so that the Zeeman splitting or broadening is more easily detected.

(4) In order to avoid systematic errors, the selected line should not be blended with neighboring ones. Moreover, it is expected that the former is rather distant from the latter.

(5) The line of interest is sensitive solely to the magnetic field and not to the changes of all other physical parameters, such as temperature. Otherwise the variations of line profiles and equivalent widths are not purely due to the presence of the magnetic field and this may further complicate the interpretation of the observational material.

Bearing these principles in mind, some astronomers have performed selections of the working lines of magnetometry. As far as we know, the earliest work was done by von Klüber (1948). In the spectral range 5000–9000 Å he chose eight lines, the main data of which are shown in Table 3.1 of this book. Among these spectral lines the most important are the three iron lines $\lambda\lambda$5250, 6173 and 6302, and they have been widely used for many years. Until the present, von Klüber's selection is still effective, but the following supplements have to be made:

(1) In recent years there have been some more comprehensive and detailed studies. For instance, Harvey (1973) compiled a large table of magneto-sensitive lines. Besides, it has been found that many generally used working lines, including the three iron lines just mentioned, suffer more or less from blending with neighboring lines.

(2) For the study of some non-magnetic phenomena, e.g. the Evershed effect, we have to exclude the influence of the magnetic field. In such cases one selects certain lines which are completely insensitive to magnetic fields, i.e. lines with $g = 0$. These lines have been tabulated by Sistla and Harvey (1970).

(3) As stated above, Fe I $\lambda5250$ is one of the three lines most widely used in solar magnetometry. Nevertheless, it is a pity that this line has been found to be rather sensitive to temperature. Concretely speaking, owing to the temperature difference of 1500 K the equivalent width of this line in the sunspot umbra is 2.42 times larger than that in the photosphere. For comparison, the corresponding ratio for $\lambda6302$ is a mere 1.09 (Ye Shi-hui, 1991). It is interesting that there are some other lines with temperature sensitivities even much higher than 5250 Å. One example is VI $\lambda6256.887$. For this line the above ratio is as large as 48!. Therefore, $\lambda5250$ deviates greatly from the fifth principle described above and so the observational material obtained with this line may be doubtful. According to Harvey and Livingston (1969) and Wittmann (1971a), the magnetic field strengths given by this line are much smaller than the intrinsic values and the difference may reach as much as 2–5 times!

The use of $\lambda5250$ in astronomical magnetometry provides a historical lesson. It reminds us that much attention should be paid to the selection of magneto-sensitive lines.

CHAPTER 4

The Magnetic Fields of Sunspots

The sunspot phenomenon, with its reduced temperature and intense magnetic field, is at present without a scientific explanation.
E.N. Parker, *Cosmical Magnetic Fields* (1979).

As we stated in Chapter 1, the earliest discovery of the cosmic magnetic field outside the Earth was achieved for sunspots by Hale (1908) at the Mt. Wilson Observatory. In the thirty more years after this, observations of this kind were carried on solely at this institution. Since 1941 and 1955, respectively, the Potsdam and Crimean Astrophysical Observatories also began to make routine measurements. In the sixties the Beijing, Purple Mountain and Yunnan Observatories of Academia Sinica successively started regular observations of sunspot magnetic fields. In the past eighty more years a large amount of observational material has been accumulated[1] and on this basis much interesting theoretical research has been performed. However, in spite of the tremendous progress, there are still many unresolved problems and puzzling contradictions in the study of sunspot magnetism.

[1]The observational data of Mt. Wilson can be found in *Publications of the Astronomical Society of the Pacific* and those of Potsdam, *Mitteilungen des Astrophysikalischen Observatorium Potsdam*. The Soviet/Russian measurement results are regularly published in *Solnechnye Dannye Byulleten*. The observational material obtained by six observatories in various countries during the International Geophysical Year was published in 1962 as Volume 23 of the IGY Annals (Stepanov *et al.*, 1962). The *Chinese Solar Geophysical Data* (Beijing Observatory) and *Monthly Solar Activity* (Yunnan Observatory) also contain the sunspot magnetic field data.

4.1 Basic Characteristics

(1) *The polarities of sunspot magnetic fields and laws of their arrangement and variation.* After many years of routine observations, Hale and his associates discovered the following regularities:

(i) Spots often appear in groups. Generally speaking, there are two large spots in a mature group. The one which is situated in the western part of the group, i.e. on the front side in the direction of solar rotation, is called the *preceding spot*, and the other spot which is at the opposite position, the *following spot*. As a rule, these two spots possess different magnetic polarities, i.e., one is N and another is S.

(ii) The solar surface can be divided into northern and southern hemispheres. In the same hemisphere almost all the preceding and following spots have respectively the same polarities, while in the other hemisphere the polarity arrangement is just the contrary.

(iii) Near the end of every *11-year solar activity cycle*, most of the spots belonging to the old cycle are located in low-latitude regions on both sides of the equator, while spots of the new cycle appear in high-latitude districts not far from the poles. It is interesting to note that in each hemisphere the spots belonging to various cycles have different polarity arrangements. This implies that after each 11-year cycle the distribution of spot magnetic polarities is reversed. Hence sunspots are said to display a *22-year magnetic cycle*.

Strictly speaking, it is not necessary that every spot group obeys this regularity. According to the statistics of Hale (1925), among the 1735 groups observed in 1908–1925 there were 41 exceptions.

The interpretation of the laws mentioned previously is concerned with the mechanism of solar activity and will be described later, in Section 11.3.

(2) *Relation between the area and magnetic field strength of spots.* The magnetic field is the most important characteristic of sunspots. All the spots possess magnetic fields. As far as we know, the strongest field ever observed is 4300 G and the weakest one about 100 G. According to the statistics of Ringnes and Janssen (1960), only about 20% of sunspots have magnetic fields more intense than 2000 G and 5% more than 3000 G. Roughly speaking, the larger a spot, the stronger its magnetic field. Some astronomers (e.g., Nicholson, 1933; Houtgast

and van Sluiters, 1948; Mattig, 1953; Ringnes and Janssen, 1960) studied the relation between areas and maximum field strengths of spots. They obtained basically the same result, which may be expressed by the following empirical formula:

$$B_m = 3700 \frac{A}{A+66} \,. \tag{4.1}$$

Here B_m is the maximum magnetic field strength of a spot (in gauss) and A is the spot area (in units of one millionth of the visible solar surface). According to this formula the strongest magnetic field strength of sunspots could be only 3700 G, and this is not the reality. So Eq. (4.1) is merely an approximate expression.

It is interesting to mention the recent work of Osherovich and Garcia (1989). They found that the relative size of the umbra of a spot is connected with the magnetic field strength at the outer boundary of the penumbra. According to their *return flux model* (Osherovich, 1982) the sunspot magnetic field is composed of two regimes. One consists of open field lines emerging from the core and extending to infinity, while the other consists of lines returning to the photosphere. The separative surface dividing these regimes is identified with the boundary between the umbra and the penumbra. Therefore, the relative quantity of return magnetic flux is closely associated with the size of the umbra compared to that of the penumbra. The essential result of their work is that small umbrae or, equivalently, large penumbrae (relative to total spot areas) are associated with large magnetic fluxes and with intense magnetic fields on the outer penumbral boundaries. Besides, their theory provides mathematical expressions for these quantities as functions of the relative size of the umbra. This will be discussed later in this section.

(3) *Relation between magnetic field strengths and heliocentric angles of sunspots*. Strictly speaking, Eq. (4.1) can be used only for spots in the central part of the solar disk. Some observations (e.g., Houtgast and van Sluiters, 1948; Cowling, 1946) show that when spots are close to the solar limb, i.e. when their heliocentric angles are large, their B_m's become smaller. For this there are two possible explanations.

(i) When spots are near the limb, the influence of scattered light is larger and so the observed magnetic field strengths diminish.

(ii) The spot magnetic field has a certain gradient. The observed radiation of spots near the limb comes from higher layers where the magnetic field is weaker.

The two causes may coexist at the same time.

(4) *Distribution of magnetic field strength in unipolar spots.* Inside the same spot the magnetic field strengths and directions of various regions may be quite different. At present, research is restricted mainly to unipolar spots. In this section we discuss only the distribution of spot magnetic field strength on the solar surface. As for the change of field intensity with height and the variation of field direction with location in the spot, these topics will be treated in Sections 6.4 and 4.2, respectively.

On the basis of observations, Hale and Nicholson (1938) found that, generally speaking, the magnetic field strength is largest at the spot center, diminishes outward, and drops to a very small value at places not far outside the penumbral boundary. Further measurements show that the field intensity does not vary much in the realm of the umbra but changes abruptly in the penumbra. In order to express the distribution of field strength in unipolar spots, many astronomers have successively proposed empirical formulae.[2] Some of these are the following ones.

(i) Hale and Nicholson (1938):

$$B(\rho) = \frac{B_0}{1+\rho^2}. \tag{4.2}$$

Here ρ is the distance of an arbitrary point from the spot center in units of the spot (penumbral) radius and B_0 is the field strength at the spot's center.

This formula was also obtained by Beckers and Schröter (1969b).

(ii) Broxon (1942):

$$B(\rho) = B_0(1-\rho^2). \tag{4.3}$$

(iii) Alfvén (1950):

$$B(\rho) = \frac{a}{\rho^3}\sqrt{1+3\sin^2\varphi}\,, \tag{4.4}$$

where φ is the angle between the magnetic vector and the solar surface and a is the moment of the magnetic dipole.

(iv) Mattig (1953):

$$B(\rho) = B_0(1-\rho^4)\exp(-2\rho^2). \tag{4.5}$$

[2]Except Alfvén's formula which is a theoretical one.

(v) According to Eqs. (4.3) and (4.5), the spot magnetic field should vanish at $\rho = 1$, and this is not true. Bumba (1960a) measured the field intensity outside the outer penumbral boundary and found it to be about 300 G. Besides, this value does not depend on the size of the spot. He proposed the following expressions:

$$B(\rho) = -\frac{1}{\rho^3}, \qquad \log B_0 = 0.27 \log a - 4.82. \tag{4.6}$$

(vi) Wittmann (1974):

$$B(\rho) = B_0(1 + \rho^2 + \rho^4 + \rho^8 + \rho^{16})^{-1}. \tag{4.7}$$

(vii) Gurman and House (1981):

$$B(\rho) = B_0(0.880 - 1.07\rho^2 + 0.416\rho^4 - 0.063\rho^6). \tag{4.8}$$

(viii) Kawakami (1983):

$$B(\rho) = B_0(1 + \rho^2)\exp(-\rho^{2.5}). \tag{4.9}$$

Finally, it is interesting to mention that Osherovich and Garcia (1989) derived the following theoretical relation between the magnetic field strengths at the spot center (B_0) and on the outer penumbral edge (B_p) as a function of the ratio of the penumbral and umbral radii (R), i.e.,

$$B_p = B_0 \frac{\sqrt{2(R^2 - 1)}}{R} \exp\left[1 - \frac{R^2}{(R^2 - 1)}\right]. \tag{4.10}$$

This formula was obtained with the *return flux theory* under the assumption that the vertical component of the spot magnetic vector vanishes on the outer penumbral boundary. As described in their work, the theoretical curve $B_p/B_0 \sim R$ calculated with the above equality agrees well with observation.

(5) *Magnetic flux of sunspots.* The total magnetic flux of a spot can be expressed as

$$\Phi = \int B_z \, \mathrm{d}A \equiv 2\pi \int_0^b B(r) \cos \delta r \, \mathrm{d}r \equiv kB_0 A, \tag{4.11}$$

where r is the distance of an arbitrary point from the spot center, b is the spot radius, δ is the angle between the magnetic line of force and the normal to the solar surface and the integration is performed for the whole spot area A. When we use Broxon's formula (4.3) and take $\delta = (r/b)70°$, it is easy to get $\Phi = 0.38B_0A$, i.e. $k = 0.38$. For Mattig's formula (4.5), we have $k = 0.28$; and for that of Hale and Nicholson (4.2), $k = 0.53$.

As stated above, in the return-flux model it is suggested that a part of the sunspot magnetic flux returns to the photosphere. According to the theoretical derivation of Osherovich and Garcia (1989), the amount of this part of the flux (Φ_{RF}) relative to the total flux (Φ_{tot}) is a function of the parameter R defined above. The mathematical expression is

$$\frac{\Phi_{\mathrm{RF}}}{\Phi_{\mathrm{tot}}} = \frac{1}{\mathrm{e}}(R^2 - 1). \tag{4.12}$$

In comparison with other solar active features, spots possess rather long lives and in the general case they can exist for several days and even tens of days. In this time interval, with the evolution of a sunspot, its magnetic flux rapidly attains its maximum and then steadily decreases approximately according to a linear law (Cowling, 1946).

4.2 Orientation

In order to obtain the complete information on the spot magnetic vector, one should determine not only its strength, but also its direction. In the following two subsections we shall discuss this problem in the 2-D and 3-D cases, respectively.

4.2.1 Angle Between Magnetic Lines of Force and Line of Sight

By measuring the Zeeman splitting one can only obtain the total magnetic intensity, while by magnetographic observations only the component along the line of sight can be obtained. In order to gain a comprehensive view of the magnetic vector, one has still to know its orientation in space. For the 2-D case, it is necessary to measure the angle γ formed by the magnetic vector $\mathbf{B}$ and the line of sight. (See the projection of $\mathbf{B}$ on the xoy-plane of Fig. 3.12). For this the following methods have been established.

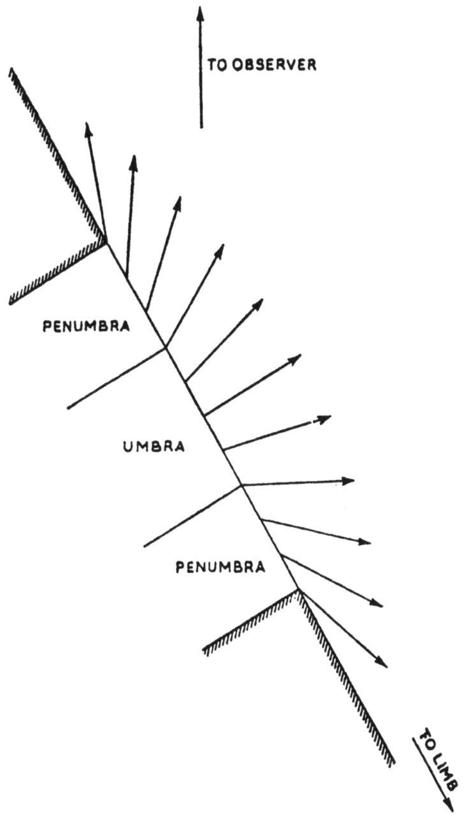

Fig. 4.1. Dividing line of polarities in a unipolar spot.

(1) Hale and Nicholson (1938) used Seares' formula to compute γ with the relative intensities of Zeeman components. According to our study (Ye Shi-hui *et al.*, 1978b) briefly described in Section 3.1.4, this method is in principle incorrect and may yield erroneous results.

(2) In accordance with the classical model, which will be described in Section 4.4, magnetic lines of force in the umbra are approximately perpendicular to the solar surface and those in the penumbra are strongly inclined. The larger is the distance from the spot center, the more inclined are the lines of force (Figure 4.1). Hence, when the spot is close to the solar limb, the magnetic polarity in the spot region nearer to the limb may change sign. In other words, it might seem to the observer that one and the same spot contained two magnetic regions with different polarities. Moreover, the closer the spot is to the solar limb, the nearer is the line dividing different polarities to the spot center. Therefore, one can determine γ with the distance of the dividing line from the spot center. The results of measurements by Hale *et al.* (1918) for

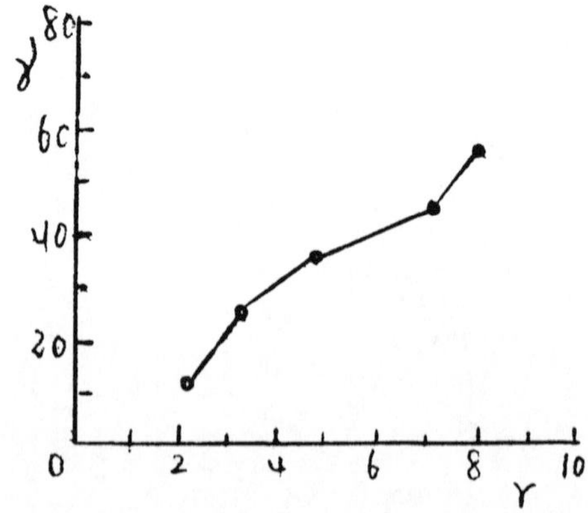

Fig. 4.2. Determination of γ with distance of polarity dividing line from the spot center.

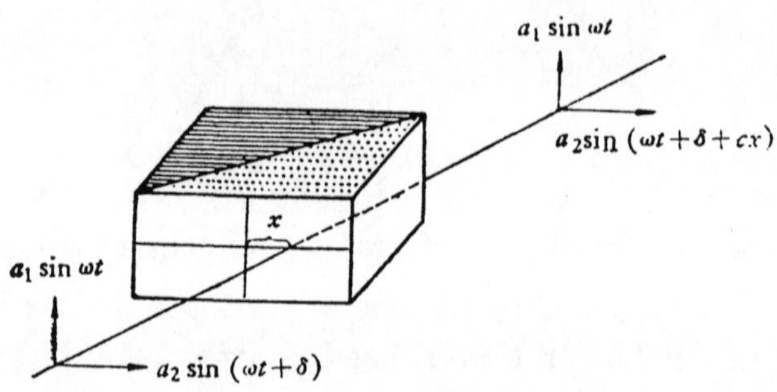

Fig. 4.3. Babinet compensator.

a lot of spots are shown in Figure 4.2, in which the penumbral radius is taken to be 10 units.

(3) Treanor (1960) devised a new method and applied it to actual observations at Oxford. Its principle is as follows. For $\gamma = 0$ (the longitudinal Zeeman effect), the two σ-components are circularly polarized. But for $0 < \gamma < 90°$, they are elliptically polarized. So it is natural that the ellipticity is related to γ. When the phase difference of the σ-components has been determined with a *Babinet compensator* (Figure 4.3), the value of γ can be obtained. To be precise, the relation between γ and the ellipticity of polarization Ψ ($= \tan^{-1}(B/A)$, A and B being the major and minor axes of the ellipse) is

$$\tan\Psi = \cos\gamma. \tag{4.13}$$

Hence when Ψ is known, γ can be readily calculated.

The Babinet compensator is composed of two quartz wedges with mutually perpendicular optical axes. The angle formed by their surfaces is α. When a beam of elliptically polarized light with components $a_1 \sin\omega t$ and $a_2 \sin(\omega t+\delta)$ passes through the compensator, the first component remains unchanged, while the second becomes $a_2 \sin(\omega t+\delta+cx)$, where $c = (2\pi/\lambda)(\mu_0 - \mu_e)$. Here μ_0 and μ_e are the coefficients of refraction of the ordinary and extraordinary light in the crystal. After passing through the compensator the light beam displays bright and dark fringes. From the interval between neighboring fringes one may determine the phase difference, δ_1. After a rotation of the compensator through angle τ, the phase difference becomes δ_2. The angle then Ψ can then be computed with the following formulae:

$$\begin{aligned} \tan 2\varphi &= \sin 2\tau \left(\frac{\tan\delta_1}{\tan\delta_2} - \cos 2\tau\right), \\ \tan 2\Psi &= \tan\delta_1 \sin 2\varphi. \end{aligned} \tag{4.14}$$

The angle γ can be evaluated by substituting the value of Ψ into Eq. (4.13).

In recent years Treanor's method has been refined and applied to the study of models of sunspot magnetic fields by Adam (1985, 1990).

(4) Nishi (1962) and Nishi and Makita (1973) broke fresh ground in this kind of work. They installed a polarizer before the spectrograph slit and then, for each magneto-sensitive line, e.g. Fe I λ6302, they obtained a set of four spectra. Two of these belong to linearly polarized light with mutually perpendicular directions of polarization, and the others to circularly polarized radiation with opposite directions of polarization. Treating these spectra with the theory of spectral line formation in the magnetic field, they simultaneously obtained the values of $B(r)$ and γ. They found that the distribution of magnetic field strength in unipolar spots agrees with Broxon's formula (4.3), but the direction of the magnetic field differs greatly from those given by other methods. The main difference is that, according to their measurements, the values of γ in the penumbra are close to 90°, i.e. the magnetic lines are almost parallel to the solar surface.

4.2.2 Azimuth

In order to fix the orientation of the magnetic vector in a 3-D space, besides the angle of inclination γ one has still to determine the azimuthal angle χ of the transverse magnetic field (see Figure 3.12). For convenience, many astronomers take this angle to be identical with the azimuth of the plane of polarization φ [$= \frac{1}{2}\tan^{-1}(U/Q)$]. So Unno's assumption of $U = 0$ implies that φ is always equal to zero. In this case the azimuth of the transverse field is thought to be fixed and the 3-D problem is reduced to a 2-D one. Nevertheless, after much controversy the existence of the Faraday rotation in sunspots has been convincingly demonstrated by Makita (1986b) with observations of the Okayama Vector Magnetograph. This means that the plane of polarization is constantly rotating and so its azimuth φ can no longer be considered to be fixed. In such a circumstance χ cannot be identified with φ and should be determined by a specific method.

Under the influence of the Faraday rotation, the azimuth φ and the parameter U cannot be taken to be always equal to zero. In this case one has to solve the following *Unno–Beckers' equations* of transfer of all the four Stokes parameters (Beckers, 1969)[3]:

$$\begin{aligned}
\cos\theta\frac{dI}{d\tau} &= (1+\eta_I)(I-B) + \eta_Q Q + \eta_U U + \eta_V V,\\
\cos\theta\frac{dQ}{d\tau} &= \eta_Q(I-B) + (1+\eta_I)Q + \rho_R U + \rho_W \sin 2\chi V,\\
\cos\theta\frac{dU}{d\tau} &= \eta_U(I-B) - \rho_R Q + (1+\eta_I)U - \rho_W \cos 2\chi V,\\
\cos\theta\frac{dV}{d\tau} &= \eta_V(I-B) - \rho_W \sin 2\chi Q + \rho_W \cos 2\chi U + (1+\eta_I)V.
\end{aligned} \tag{4.15}$$

The parameters which represent magneto-optical effects (including the Faraday rotation) are

$$\begin{aligned}
\rho_R &= \frac{\eta_0}{H_0(0,a)}\cos\gamma[F(v-v_H,a) - F(v+v_H,a)],\\
\rho_W &= \frac{\eta_0}{H_0(0,a)}\frac{\sin\gamma}{2}[F(v-v_H,a) + F(v+v_H,a) - 2F(v,a)],
\end{aligned} \tag{4.16}$$

where $H(0,a)$ is the Voigt function and

$$F(v,a) = \frac{1}{2\pi}\int_{-\infty}^{\infty}\frac{u}{u^2+a^2}\,e^{-(u-v)^2}\,du. \tag{4.17}$$

is the *Faraday-Voigt function*. Moreover,

$$\eta_I = \frac{\eta_p}{2}\sin^2\gamma + \frac{\eta_b+\eta_r}{4}(1+\cos^2\gamma),$$

[3]There are some inexactitudes in the original form of the equations given by Beckers and they have been corrected here.

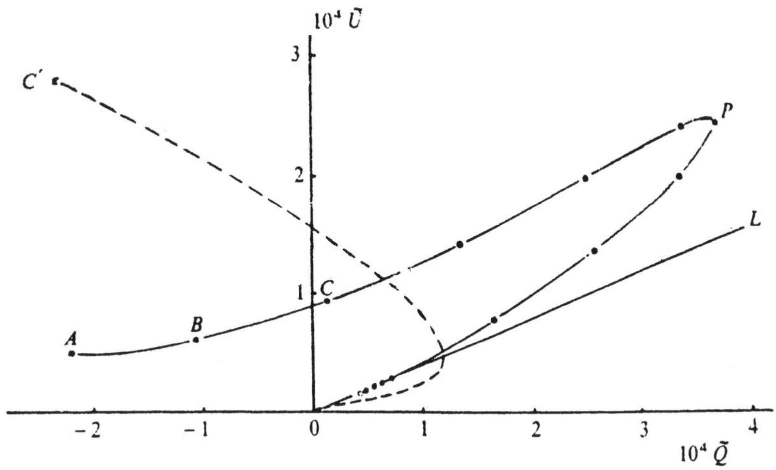

Fig. 4.4. Change of φ from line center to wing.

$$\begin{aligned} \eta_Q &= \left(\frac{\eta_p}{2} - \frac{\eta_b+\eta_r}{4}\right)\sin^2\gamma\cos 2\chi, \\ \eta_U &= \left(\frac{\eta_p}{2} - \frac{\eta_b+\eta_r}{4}\right)\sin^2\gamma\sin 2\chi, \\ \eta_V &= \frac{\eta_r-\eta_b}{2}\cos\gamma. \end{aligned} \tag{4.18}$$

Choosing a certain value for χ, e.g. 10°, adopting Kneer's model of sunspot umbra and taking a series of values for B, θ and γ, we (Ye Shi-hui and Jin Jie-hai, 1987) performed numerical solutions of the system of Eqs. (4.15) and calculated the following quantities defined for a wavelength range $\delta\lambda = 0.01$ Å as

$$\begin{aligned} \tilde{Q} &= \int_{\lambda}^{\lambda+\delta\lambda} Q\,\mathrm{d}(\Delta\lambda), \\ \tilde{U} &= \int_{\lambda}^{\lambda+\delta\lambda} U\,\mathrm{d}(\Delta\lambda), \end{aligned} \tag{4.19}$$

where $\Delta\lambda = \lambda - \lambda_0$ and λ_0 is the wavelength of the line center. Our calculations were made for various spectral regions starting from the line center ($\Delta\lambda = 0$) and moving consecutively to the red wing of Fe I λ5250 until $\Delta\lambda \simeq 0.20$ Å. With these results we constructed the *azimuth diagram* shown in Figure 4.4. (This diagram was first proposed by Makita and the dotted line is the curve suggested by him). The point A represents the line center. For point B, $\Delta\lambda = 0.01$ Å; for C, $\Delta\lambda = 0.02$Å, etc. The angle $\angle$ QOA is equal to twice the azimuth of the plane of polarization of the radiation in the very center of the line, i.e. 2φ ($\Delta\lambda = 0$). This is because, as we mentioned at the beginning of this subsection, φ is equal to one half of the arctangent of the ratio U/Q. From Figure 4.4 we have $\varphi(\Delta\lambda = 0) = 84°$, far larger than the value of χ chosen by us, i.e., 10°. This means that due to the action of the Faraday rotation the observed azimuth of the plane of polarization is much larger than the intrinsic azimuth of the transverse field. Therefore, the former cannot be identified with the latter. However, for points B, C, etc. the difference between these two angles successively diminishes. Beyond the turning point P, the curve connecting A, B, C, ... approaches the asymptote OL. It is interesting that this straight line forms an angle of 20° with the horizontal axis, which is just equal to 2χ. This implies that the effect of Faraday rotation on the radiation from far wings of the line is negligible. Nevertheless, these spectral regions are magnetically insensitive and seldom used in magnetometry. So the better choice is to take measurements in the central part of a magneto-sensitive line and then try to eliminate the influence of the Faraday rotation. For this we suggest obtaining numerical solutions of Unno-Beckers' equations for the spectral line and wavelength region used, and then constructing theoretical azimuth diagrams. Computations should be done for a series of values of both B and γ. When these two quantities are determined from independent observations, one may decide which curve on an azimuth diagram can be used to derive χ from the directly measured φ. In this way it is possible to get the accurate azimuth of the transverse field, free from the influence of the Faraday rotation.

4.3 Fine Structures

It has become increasingly evident, for a long time now, that various solar active features, as well as the undisturbed solar atmosphere, are highly inhomogeneous. In other words, they possess very complicated fine structures. The granules, moustaches, chromospheric spicules, etc. can be taken to be evidence. However, it is a pity that the scale of fine structures is close to and, quite possibly, even much smaller than the limits of resolution of the best existing instruments as well as the atmospheric seeing ($\simeq 0.5$ arcsec). So it is expected that in the near future still finer structures will be discovered. In what follows we enumerate a series of facts which show that sunspot magnetic fields

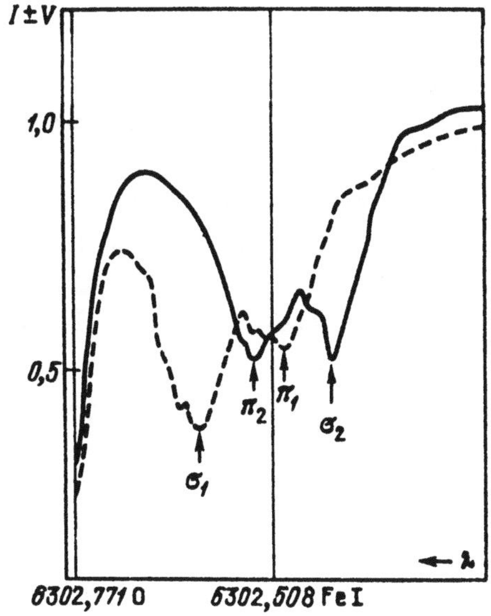

Fig. 4.5. π-component splitting.

also have very delicate structures.

(1) *Splitting of the π-component.* When the magnetic field of a spot umbra is observed with a Zeeman triplet, between the two σ-components appearing in the longitudinal case, there are sometimes two other components π_1 and π_2 with opposite directions of circular polarization (Figure 4.5). This phenomenon, called the *π-component splitting*, was first noted by Severny (1959) and studied by many others (e.g., Mogilevskii *et al.*, 1967; Beckers and Schröter, 1969a; Grigoriev and Katz, 1972; Moore, 1981). Its fundamental characteristics are as follows:

(i) The amount of the π-component splitting is about 10–20 mÅ.

(ii) This quantity does not depend on the magnetic polarity of the spot, but increases with its field strength.

(iii) The sense of splitting of the π-component is always opposite to that of the σ-components.

(iv) The distance of a split π-component from the line center is about ten times smaller than that of the corresponding σ-component.

A reasonable explanation of the π-component splitting is that inside an umbral magnetic field of about 2000–4000 G is imbedded a small-scale field of approximately 200–400 G with a polarity opposite to that of the main field. Hence the seemingly split π-components are in reality the σ-components of the small-scale field. However, some astronomers (Rachkovskii, 1962, 1972; Beckers, 1969; Staude, 1970b, 1973; Wittmann, 1971b, 1972) interpreted the before-mentioned phenomenon in terms of magneto-optical effects. It is interesting to note that this explanation also requires that the magnetic field is inhomogeneous. (A detailed account is given in Obridko, 1985.)

(2) *Magnetic field strengths measured with different spectral lines.* Two kinds of lines, i.e., those of Fe II, Sr II and C II as well as those of Ti I and Fe I, can appear simultaneously in the spectrum of one and the same spot. Lines of the first group are formed in conditions similar to those of the photosphere, while the temperature in the region of formation of lines of the second group should be 1500–2000 degrees lower. This led Makita (1963) to the hypothesis of *two-component structure* of the spot umbra. But Zwaan (1965) suggested that Fe II lines in the umbra simply belong to the scattered light of the photosphere and penumbra. This idea has been negated by Obridko (1985), because direct measurements in Fe II $\lambda\lambda$4924 and 5048 do display Zeeman splittings. Moreover, their splittings are maximum at the same places as the Ti I and Fe I lines. This means that the radiation of Fe II lines really comes from the umbra. It is important to find that the magnetic field strengths given by Fe II lines are much smaller than those of Ti I and Fe I. The mean value of the ratio $B_{\text{Fe II}}/B_{\text{Ti I}}$ is 0.55.

(3) *Inversion of the V-profile.* It is often observed in spot umbra that the profile of the Stokes parameter V has an inversion near the line center (Figure 4.6). The inverted maxima are 5–10 times lower than the primary ones and are much nearer to the line center. The inversion cannot appear if the umbral magnetic field is approximately homogeneous (Beckers and Schröter, 1969a). The sense and separation of these two smaller extrema correspond to an opposite-polarity field with strength about one order of magnitude weaker than the primary one. Besides, Beckers and Schröter (1968, 1969a) proposed that the reversed weak field is located in umbral dots.

The V-profile inversion can also be produced by the magneto-optical effect. A discussion of this alternative can be found in Moore (1981).

(4) *Direct observations.* Besides the interpretation of the above phenomena, evidence is also provided by the following observations:

(i) *Transverse fields in umbrae.* Evershed (1944) first observed that even in the umbra there are transverse magnetic fields. This is in

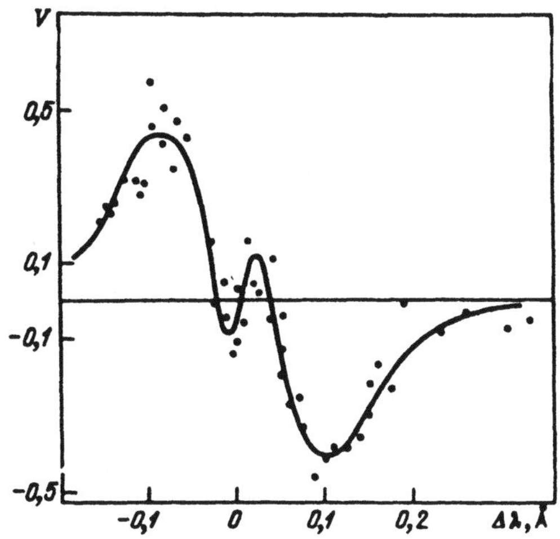

Fig. 4.6. *V*-profile inversion.

direct contradiction with the *fan-shaped model*[4] (Hale and Nicholson, 1938.) and many other sunspot magnetic field models.

(ii) *Umbral dots.* As measured by some astronomers (e.g., Zwaan, 1968; Zwaan and Burman, 1971; Kneer, 1973), the magnetic fields of umbral dots are much weaker than those of the intervening darker regions of umbrae. But the observation of Lites *et al.* (1989) shows that the difference is not large. The recent measurement of Pillet and Vázquez (1993) with the Gregory Coudé Telescope of the Teide Observatory on the Canary Islands leads to the following result. At places of clusters of umbral dots and light bridges the magnetic field is weaker than the intermediate districts in the umbra and the difference is less than 200 G.

(iii) *Penumbral grains.* Their field strengths are also less than those of the dark surroundings (Beckers and Schröter, 1969b).

Since the sizes of both dots and grains are quite small ($\lesssim 0''.5$), the inequality of field strengths between them and the neighboring regions implies that sunspot magnetic fields are highly inhomogeneous. But the magnitude of the differences measured up to now might be not quite certain.

(iv) *Patterns of field strength scannings.* When the 1-D distribution

[4]See Section 4.4.1

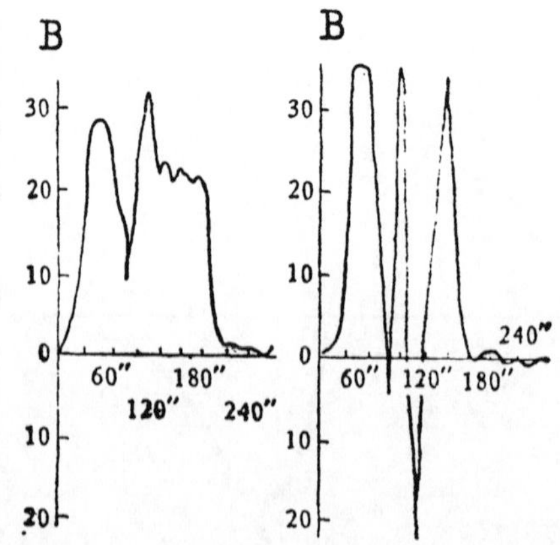

Fig. 4.7. Fine structures of a seemingly unipolar spot (B in units of 10^2 G).

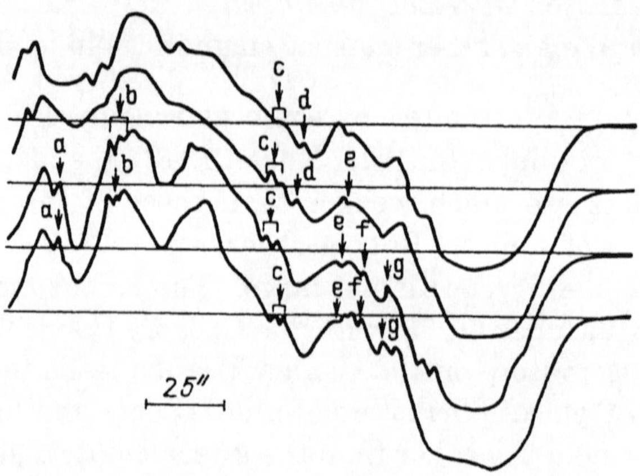

Fig. 4.8. "Pleats" in spot magnetic fields.

of the magnetic field is obtained by scanning across a spot with a low spatial resolution (height of slit = 27″), the spot field seems to be unipolar (see the left part of Figure 4.7). However, if a higher resolution (5″) is used, the same spot becomes a multi-polar one (right side of the figure). Therefore, strictly speaking, unipolar spots practically do not exist.

The scannings of spot magnetic fields with the Crimean solar magnetograph show the existence of very delicate structures, called as "pleats". Some examples are presented in Figure 4.8. The horizontal lines represent null field strengths. (A detailed account can be found in Severny, 1988.)

(v) *Penumbral filaments.* As can be readily seen on high-resolution white-light images of sunspots, the penumbra contains many dark and bright filaments, which are approximately in the radial direction and have lengths and widths of about 5000 and 300 km, respectively. For many years astronomers have known that the magnetic field intensities of those two kinds of filaments differ greatly from each other. This is another evidence of the inhomogeneity of spot magnetic fields. Nevertheless, it is still not clear which of them possesses stronger magnetic fields. Some observers, e.g. Beckers and Schröter (1968), found that the field of dark filaments is significantly more intense than that of bright ones. (In passing, we should mention that a similar situation exists for the dark granules of umbra (Obridko, 1968; Liamova, 1971; Guseinov, 1974 a, b)). However, from analysis of the observations with the Huairou Magnetic Field Telescope Liu Jianchiang *et al.* (1991) claimed that the bright penumbral filaments have more intense magnetic fields than the dark ones, showing a difference of about 300 G. This is in need of further investigation.

4.4 3-D Models

It follows from the preceding section that the magnetic field of sunspots is quite complicated and so one could hardly set up a precise model which might represent the detailed structures. However, from the macroscopic point of view some broad configuration is both possible and useful, and much work has been done in this field.

4.4.1 Fan-Shaped Model

On the basis of observations of many years Hale and Nicholson (1938) established the fan-shaped model, which can be expressed by Eq. (4.2) together with the following formula:

$$\gamma(\rho) = a\frac{\pi}{2}\rho, \tag{4.20}$$

where $\rho = r/b$, γ is the inclination of line of force to the normal (see Figure 4.9) and a is an adjustable coefficient. The two authors took $a = 0.75$.

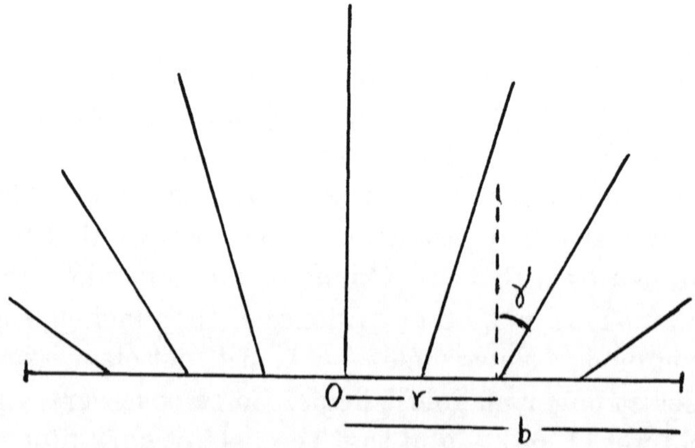

Fig. 4.9. Fan-shaped model.

This empirical model can be readily interpreted theoretically. The hydrostatic equilibrium of a spot on an arbitrary level z may be expressed by the condition

$$p_i(z) + \frac{[B(z)]^2}{8\pi} = p_e(z), \tag{4.21}$$

where p_i and p_e are gas pressures within and outside the spot. As pointed out by Cowling (1957), at sufficiently high levels the magnetic pressure exceeds the gas pressure of the surroundings and the magnetic field of the spot pushes aside the exterior gas. In this situation the magnetic lines of force have the shape of a fan.

Treanor's (1960) observations with the Babinet compensator confirmed the fan-shaped distribution of spot magnetic lines of force. But he found that about 40% of the total amount of magnetic flux is contained in the unit solid angle around the spot axis, while Hale and Nicholson's formulae (4.2) and (4.20) should yield a corresponding value of only 20%.

Bumba (1960a) improved the fan-shaped model by assuming that magnetic lines of force are not straight but curved. Then he satisfactorily explained the observed behavior of the magnetic field close to a spot with a polarity opposite to that of the proper spot and some morphological peculiarities of the Ca II H and K lines in the chromospheric flocculi above sunspots.

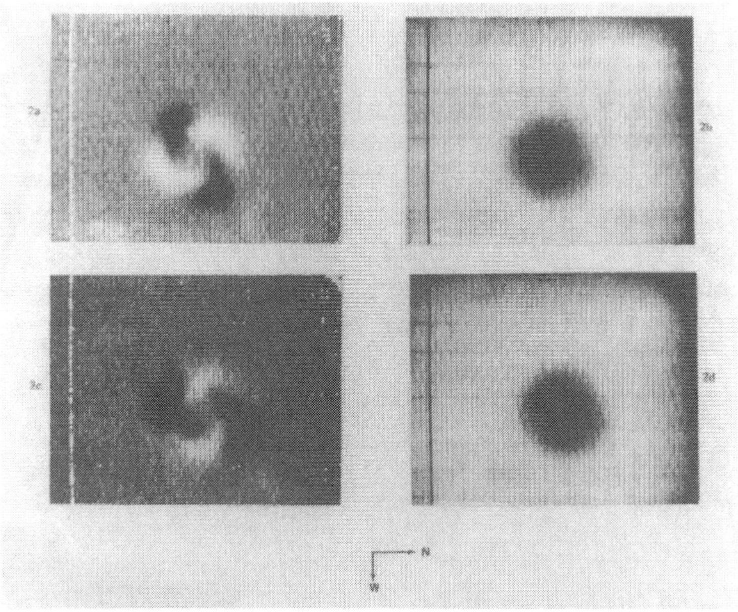

Fig. 4.10. Monochromatic $\tilde{Q}$ and $\tilde{U}$ images of unipolar spots (2a and 2c) and corresponding white-light images (2b and 2d).

4.4.2 Configuration of Magnetic Field Lines on the Spot Surface Plane

According to the fan-shaped model, magnetic field lines of spots are expected to exhibit a radial distribution on the surface plane. However, the vector magnetograph of MSFC yields somewhat strange $\tilde{Q}$ and $\tilde{U}$ monochromatic images of unipolar spots in Fe I $\lambda5250$. To be precise, $\tilde{Q}$ and $\tilde{U}$ are the *signals of linear polarization* defined as follows:

$$\tilde{Q} = \frac{\int Q(\lambda) f(\lambda)\,\mathrm{d}\lambda}{\int I(\lambda) f(\lambda)\,\mathrm{d}\lambda}, \qquad \tilde{U} = \frac{\int U(\lambda) f(\lambda)\,\mathrm{d}\lambda}{\int I(\lambda) f(\lambda)\,\mathrm{d}\lambda}, \tag{4.22}$$

where $f(\lambda) = C\exp\{-[(\lambda - \lambda_0)/(\Delta\lambda_F)]^2\}$ is the transparency function of the birefringent filter and $\Delta\lambda_F = 125/(2\sqrt{\ln 2}) \approx 75$ mÅ. It is interesting to note that such images (e.g., 2a and 2c in Figure 4.10) have a conspicuously spiral form (Hagyard and Cumings, 1975). In striking contrast, the white-light pictures of the same spots have a quite simple and round form. So it was once thought that the magnetic field lines in spots are spiral. Nevertheless, Landi Degl'Innocenti (1979) pointed out that if magneto-optical effects are taken into account, a purely radial field may give rise to spiral $\tilde{Q}$ and $\tilde{U}$ monochromatic images. We (Ye Shi-hui and Jin Jie-hai, 1986) justified this argument with theoretical calculations. We found that the radial or fan-shaped model, expressed by Eqs. (4.2) and (4.20), does lead to theoretical $\tilde{Q}$ and $\tilde{U}$ spot images which are spiral and similar to the observed ones, while Hagyard's *et al.* (1977) spiral model of distribution of lines of force produces irregular images, quite unlike the observation. Therefore, one may conclude that the radial model is closer to reality.

4.4.3 POSSIBLE TWISTING OF MAGNETIC FIELD LINES

According to the fan-shaped model, the azimuth of the transverse field (χ) does not vary with depth, i.e. the field lines of spots are not twisted. However, Severny (1964a) claimed that there is a "strong variation of field orientation with depth". In order to explain the observation of broad-band circular polarization of spots, Makita (1963) assumed a right-handed differentially twisted magnetic field. Hence it is necessary to decide whether the *twisting of field lines* really exists. If this is so, one should further determine the amount of twisting. For this we (Ye Shi-hui and Jin Jie-hai, 1990) have proposed the following diagnostic method.

For a fixed point in a spot and for a certain wavelength interval of a magneto-sensitive line, one may measure $\tilde{Q}$ and $\tilde{U}$ and then calculate the azimuth of the plane of polarization $\varphi = \frac{1}{2}\tan^{-1}(\tilde{U}/\tilde{Q})$. If the wavelength interval of observation is moved successively from line center to a wing, one may draw the azimuth diagram $\tilde{U} \sim \tilde{Q}$. According to our theoretical computations, if field lines are strongly twisted, the curve on the diagram contains loop structures (e.g., Figure 4.11). In the opposite case, it is comparatively simple and similar to a parabola (Figure 4.12).

We (Ye Shi-hui, 1992) applied this method to the observed umbral Stokes profiles of Mg I λ5172 published by Lites and Skumanich (1988) and obtained an azimuth diagram which has no loop structures. This implies that the magnetic lines of force of the observed spot are not strongly twisted, or even that they do not have any twisting at all.

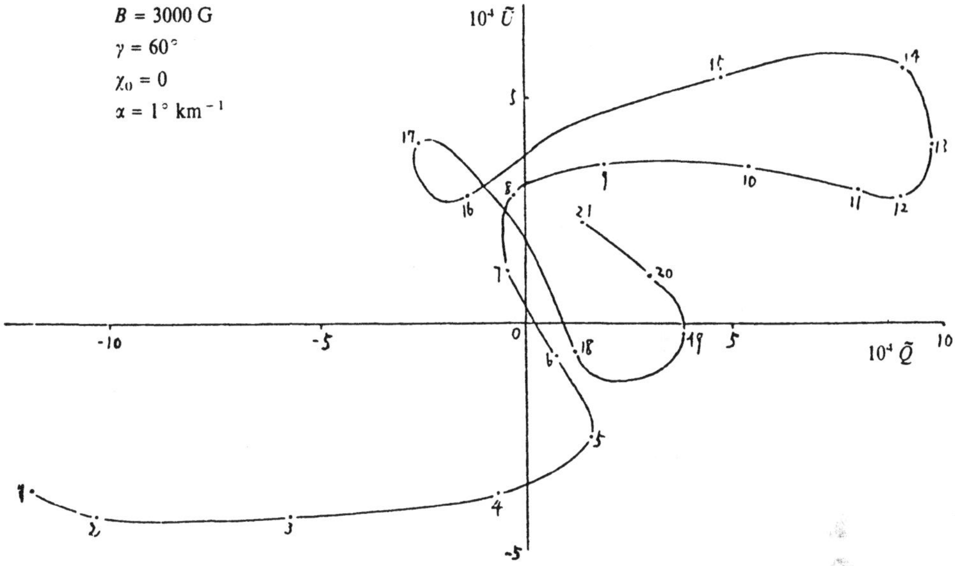

Fig. 4.11. Theoretical azimuth diagram for strongly twisted magnetic field lines (α is the coefficient of twisting).

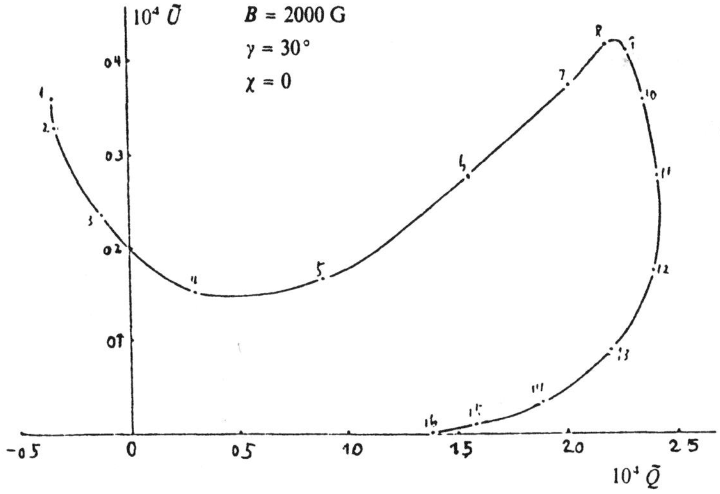

Fig. 4.12. Azimuth diagram without twisting of magnetic field lines.

4.4.4 Inclination of Magnetic Field Lines

When the curves of equal intensities are traced out, the monochromatic $\tilde{Q}$ and $\tilde{U}$ images of unipolar sunspots exhibit an interesting configuration. Each image consists of four quadrants and the neighboring ones have alternate plus and minus signs. Moreover, each quadrant contains a set of concentric closed curves. Figure 4.13 is the $\tilde{U}$ image of a large spot observed in Fe I $\lambda 5324$ at the Huairou Station. Figure 4.14 is the longitudinal magnetogram of the same spot. It is impressive to note that these two images, belonging to one and the same spot, have drastically different shapes. Observations in Fe I $\lambda 5250$ with the vector magnetograph of Marshall Space Flight Center yield the same result.

The above phenomenon can be interpreted in terms of magneto-optical effects (Ye Shi-hui and Jin Jie-hai, 1993a, b). It is interesting to find that the theoretical $\tilde{Q}$ and $\tilde{U}$ spot images are concerned not only with the magnetic field strength B,[5] but also with the inclination angle γ. In the fan-shaped model this angle is given by Eq. (4.20). So the theoretical $\tilde{Q}$ and $\tilde{U}$ spot images are connected with the coefficient a. By means of numerical solutions of Unno-Beckers' equations (4.15), we can construct theoretical images with different values of a. We discovered that for $a = 1$ the theoretical images agree rather well with the observed ones. But when a becomes smaller, e.g. $a = 0.833$ and 0.3, the agreement is worse (see Figure 4.15). This is shown by the fact that for smaller a more outer contours in the theoretical images are no longer closed, and this is in conflict with observations. Therefore, if the fan-shaped model is adopted, it seems better to take $a = 1$.

Of course, the classical fan-shaped model is not the unique sunspot model we can use. The return-flux model of Osherovich (1982) may be considered to be more reliable, since its theoretical distribution of the angle γ across the spot agrees satisfactorily with the observations of Beckers and Schröter (1969a) and Wittmann (1974). We used Osherovich's theoretical curves showing the changes of γ and B_0/B with ρ and computed the theoretical $\tilde{Q}$ and $\tilde{U}$ images of unipolar spots. Figure 4.16 presents an example, which is a $\tilde{Q}$ image calculated with $B_0 = 3000$ G. It is of some interest to find that the result of numerical simulation with the return-flux model seems to be better than that with the fan-shaped model. Maybe this implies that the former is closer to reality than the latter. Thus the monochromatic spot images in linearly polarized radiation can be used to check the regularities of change of γ with ρ.

[5]This has been demonstrated in our previous work (Ye Shi-hui and Jin Jie-hai, 1986).

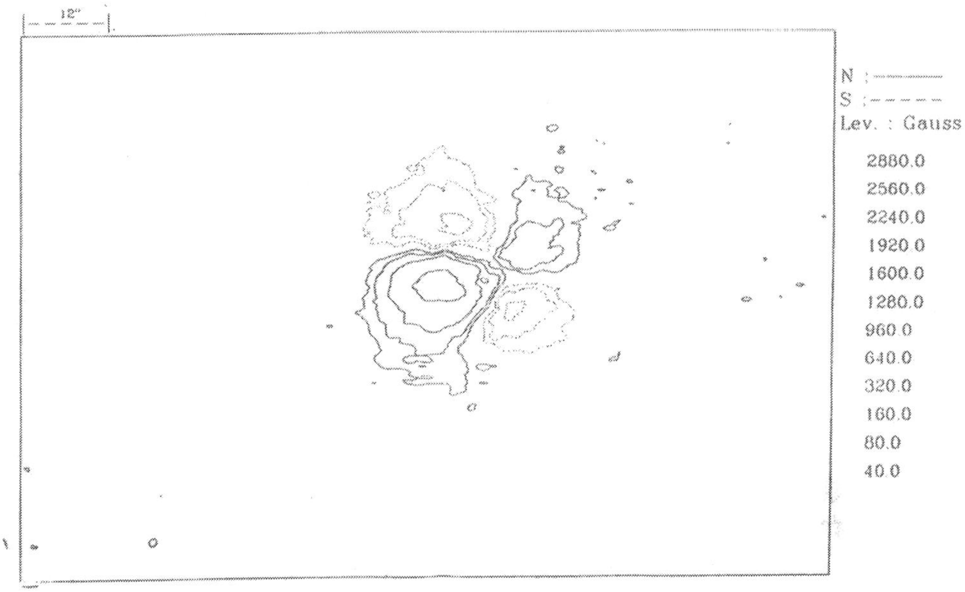

Fig. 4.13. Monochromatic $\tilde{U}$ image of a unipolar spot of April 28, 1988 with heliographic coordinates E12 and S18.

Fig. 4.14. Longitudinal magnetogram of the same spot as in Figure 4.13.

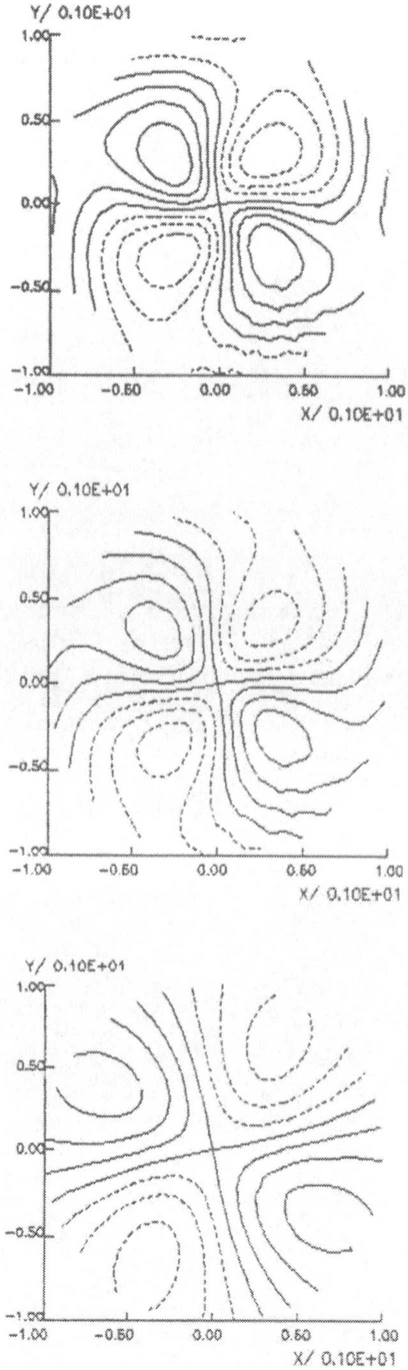

Fig. 4.15. Theoretical $\tilde{Q}$ images computed for fan-shaped models with $a = 1$, 0.833 and 0.3 (from top to bottom).

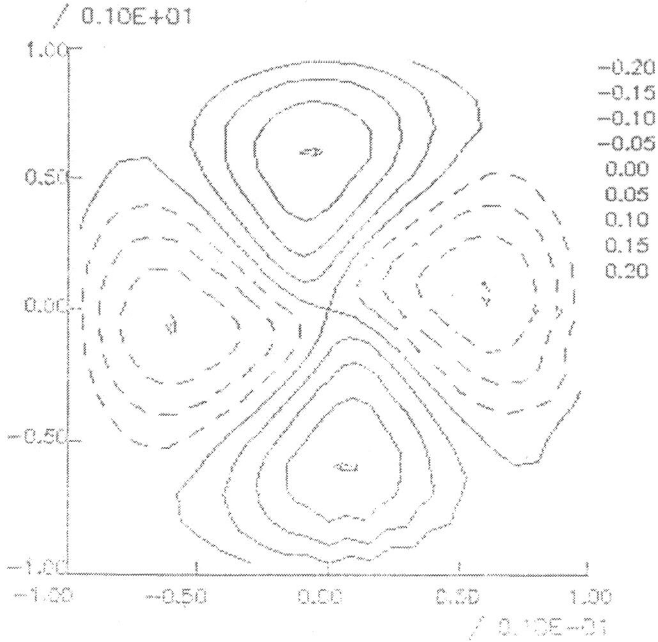

Fig. 4.16. Theoretical $\tilde{Q}$ image constructed with Osherovich's return-flux model.

4.5 Sunspot Magnetic Fields and the Evershed Effect

Matter is in constant motion, everywhere in the universe. The situation is, of course, the same in sunspots. Since spots are embedded in strong magnetic fields, their internal motions may be influenced by the presence of magnetism.

(1) *A brief historical account*. More than eighty years ago Evershed (1909, 1910) discovered from displacements of spectral lines that there are constant flows of matter in penumbrae, mainly along spot radii. More precisely, the matter moves *outward* on the solar surface with a velocity of about 2 km s^{-1}. But at the higher atmospheric level where H_3 and K_3 are formed, matter flows inward. This phenomenon is called the *Evershed effect*.

Soon afterwards, St. John (1913) performed careful measurements in 506 spectral lines, including Ca II H and K. He found that the direction of movement in the penumbra depends on line intensities. Outward flows are exhibited by weak lines with Rowland intensities of about 10 or even less. For lines with Rowland intensities around 20, no significant motion can be detected. As for stronger lines, mass motions are inward. Besides, for spectral lines of equal intensities the displacements are proportional to wavelengths. This justifies the reasoning that the Evershed motions can be interpreted with the *Doppler effect*.

Since the thirties many astronomers (e.g., Abetti, 1932; Kinman, 1952, 1953) continued the study of mass motions in spots. Their work basically agreed with Evershed's conclusion that, near the spot bottom, matter flows from the interior to the surroundings, while at upper levels it flows from the exterior into the spot. However, there is much controversy among different authors. On this the reader may consult Section 4.4.2 of Bray and Loughhead (1964). Here we only wish to emphasize that the Evershed effect is concerned with spot magnetic fields. This can be demonstrated with both magnetograms and dopplergrams, which are obtained simultaneously by a magnetograph. Using such observational data, Bumba (1960b, c) discovered the following phenomena:

(i) *Matter flows along magnetic lines of force.* For unipolar spots, movements take place on fan-shaped planes, which are inclined to the solar surface. As we have mentioned, in the fan-shaped model, as improved by Bumba, the magnetic field lines are curved and so matter flows from the umbra to the penumbra.

(ii) *In the penumbra the spectral line is not displaced as a whole.* Bumba found that there is merely a small displaced branch, which he called a "flag" (Figure 4.17). This can be interpreted as follows. The Evershed movements do not take place everywhere in the penumbra. The matter flows outward only along penumbral filaments.

Moreover, Beckers (1968a) discovered that Evershed motions are concentrated in dark filaments. This has been confirmed by Beckers and Schröter (1969b), Abdusamatov and Krat (1970), Stellmacher and Wiehr (1980), Wiehr *et al.* (1984) and Küveler and Wiehr (1985). So it seems to be generally recognized that matter flows in the penumbra along the magnetic field lines which are frozen in the spot plasma. This motion may be thought to be a laminar flow, which is formed under the joint action of the magnetic force, gravity, gas pressure and some other factors.

The study of the Evershed effect continues up to the present. Wiehr and Stellmacher (1989) made high resolution spectroscopic observations at Tenerife and found that there might be no correlation between Evershed movements and brightness structures. The observation of Kitai *et al.* (1993) at Hida also shows that the fine details of the brightness fields do not match well with those of the velocity fields of spots.

(2) *Theoretical interpretation.* From the theoretical point of view

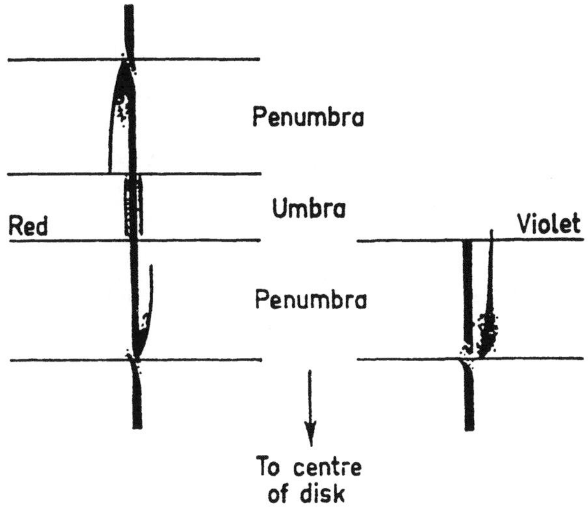

Fig. 4.17. The "flag" phenomenon in a spot spectral line.

the Evershed motion is both natural and necessary. As remarked by Obridko (1985), the expansion of magnetic lines of force expressed by Eq. (4.21) is insufficient to explain the formation of the penumbra and so a supplementary mechanism is needed. According to his calculations, a small deviation from the hydrostatic equilibrium is sufficient to produce strong pressing on the magnetic field lines and to force them to lie in the horizontal direction. Concretely speaking, he found the following inequality

$$\tan\gamma > 2 \times 10^{-3}\delta g, \tag{4.23}$$

where g is the gravitational acceleration on the solar surface. So even a deviation of ten percent from the hydrostatic equilibrium (i.e. $\delta g = 3 \times 10^3\ \mathrm{cm\,s^{-2}}$) may give rise to $\tan\gamma > 6$ or $\gamma > 80°$. This means that the magnetic lines of force are almost parallel to the solar surface.

(3) *The realm of Evershed motions.* Does the Evershed flow stop on the spot boundary? Some authors, e.g. Wiehr and Degenhardt (1992), say that the flow ends on the outer penumbral edge, while some others, like Börner and Kneer (1992), claim that it continues beyond this boundary. Solanki *et al.* (1992) analyzed the Stokes I- and V-profiles of the infrared lines Fe I 1.5648μ and 1.5653μ and found that magnetic field lines can stretch to outside the visible contours of sunspots in the form of low-lying superpenumbral canopies. Moreover, they have

discovered that the Evershed flow on the canopy can go far beyond the penumbra, while the flow beneath the canopy stops on the penumbral boundary. In other words, the magnetic lines of force, which form canopies, carry the Evershed flows to the outside of spots. This again shows the close connection between the Evershed effect and sunspot magnetic fields.

To conclude, we would say that the Evershed effect seems to be more complicated than might have been expected.

4.6 Magnetic Fields of Sunspot Groups and Complexes of Activity

The contents of the preceding sections of this chapter are mainly concerned with individual sunspots. However, as is well known, spots often appear in groups. It is therefore necessary to study the magnetic fields of spot groups.

4.6.1 Magnetic Types of Spot Groups

Up to the present there are three kinds of classification of sunspot groups, which may be called (1) the Mt. Wilson magnetic classification; (2) the Zürich evolutionary classification, and (3) McIntosh's three-dimensional classification. Because intense magnetism is the fundamental characteristic of sunspots, any classification into groups is naturally connected with their magnetic behavior.

In the first classification, spot groups are divided according to their magnetic polarities into three classes: α, β and γ. They correspond, respectively, to *unipolar*, *bipolar* and *multipolar* groups. There are also some subclasses. If the center of gravity of the accompanying flocculus is located ahead of or behind the spot (in the sense of solar rotation), a letter p or f is attached to the signature of class α. This specification has some physical significance. According to Howard's (1959) observation, flocculi are basically bipolar in magnetism. So one may consider that the magnetic lines of force coming from a unipolar spot can go back to the neighboring flocculus region. Just in this sense, Hale believed that there are "hidden spots" in flocculi. A bipolar group is called βp or βf, if the preceding spot is much larger or smaller than the following one. According to the statistics of Hale and Nicholson (1938), the percentage of spot groups of different classes is given in Table 4.1.

If two or more umbrae with different polarities exist in the same penumbra, the spot group which contains such an umbra is designated as $\beta\gamma$ or δ. Although these groups as well as those of class γ, are quite few, they produce the greater part of large solar flares, in particular proton flares.

The Zürich Observatory classification consists of nine categories, given in Table 4.2.

This classification traces the evolution of a fully developed group and is closely connected with its magnetic behavior.

The 3-D classification proposed by McIntosh (1990) uses three indices, i.e., Z, p and C. Z is the Zürich class, p is the type of the main spot and C is the degree of compactness of the spots in a group. The outstanding merit of this system is its high correlation with solar flares. It is worth to pointing out that this classification also mainly relies on the magnetic properties of spot groups.

Finally, there is a special class of sunspot groups which violate Hale's law of distribution of magnetic polarities (Tang, 1982). They occupy about 3% of the total number of spot groups.

4.6.2 MAGNETIC FIELD CONFIGURATIONS OF COMPLEX SPOTS AND SPOT GROUPS

The models of spot magnetic fields introduced above (e.g. the fan-shaped model of Hale and Nicholson and the return-flux model of Osherovich) belong basically to unipolar spots. The overall distribution of magnetic lines of force in these models is rather simple and regular. This is so even for bipolar spot groups. As stated by Severny (1964b), their magnetic field lines display a configuration like that of the iron filings on a glass plate placed above a magnetic bar. However, the situation is drastically different for complex spots; in particular spots of the δ-type. According to Patty (1981), their complicated magnetic structure can be interpreted in terms of the combination of a large longitudinal gradient with a strongly sheared transverse component.

As for the magnetic field configuration of spot groups, it cannot of course be expressed with the simple models mentioned above. However, there are some basic principles which may be believed to be valid for any group. First, magnetic fields exist before the birth and after the disappearance of spot groups and are basically bipolar. Second, the Babcocks' (1955) measurements show that the absolute values of magnetic fluxes with different polarities are approximately equal to each other. Also, Bumba (1963) found that the overwhelming part of magnetic field lines are concentrated in narrow tubes, which connect individual spots; only a small portion is scattered to the wide space outside spots. It is interesting to note that the distribution of the magnetic field does not necessarily agree with the optical image of the same spot group. Local strong magnetic regions (the so-called "magnetic hills") do not always appear at the site of spots. Sometimes, in the vicinity of "magnetic hills" of several hundred gauss there is no spot at all. These facts imply that the situation for multipolar groups is very complicated.

4.6.3 MAGNETIC FIELDS OF COMPLEXES OF ACTIVITY

Magnetic relations exist among not only neighboring spots, but also separate active regions (Stepanov and Gopasyuk, 1962). Several active regions, which are connected by a common magnetic field, are said to form a *complex of activity* (Obridko, 1985). This is a very dynamic feature. Once born, it expands with a speed of 10°–20° per day or 120 $\mathrm{m\,s^{-1}}$. The existence of two complexes on the Sun increases the degree of activity of each of them, especially on their interacting borders. A large complex may exist in 10–15 solar rotations, span both the northern and southern hemispheres and produce a series of proton flares.

The magnetic flux of a large active region may attain $5\text{–}8 \times 10^{22}$ Mx. The amounts of fluxes of opposite polarities are only approximately equal. In the ascending phase of activity the unbalanced part can occupy 16% of the total amount of fluxes, and in the descending phase it can be 30%. The total net quantity of magnetic fluxes of all the spots, which exist at the same time on the Sun, often differs from zero. Also, the polarity of the net flux is always opposite to that of the general magnetic field of the Sun. Therefore, one may say that the magnetic fields of active regions are in some way connected with the general field. This justifies Babcock's (1961) point of view.

4.7 Influence of Magnetic Fields on Sunspot Spectra

Magnetic fields of several thousand gauss may have some influence on the spectra of sunspots. More than half a century ago, ten Bruggencate and von Klüber (1939) suggested that magnetic fields can widen absorption lines in the spot spectrum, give rise to the so-called "pseudo-Doppler effect" and deform the curve of growth. It is well known that the middle part of a curve of growth is a flat transition region. According to the reasoning of these two authors, just here appears the main deformation. The stronger is the magnetic field, the higher becomes this section of the curve of growth.

How does the magnetic field change the equivalent width (W_λ) of a spectral line? Hale and Seares (1913) attacked this problem, yet their work is in principle incorrect. As stated in Section 3.1.4, the formulae which are obtained for emission lines, i.e. for the *direct* Zeeman effect, cannot be applied to absorption lines (*inverse* Zeeman effect). Besides, they treated only two special cases: $\gamma = 0°$ and $90°$. In the fifties, Unno (1956) and Stepanov (1958a, b) studied this question with their respective theories of the inverse Zeeman effect. Both recognized that the *magnetic intensification* of spectral lines really exists, but they got different results. To be precise, Unno's value of intensification ($\Delta W_\lambda / W_\lambda$) is less than 10%, while ten Bruggencate and von Klüber found it to be up to 80%. Stepanov obtained intermediate values. We (Ye Shi-hui *et al.*, 1978a; Ye Shi-hui, 1991) studied this problem with our numerical solution of the equations of transfer of the Stokes parameters and drew the following conclusions.

(1) ΔW_λ depends on both B and γ and increases with each of these two quantities (see Figure 4.18).

(2) When $\gamma = 0°$, $\Delta W_\lambda = 0$. This implies that the pure longitudinal Zeeman effect does not cause magnetic intensification.

(3) For $\gamma > 60°$, ΔW_λ gradually becomes saturated. This does not agree with Unno's prediction that ΔW_λ is maximum at = 55°.

(4) The larger B or γ, the more conspicuous is the effect of saturation.

TABLE 4.1
Distribution of various magnetic types of sunspot groups (%)

class	α	αp	αf	β	βp	βf	$\beta\gamma$	γ
percentage	15	28	3	11	29	10	3	1

TABLE 4.2
The Zürich Observatory classification scheme

A	a single pore,
B	two pores with the bipolar configuration,
C	bipolar group in which one major spot has penumbra,
D	bipolar group, of which both the main spots possess penumbrae,
E	complicated bipolar group with many small spots and with a span in longitude exceeding 10°,
F	very large bipolar or complex group with a longitude span larger than 15°,
G	bipolar group without small spots,
H	unipolar spot with a penumbra of diameter larger than $2^{\circ}\!.5$,
J	unipolar spot with penumbra smaller than $2^{\circ}\!.5$.

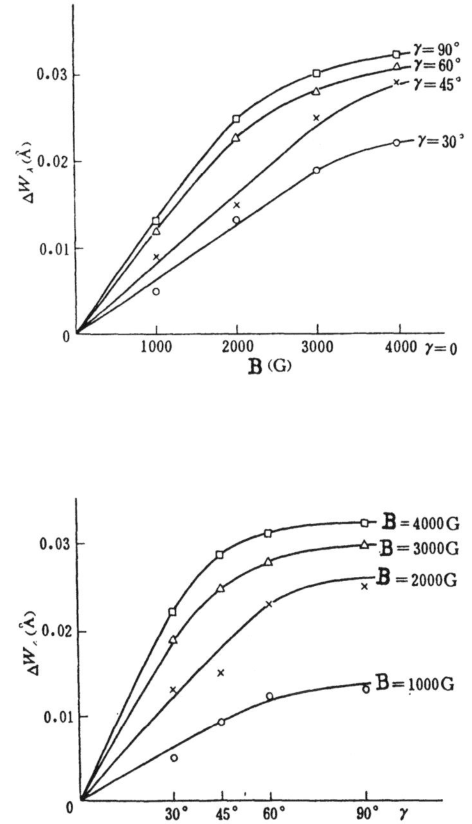

Fig. 4.18. Magnetic intensification of Fe I λ6302.

CHAPTER 5

Background and Local Magnetic Fields on the Solar Surface

If the Sun had no magnetic field, it might be a quiet, "classical" star – if such stars exist, with no corona, chromosphere, sunspots, or solar activity.
H. Zirin, *Astrophysics of the Sun* (1988).

In the last chapter we described the strongest magnetic fields on the solar surface: those of sunspots. In fact, it is not only spots that possess magnetism. Magnetic lines of force pervade various parts of the solar disk and every layer of the solar atmosphere. Strictly speaking, there is no region which is a real vacuum for magnetism. Starting here, we shall devote three chapters to the solar magnetic fields outside spots. Roughly speaking, this chapter introduces the fields of different parts of the solar surface in the transverse direction and emphasis is laid on active features. Then, in the vertical direction, Chapter 6 will discuss the magnetic fields in various layers of the solar atmosphere with stress on quiet regions. The overall characteristics of the solar magnetism will be treated in Chapter 7. This kind of division is, of course, crude. Because various regions of the Sun may be interlocked and many phenomena closely connected, the contents of these chapters will inevitably have some overlaps and repetitions.

5.1 Solar Background Magnetic Field

It was stated in Section 4.3 that sunspot magnetic fields possess very fine structures. Now we want to emphasize that the situation is similar for magnetic fields outside spots as well as active regions. High-resolution videomagnetograms of separate undisturbed districts as well as the whole quiet Sun consist of innumerable tiny *magnetic elements* or *flux tubes*, and the upper panel of Figure 3.18 and Figure 3.10 are excellent examples. This may be called the *solar background magnetic field*. Such a field is quite weak, but due to its wide space it may occupy a significant portion in the total amount of solar magnetic flux. In recent decades many solar physicists pay much attention to the investigation of the background magnetic field and some achievements will be described in what follows.

5.1.1 "HIDDEN" MAGNETIC FIELDS

The study of the background magnetic field is even more difficult than that of the spot field. This is because, first, the former is two or three orders of magnitude weaker than the latter and so is harder to measure. Second, the background field is highly intermittent. This means that although it is weak as a whole, there are in its ample space quite intense magnetic elements or flux tubes with strengths of a few kilogauss (kG). The background field may be thought to contain both kG elements and intrinsically weak "hidden" fields. The mean observed field strength $\langle B\rangle$ is related to the intrinsic strength B by the following expression:

$$\langle B\rangle = \alpha B, \tag{5.1}$$

where α is called the *filling factor*. For the background field, α is small; about 0.01 or maybe even less. By contrast, the spot field has a filling factor almost equal to unity. This implies that the spot is full of flux tubes. Hence in the study of the background field we have to determine not only $\langle B\rangle$, but also α. This, of course, makes the task more difficult. Third, the main trouble arises from the so-called *mixed-polarity unresolved fields*. When extremely small magnetic fields with different polarities are located in one and the same resolved region, their fluxes cancel one another to a great extent and so the net magnetic flux over the resolved element is practically equal to zero. In this situation the field strength can hardly be measured by the commonly used longitudinal Zeeman effect.

In spite of the difficulties enumerated above, it is still possible to measure the background and, in particular, the "hidden" magnetic fields. As summarized by Stenflo (1992a) the following methods can be adopted: (1) Zeeman broadening of unpolarized line profiles, (2) Hanle-effect depolarization, (3) transverse Zeeman effect, and (4) direct observations with instruments of highest spatial resolution. The

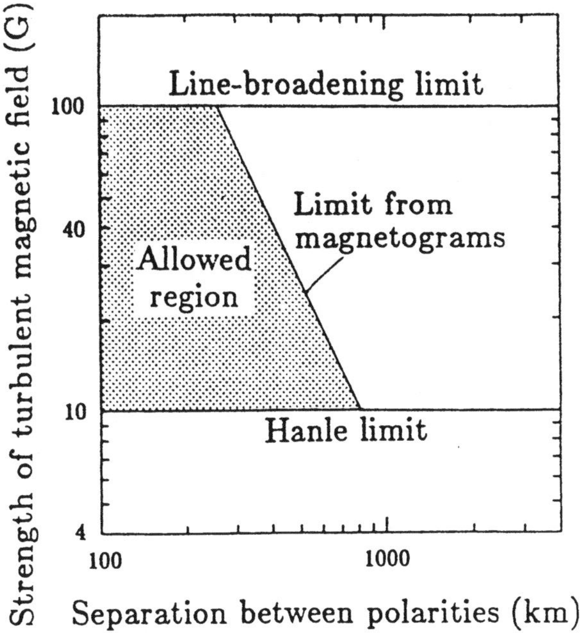

Fig. 5.1. Observational constraints of "hidden" or "turbulent" magnetic fields.

realms of applications of these methods are schematically shown in Figure 5.1. The shaded region might be accessible for the Large Earth-Based Solar Telescope (LEST) now under construction. Due to lack of space we can merely mention the following main results of investigation with these methods: (1) The strength of the "hidden" magnetic field is likely to be in the range 10–100 G. (2) The distribution of field vectors is almost isotropic. (3) The characteristic scale size is smaller than $1''$.

5.1.2 Infrared Diagnostics

Quite recently, the intrinsically weak magnetic field was measured by Stenflo (1992a). He used the *abnormal V-profiles* of the infrared line pair Fe I 1.5648 and 1.5653 μm, which have Landé factors 3.0 and 1.5, respectively. These profiles are double-peaked (see the left part of Figure 5.2). The solid curve represents the observed profiles, while the dotted line denotes the synthetic profiles produced with a model of two discrete magnetic components belonging to one and the same resolved magnetic element. The separate V-profiles of these two components are shown schematically on the right side of the figure. By means of theoretical simulation, Stenflo found that one component has a strong

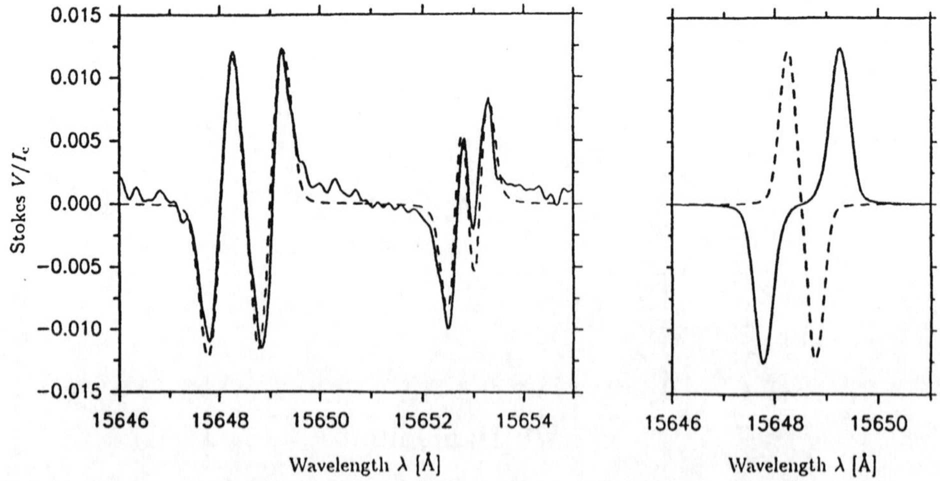

Fig. 5.2. Double-peaked V-profiles of Fe I 1.5648 and 1.5653μm.

field strength of 1.4–1.6 kG, while the other may be quite weak: 0.4 kG. It is evident that the latter belongs to the intrinsically weak "hidden" field. Moreover, it has to be noted that the two components possess opposite magnetic polarities. As for the filling factor, it is approximately proportional to the average of the red and violet wing areas of V/I_c, where I_c is the intensity of the neighboring continuum.

In passing, it is worthwhile mentioning that the double-peaked V-profiles shown in Figure 5.2 have also been observed for Fe I λ6302 across a flaring neutral line (Skumanich and Lites, 1991). For the interpretation it is also necessary to assume that inside one and the same resolved element there are magnetic fluxes with opposite polarities and that they cancel each other.

5.1.3 Correlation of the Background Field with Other Magnetic Structures

It is interesting to find that all the resolved magnetic features of various scales, such as sunspots, *ephemeral active regions* and *intranetwork fields* (these magnetic structures will be defined and described in the following section), are bipolar in nature. Moreover, they seem to form a continuous sequence of emerging flux regions. As remarked by Zirin (1987), the observed flux emergence rates of active regions, ephemeral active regions and intranetwork fields are roughly in the proportions

$$10^{20}\ \mathrm{Mx/day} : 10^{22}\ \mathrm{Mx/day} : 10^{24}\ \mathrm{Mx/day} = 1 : 10^2 : 10^4,$$

while their geometrical scales are in the approximate proportions

$$75'' : 15'' : 3'' = 25 : 5 : 1.$$

This seems to tell us that the smaller the magnetic feature, the faster the rate of flux emergence $\mathrm{d}\Phi/\mathrm{d}t$. By use of the relation between this quantity and the rate of change of the mean magnetic field strength, i.e.

$$\frac{\mathrm{d}\overline{B}}{\mathrm{d}t} = \frac{1}{4\pi R_\odot}\frac{\mathrm{d}\Phi}{\mathrm{d}t}, \tag{5.2}$$

the following empirical expression can be derived:

$$\frac{\mathrm{d}\overline{B}}{\mathrm{d}t} = 350(d'')^{-2.8}\ \mathrm{G/day}. \tag{5.3}$$

Here d'' is the geometrical size in arcseconds. Substituting the known value of d'' into this equality, we have for active regions $\mathrm{d}\overline{B}/\mathrm{d}t$ = 8 G/11 years. This is of the right order of magnitude according to the *Babcock–Leighton model of the solar activity cycle* (see Section 11.3). For intranetwork fields, the rate is 16 G/day, which is in nearly four orders of magnitude larger than that of active regions. If we could extrapolate the above sequence to the background field and take its scale to be $0\overset{''}{.}1$, then we might get $\mathrm{d}\overline{B}/\mathrm{d}t = 150$ G/min! Such a rate seems too large to be possible and should be checked in further research. However, it is in any case reasonable to think that the replenishment of the background-field pattern could be much faster than resolved magnetic features and this might be of vital importance for the evolution of solar magnetic fields. Therefore, as pointed out by Stenflo, there may be a close connection between the small-scale dynamics of magnetic features and the global solar-cycle evolution.

5.2 Small-Scale Magnetic Fields

Nature might be thought to be a skillful architect, who could construct various colorful structures with magnetic elements or flux tubes on the solar surface. When a magnetic structure is larger or smaller than a mature active region, which has a linear dimension of about 10^4 km, it may be called a large- or small-scale magnetic field. This section is devoted to the characteristics and evolution of various kinds of small-scale fields. Some selected topics will be discussed in the following.

(1) *Magnetic field strengths and sizes of flux tubes outside sunspots.* Flux tubes have different characteristic sizes in various magnetic regions. Sunspots possess the largest flux tubes, with diameters of one arcsec or even larger. These flux tubes appear as dark photospheric features. Nevertheless, the flux elements outside spots are much smaller and form objects which are brighter than the photosphere. Magnetic fluxes on the surface are believed to emerge from the solar interior, and under the action of turbulent motions they are shredded into discrete fragments. This process cannot be carried on infinitely to smaller and smaller scales and should be balanced by an opposite one. This is the *instability mechanism of convective collapse*, which causes weak, vertical flux tubes to gather together. Such a mechanism is restrained by the following two prerequisites. First, the magnetic pressure within a flux tube should not be stronger than the exterior gas pressure, which pushes flux tubes to approach one another. According to Spruit and Zweibel (1979), this requires that $\beta(\equiv 8\pi P/B^2) \lesssim 1.8$. For the photosphere this condition leads to $B \geq 1.3$ kG. Second, the flux tube must be large enough so that it is optically thick and thermally insulated from its surroundings. For the photosphere this scale is expected to be 100 km. These two critical values agree well with those given by diagnostic methods.

(2) *Active regions and spots.* In active regions, especially in spots, magnetic flux tubes emerging from the solar interior are highly concentrated. In other words, their filling factors are much larger than in undisturbed regions. By definition the filling factor α cannot exceed unity. Now we should stress that even in spots α is not very close to this critical value. The following fact may be considered to be a possible illustration. In spot umbrae, which are much "darker" than the photosphere, there are comparatively *bright dots*. According to Parker (1979b) and Spruit (1981), they are the field-free gas that rises from the solar interior to the surface among separate flux tubes. If this interpretation is plausible, the magnetic field in umbral dots should be much weaker than that in umbrae. Due to the influence of stray light, the direct justification is difficult and so there is controversy among various authors. Kneer (1973) found that the field strength in dots is about twice as small as in surrounding umbrae. However, Zwaan *et al.* (1985) and Lites and Scharmer (1988) claimed that the difference is not so large. This is in need of further investigation.

(3) *Magnetic networks.* Under the action of turbulent motions the magnetic flux of old decaying active regions is divided into very small portions. It is interesting to find that these flux elements are not ran-

domly distributed on the solar disk, but form *network patterns* (Simon and Leighton, 1964) with a characteristic size of about 30,000 km. Magnetic networks coincide in position with the networks of two other kinds. One of them is the network on monochromatic images of various spectral lines, especially those formed in the chromosphere-corona transition zone where the coronal heating takes place. This shows that the magnetic field is important in the heating and energy balance of the solar outer atmosphere. The other is the boundaries of supergranules which are known to be the patterns of velocity fields. This has been repeatedly demonstrated by observations at Kitt Peak (Livingston *et al.*, 1976a, b), BBSO (Haimin Wang, 1988a, b; Haimin Wang and Zirin, 1987, 1988) and Huairou Station (Li Jing *et al.*, 1989). An example is presented in Figure 5.3, which is a Kitt Peak magnetogram superposed on the corresponding dopplergram. The white and dark contours represent the receding and approaching motions, respectively, while the light and dark gray levels embody the magnetic fields with positive and negative polarity. With the BBSO observational material Haimin Wang and Zirin calculated the coefficient of cross-correlation and quantitatively described the close relation between the supergranule boundary and magnetic networks. Besides, Haimin Wang found from the autocorrelation curve that the average size of network magnetic elements is approximately 5700 km and that the element possesses a magnetic flux more than 5×10^{18} Mx.

When a magnetic network is relatively strong, complete and of basically one and the same polarity, it is called an *enhanced network*.

The magnetic fields inside a network are called *intranetwork fields* (IN). They were first discovered by Livingston and Harvey (1975). These fields also consist of discrete magnetic elements, which have fluxes as small as 5×10^{16}–1×10^{17} Mx and lifetimes 5–90 min. Their polarities are mixed and independent of the dominant polarity of the surrounding network. Although the strength of network fields can attain a few kG, the IN fields may be intrinsically weak. Tarbell *et al.* (1979) found an upper limit of 50 G for their strength. As discovered by Zirin and Haimin Wang, the IN elements often move toward network boundaries. So, if the network is formed through the concentration of IN elements, 10 to 100 such elements with the same polarity should be collected. The probability of one element meeting another with the same or the opposite polarity is equal to 50% in each case. Hence the probability of an IN element gathering together with 10 to 100 other elements of the same polarity is quite small. This implies that the magnetic network is the product of a complicated process in which numerous magnetic

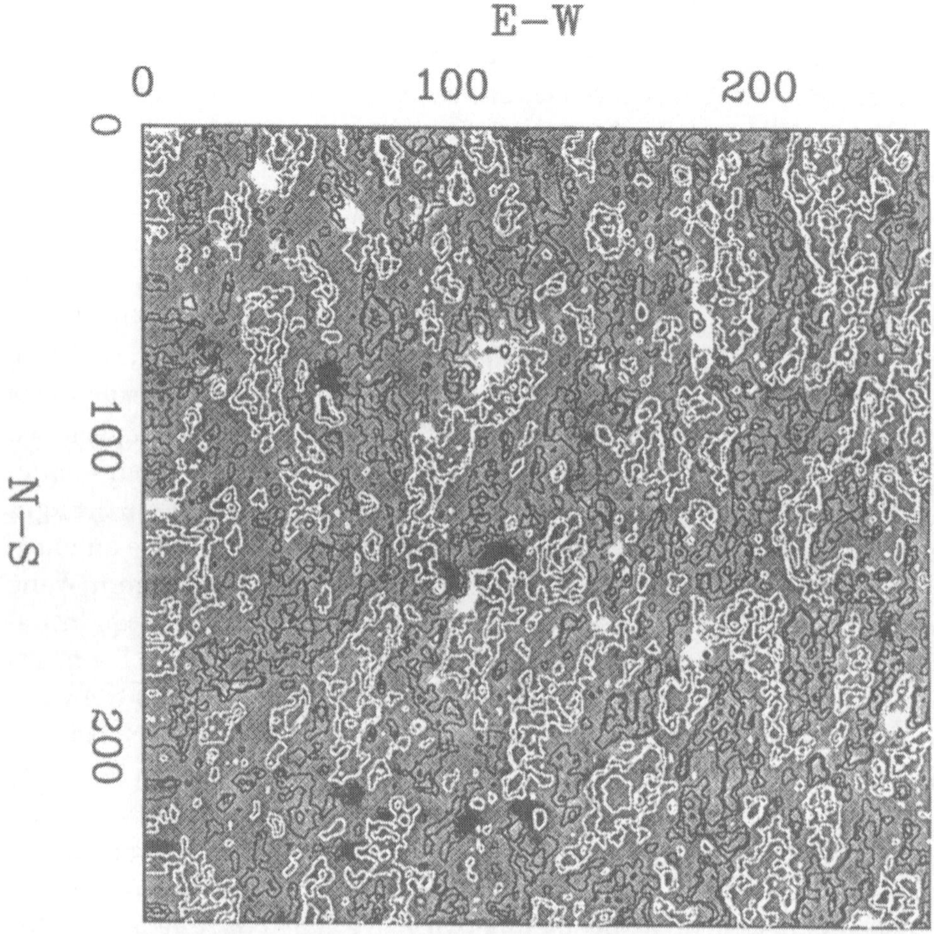

Fig. 5.3. Magnetic networks and supergranules.

elements incessantly interact with one another.

(4) *Ephemeral regions (ERs).* An ER consists of a pair of small and short-lived bipolar flux elements. It appears as a small bright area in the monochromatic images of chromospheric lines and X-rays. It is brighter than networks, but fainter than active regions. ERs exist for only about one day and do not have sufficient flux to form spots. Harvey and Martin (1973) found that there are many ERs all over the solar surface, even in the polar regions, and that they do not obey the Hale–Nicholson law of magnetic polarity. There may be some correlation between the number of ERs and the solar activity cycle. Evidence is provided by Harvey *et al.* (1975), who found that the average ER numbers in 1970 and 1973 were 373 and 179, respectively. The average

flux of an ER is about 10^{20} Mx, while that of a common active region (AR) is 4×10^{22} Mx. Because ERs are much more than ARs, the total flux carried to the solar surface by ERs may be several times more than in all spots (Zirin, 1988). So the cancellation and emergence of ERs can be considered to be the main sources for the loss and replenishment of magnetic flux of the quiet Sun.

(5) *Emerging flux regions (EFRs).* The emergence of magnetic flux constantly takes place on the solar surface. The typical process of evolution of an EFR consists of the following steps: the appearance of an initially small and bright bipolar plage, the growing separation of the poles, the formation of an arch filament system and the coalescence of magnetic elements into spots. Several EFRs can form a large AR.

EFRs obey the law of Hale and Nicholson, and so they are very probably associated with the global magnetic fields involved in the solar activity cycle. Such fields are expected to be deeply rooted in the solar interior. As estimated by Zirin (1988), their strength increases with depth faster than the density and at $R = 0.8\ R_{\odot}$ it may be as strong as 10^4 G. Therefore, from the study of EFRs one might get some information about the invisible controlling magnetic structure in the solar interior.

(6) *Evolution of small-scale fields.* The various magnetic structures described above may be divided into two categories: (i) *discrete bipolar regions*, including entire series of ARs from large ones containing spots to very small ERs; and (ii) *mixed-polarity IN fields.* There are some remarkable differences between these two groups. Almost all bipolar regions observe Hale–Nicholson's law of polarity, but the IN fields do not. Besides, bipolar regions are approximately symmetric to certain preferred latitudes, while the IN fields are not . However, the appearance and disappearance of magnetic structures of all kinds are practically the same. To be precise, the appearance is realized through the emergence of new flux from the solar interior and the disappearance through (i) cancellation of closely spaced flux elements with opposite polarities, (ii) fading *in situ* without observed dispersion, and (iii) fading with dispersion.

According to the study of Livi *et al.* (1985) using BBSO videomagnetograms, route (i) is the dominant one. As an example, four sites of magnetic flux cancellation are marked with incomplete rectangles in Figure 5.4. The process of cancellation at each site can be traced from the digital magnetograms obtained at consecutive moments for one and the same solar region. Besides, these authors stated that the cancellation is often preceded by the transport of flux fragments towards each other. Furthermore, the transport may be associated with the following types of motion: (i) the convective flow of IN fields from cell interiors to boundaries, (ii) the growth and separation of the poles of ERs, (iii) the migration of network presumably along the boundaries of network cells, and (iv) the direct motion of opposite polarity features towards one another. Moreover, the cancellation is generally accompanied by a gradual increase of the gradient of the magnetic field close to the site of cancellation. This can be thought to be a natural consequence of the migration and coalescence of small flux fragments.

5.3 Magnetic Fields of Active Regions

Solar active features often appear together in some specific districts, called *active regions*. They are characterized by *local magnetic fields* that are much stronger than the fields of quiet regions. Astronomers began to study the AR magnetic fields more than seventy years ago, but most earlier works were devoted to certain separate aspects of individual ARs. In recent years Howard (1989, 1990, 1991a–f) carried out systematic analyses of a series of statistical characteristics of AR magnetic fields with the daily magnetograms accumulated on Mt. Wilson in 1967–1990. Observations were made in Fe I λ 5250 for the whole solar disk. Magnetic measurements were averaged in the initial data reduction step into "coarse arrays" with 34 equal steps in the sine of the latitude (φ) by 34 equal steps in the sine of the central meridian distance (CMD). Each resolution element has an area of 1.68×10^{19} cm^2 or 550 millionths of a hemisphere. A magnetic region is defined to be an accumulation of contiguous resolution elements with absolute field strength of 10 G or greater. It must consist of at least two elements with different magnetic polarities. Altogether 8976 regions were investigated, which appeared in the period from January 1, 1967 to April 9, 1990. Some selected topics of interest will be briefly described in the following.

(1) *Magnetic polarity orientation.* The average angular distance between magnetic polarities in an AR is about 7°. The line connecting them has a certain orientation, which can be expressed by the angle from the positive polarity toward the negative one, with 0° being the vector pointing west and 90° the vector pointing north. The relation between the polarity orientation and the average region flux is shown in Figure 5.5. It is interesting to find that near the normal orientations, i.e., 0° and 180°, the average region flux is small, but it rises rapidly on either side and then decreases gradually. The relation of the polarity

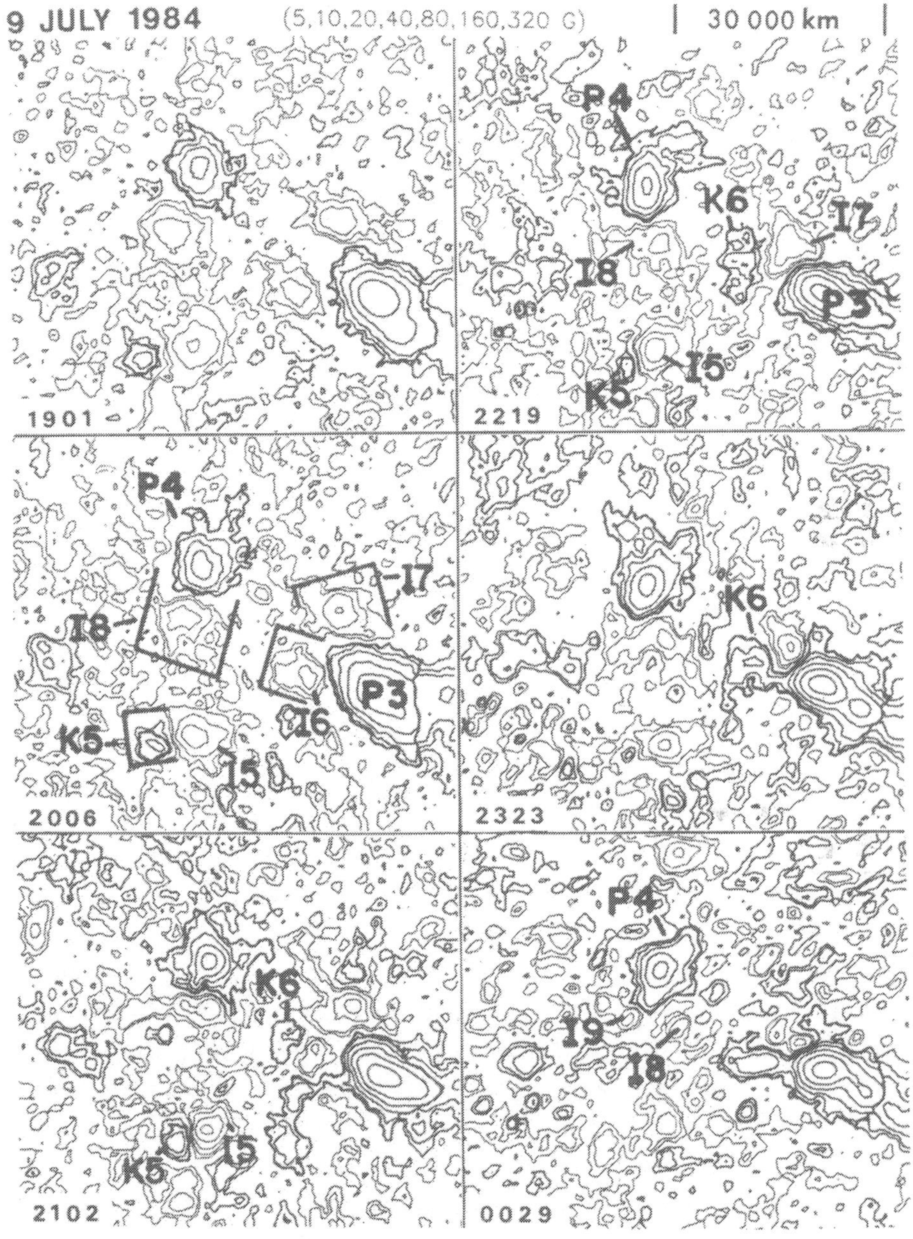

Fig. 5.4. Cancellation of small-scale magnetic fields.

orientation with the region area can be expressed by a very similar plot. This means that smaller regions tend to show larger deviations on average from the usual polarity orientation.

(2) *Tilt angle of magnetic axis*. This quantity is defined as the angle

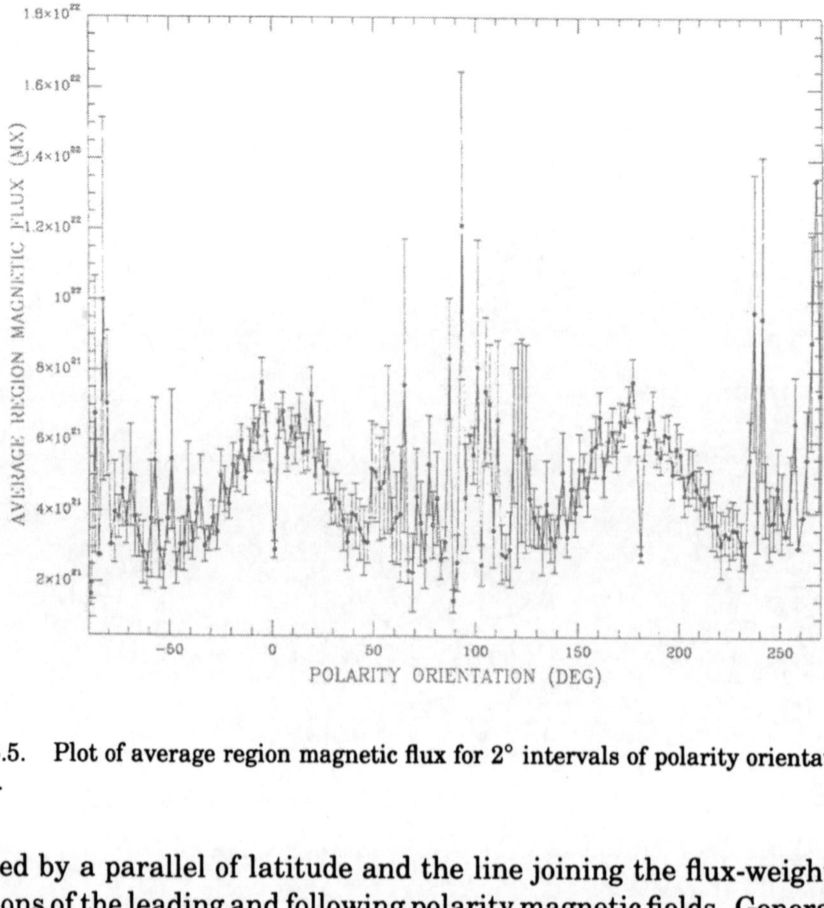

Fig. 5.5. Plot of average region magnetic flux for 2° intervals of polarity orientation angle.

formed by a parallel of latitude and the line joining the flux-weighted portions of the leading and following polarity magnetic fields. Generally speaking, the leading fields are nearer to the equator and in this case the tilt angle is said to be positive. Howard determined the daily variations of the tilt angle and found that regions with positive tilt angles show negative daily changes, viz., the tilt angle decreases and *vice versa*. Besides, the larger the tilt angle of either sign, the larger the average daily change. These relations are illustrated in Figure 5.6.

(3) *East–west inclination of magnetic field lines*. The inclination angle α can be determined with the east–west flux ratio defined as

$$\beta = \frac{|F_E| - |F_W|}{|F_E| + |F_W|}, \tag{5.4}$$

where F_E and F_W are the magnetic fluxes measured at some locations equidistant (east and west) from the central meridian, respectively. Taking a to be the value of CMD, we have the following relation between

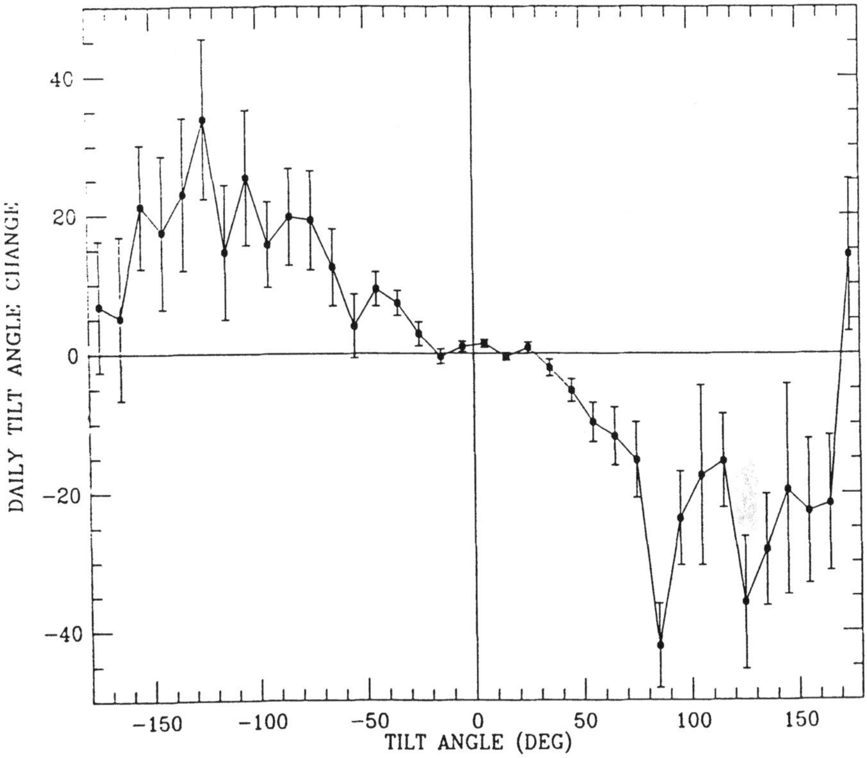

Fig. 5.6. Average daily tilt angle changes in deg day^{-1} for various tilt angles.

α and β:

$$\alpha = \tan^{-1} \frac{\beta}{\tan a} \,. \tag{5.5}$$

From data of 8422 regions the connection between the average magnetic flux per region and the CMD has been derived. This can be seen in Figure 5.7 in which the negative CMD values belong to the west side of the solar disk and positive ones to the east. From the asymmetry of the curve it has been found that the average inclination of a whole region is such as to trail the solar rotation (namely, to incline toward the east) by about 1$^\circ\!\!.$9. Moreover, Howard discovered that the leading and following polarities of AR magnetic fields tilt toward each other by about 16°. Besides, the average inclinations of growing and decaying regions are different. As for the relation with the activity cycle, the situation seems to be more complex. For the leading portions of ARs no remarkable variation of the inclination throughout the cycle is found,

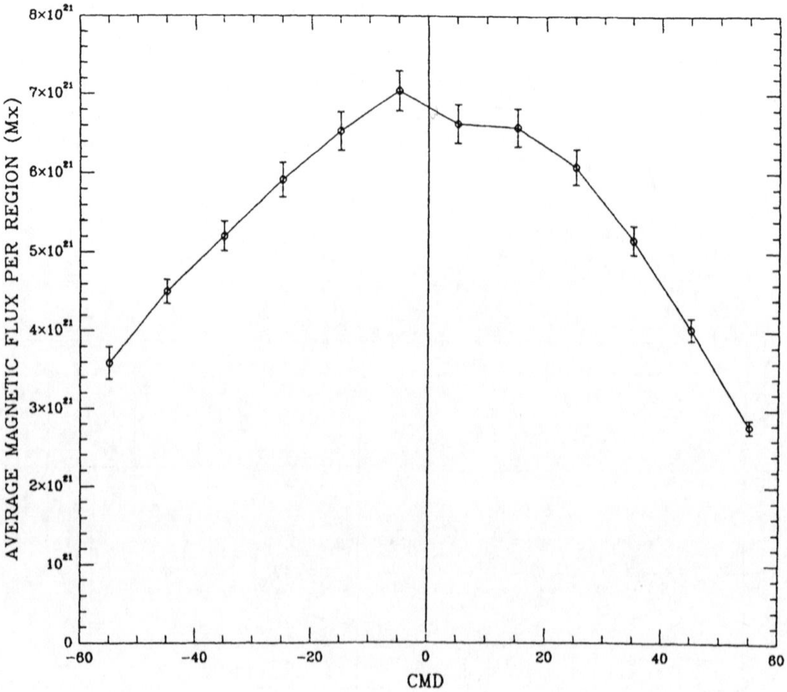

Fig. 5.7. Change of average magnetic flux with CMD.

but the following magnetic fields show a steady decrease from about 10° (leading the rotation) during the rise to maximum, to 3° at minimum.

(4) *Growth and decay of magnetic flux.* The total magnetic flux ($F^T = |F^+| + |F^-|$) for each region on each day is derived from the sum of the absolute field values for each polarity in every region with field strength larger than the lower limit (10 G). These values, multiplied by the area of the resolution element and summed, yield the total magnetic flux of the region. The average daily percentage flux change has been found to be a function of the region size (in units of 1.68×10^{19} cm^2). This is shown in Figure 5.8. The increasing and decreasing flux changes are plotted separately. It is worthwhile noting that the daily percentage flux growth is approximately twice as great as the percentage flux decay. Besides, relationships are found between daily flux change and hemisphere, cycle phase, leading and following polarities, rotation rate, etc. In general, both the growth and decay of flux are faster in more active situations: at a cycle maximum or in the more active hemisphere. Moreover, leading polarity fields grow and

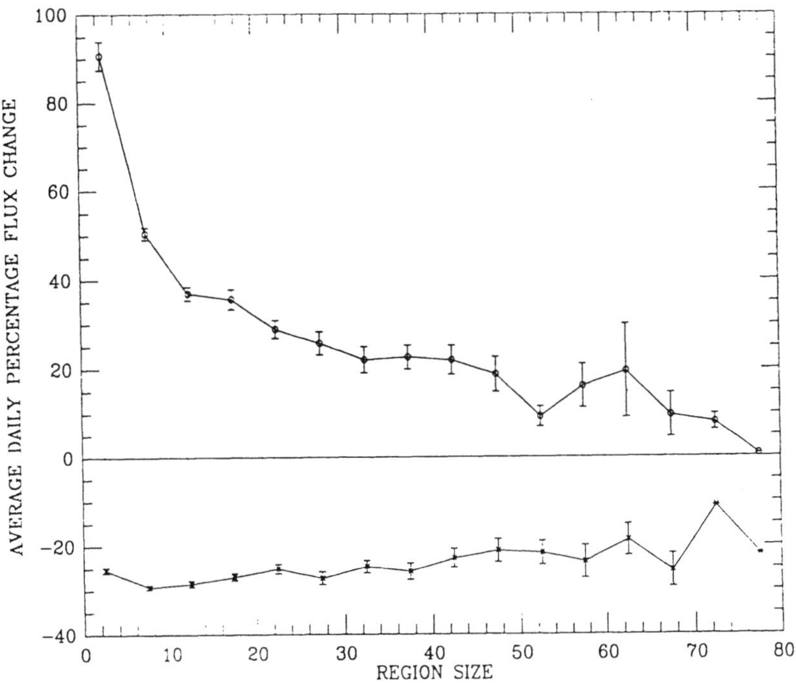

Fig. 5.8. Average percentage daily magnetic flux change for regions of different sizes.

decay more rapidly than following polarity fields.

(5) *Rotation rate.* It is well known that the Sun rotates in a differential fashion. In other words, the angular velocity (ω) is a function of the latitude (φ). The average rotation rate can be expressed as $\omega = 13.07 - 1.83 \sin^2 \varphi$. The rate of rotation of AR magnetic fields has been found to be smaller than the average rate of the undisturbed atmosphere, and the difference is called the *rotation residual.* As shown in Figure 5.9, this quantity varies with the size of the AR. Moreover, it is concerned with the phase of the activity cycle. From Figure 5.10 it is readily seen that around 1976 and 1986, i.e. in the two recent solar minima, the rotation residual attained peak values.

(6) *Meridian drift.* For a long time the meridian motions of the solar material have been measured with the Doppler effect and various tracers. Due to the low amplitude of these motions, such measurements are very difficult. A better method is to observe the magnetic flux weighted positions of ARs. In this way Howard discovered that the daily meridian drift is a function of the latitude (see Figure 5.11). The

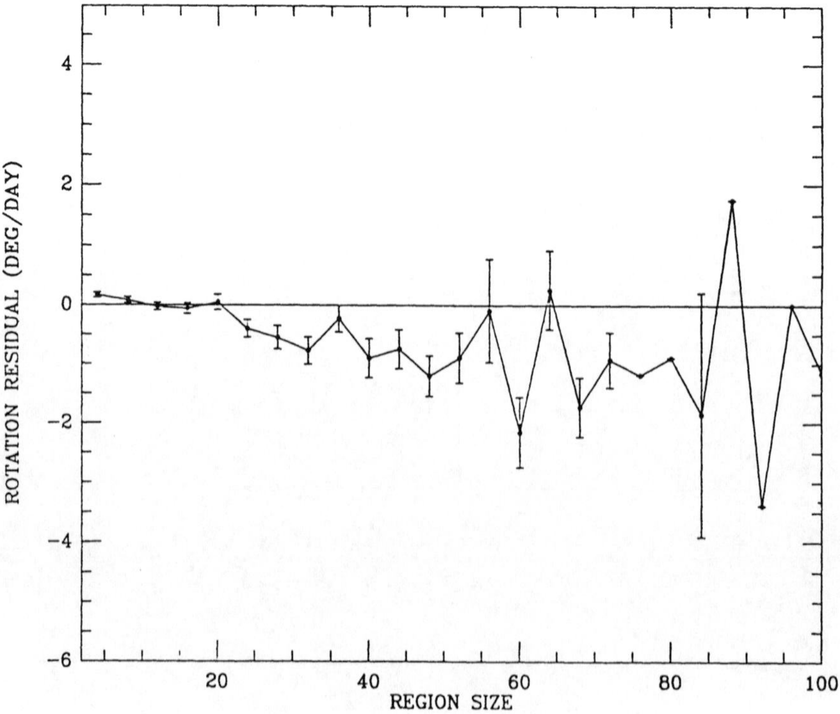

Fig. 5.9. **Residual rotation rate plotted against region size.**

drift is expressed in units of deg day^{-1} ($\simeq$ 140 m s^{-1}) and the positive drift is poleward in both hemispheres. The figure tells us that the amount of daily drift decreases with the increase of latitude. Besides, the meridian motions might be concerned with the phase of the solar cycle. But the Mt. Wilson magnetograms cover only about two cycles and are insufficient to provide a reliable conclusion.

In the above we have made a concise account of some of the characteristics of the AR magnetic fields revealed by Howard. Most of them can hardly be foreseen by intuition and are very probably concerned with the mechanisms of formation and evolution of solar ARs. For their thorough understanding and interpretation much still remains to be done.

5.4 Large-Scale Magnetic Fields

On the solar surface there are some magnetic structures which are larger than ARs, and they are called *large-scale fields*. The complexes

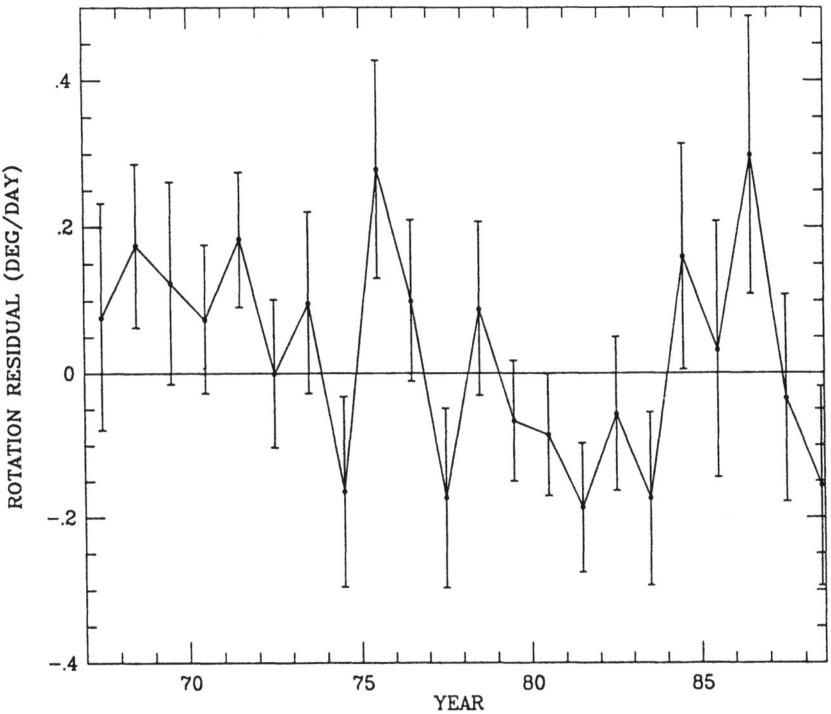

Fig. 5.10. Rotation residual plotted against time.

of activity introduced in Section 4.6.3 possess such fields. Their observation and interpretation are necessary for an understanding of the evolution of solar magnetism and the nature of the solar activity.

(1) *Arrays of large-scale magnetic fields.* When the magnetograms of consecutive days within a period of solar rotation (about 27 days) are projected on a plane and combined together, they form the so-called *synoptic charts*, which may be conveniently used for the inspection of large-scale fields. An example is provided by Figure 5.12 in which the districts surrounded by solid lines are magnetic fields with positive or N polarity, while those by dotted lines represent the negative or S polarity. In order to reduce the influence of differential rotation, one may extract from succedent charts the parts belonging to one and the same latitude interval and put them together. For instance, Figure 5.13 shows the magnetic field configuration in the latitude range $+20°$ to $-20°$ in 16 consecutive periods of solar rotation, beginning from the 1726th. The gray districts represent fields of N polarity and the dark ones S polarity. It is evident that in the neighboring stripes magnetic fields of both

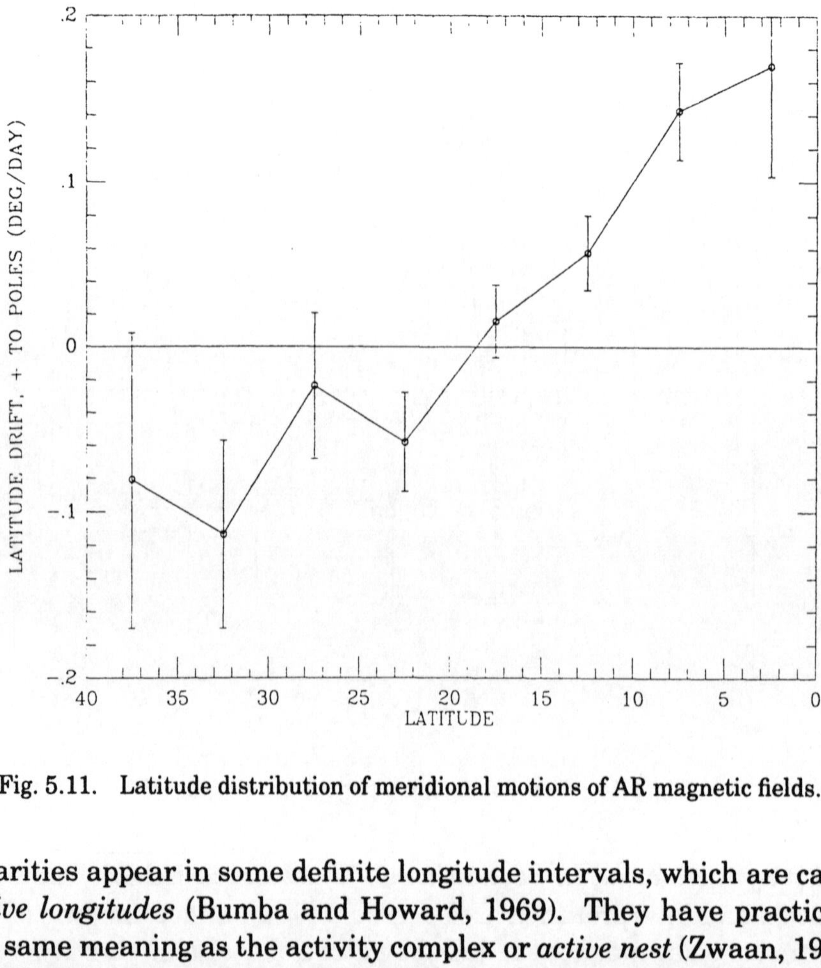

Fig. 5.11. Latitude distribution of meridional motions of AR magnetic fields.

polarities appear in some definite longitude intervals, which are called *active longitudes* (Bumba and Howard, 1969). They have practically the same meaning as the activity complex or *active nest* (Zwaan, 1987).

(2) *Resonant modal structure of solar magnetic fields.* In order to interpret the complex pattern of the large-scale field and its evolution, Stenflo and his associates (Stenflo, 1984, 1986b, 1988; Stenflo and Gudel, 1987; Stenflo and Vogel, 1986; Stenflo and Weisenhorn, 1987) represented the pattern with spherical harmonics, which for a spherical surface are the orthogonal functions. When a wave equation is used to describe perturbations in a spherical system, the spherical harmonics are the eigenfunctions corresponding to various eigenvalues, i.e., resonant frequencies. The modal structure can be expressed by the coefficients of spherical harmonics C_l^m, where l and m may be called "quantum numbers". Using the power spectrum analysis of the time series of these coefficients, the resonances in the modal structures can be found. Stenflo *et al.* used the Mt. Wilson and Kitt Peak synoptic maps of 1959–1984 and chose modes with $m = 0$, which are symmetric

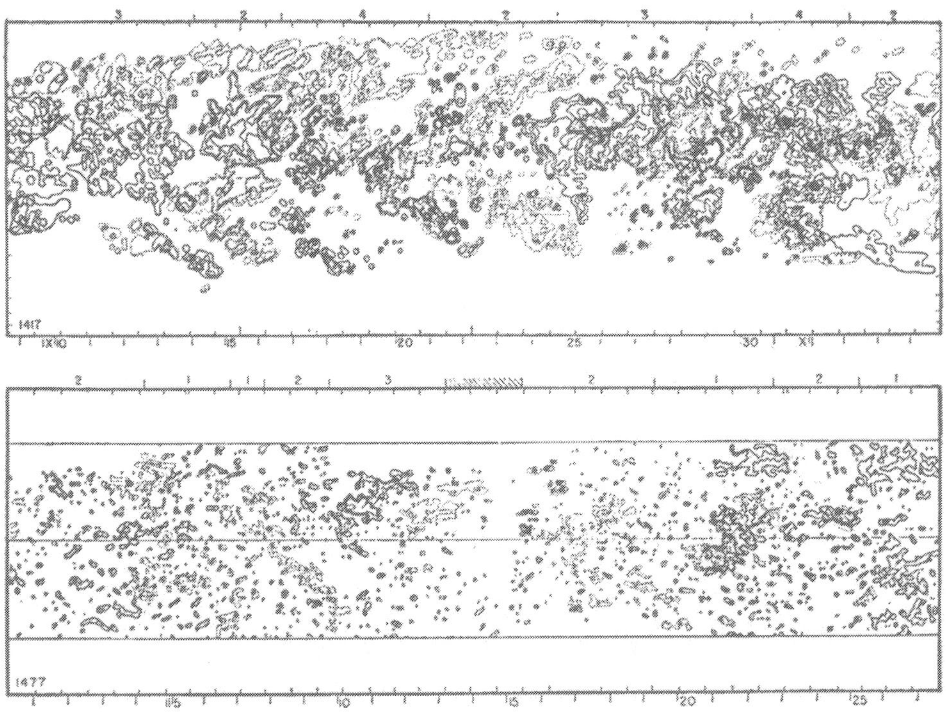

Fig. 5.12. Magnetic synoptic chart for solar rotation No. 1417 (August 1959).

around the rotational axis and independent of the longitude system. They discovered that the behavior of the power spectra of C_l^0 is different for odd and even l's. The odd l's represent field patterns which are anti-symmetric with respect to reflections on the equatorial plane, while the even l's are symmetric patterns. For the odd modes, almost all the powers are concentrated around the resonant frequency, which corresponds to the 22-year magnetic cycle (see left side of Figure 5.14).

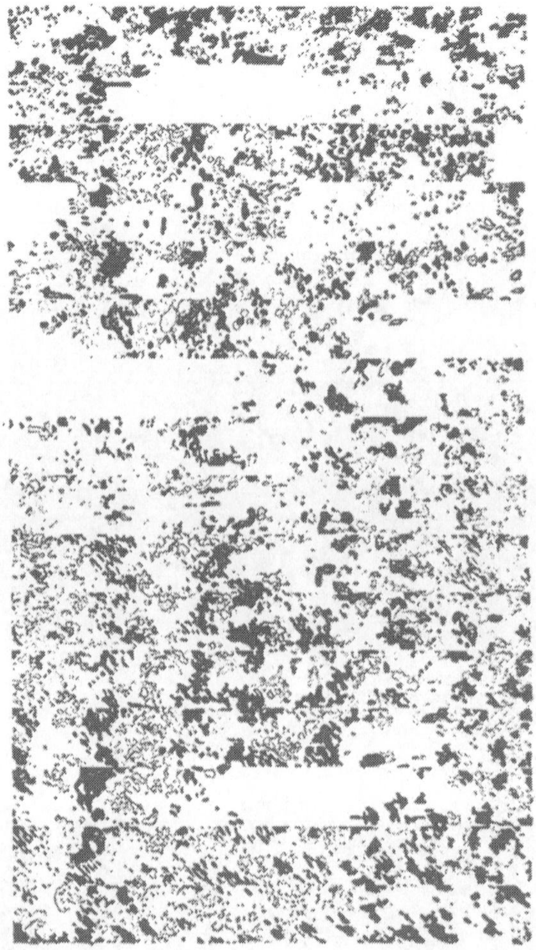

Fig. 5.13. Arrays of large-scale fields near the solar equator.

For the modes of even parity, a smooth curve can be drawn through the maxima of power spectra (right side). For $l = 2$, the resonant period is about 14 years, and for $l = 14$ it is 1.4 years. This l-dependence is reminiscent of the p-mode oscillations of the Sun. The interpretation of these diagrams may provide information on the origin of the solar magnetism and activity and also on the magnetic field in the solar interior.

(3) *Theory of the evolution of large-scale fields.* As pointed out by Leighton (1964), the random motions of the flux elements in supergranular patterns may cause the diffusion of large-scale magnetic fields.

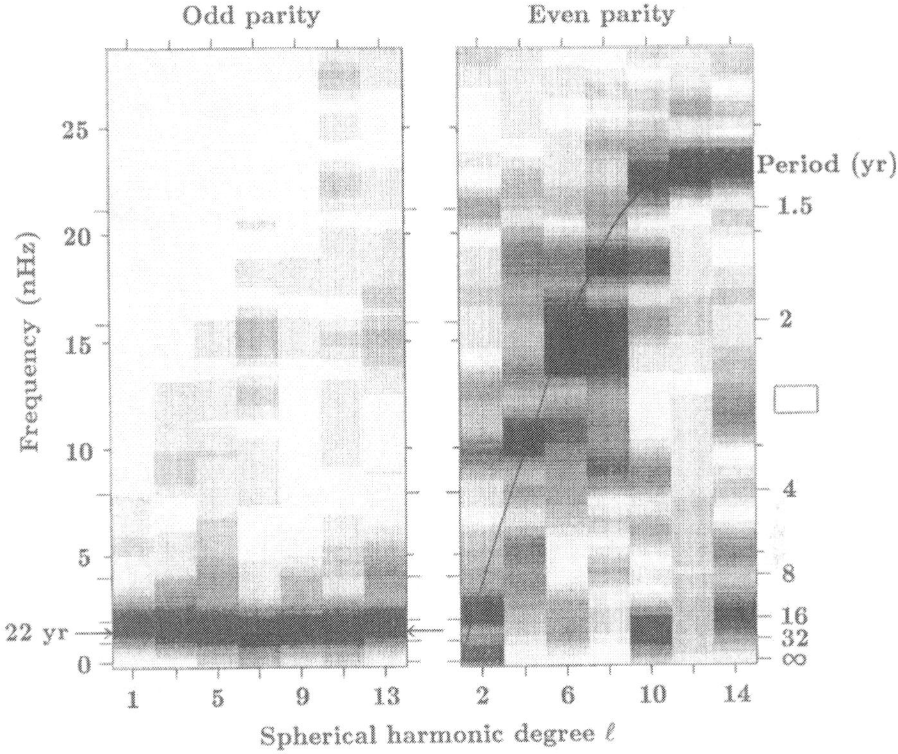

Fig. 5.14. Power spectra of harmonic coefficients as function of frequency in nHz and spherical harmonic degree l (for $l \leq 14$).

This can be described by the following *transport equation*:

$$\frac{\partial B}{\partial t} + \frac{1}{R \sin\theta} \frac{\partial}{\partial t} [B v(\theta) \sin\theta] + \cos\theta \frac{\partial B}{\partial \phi}$$
$$= \frac{k}{R^2} \left[\frac{1}{\sin\theta} \frac{\partial}{\partial \theta} \left(\sin\theta \frac{\partial B}{\partial \theta} \right) + \frac{1}{\sin^2\theta} \frac{\partial^2 B}{\partial \phi^2} \right], \tag{5.6}$$

where $B(0, \phi, t)$ is the radial component of the photospheric magnetic field, θ is the colatitude, R is the solar radius, $v(\theta)$ is the meridional flow velocity, and k is the rate of diffusion. This equation was solved both analytically and by numerical simulation (Leighton, 1964; Schatten *et al.*, 1972; Sheeley *et al.*, 1987; De Vore and Sheeley, 1987). According to De Vore (1987), in the absence of bipolar sources of flux the large-scale photospheric field decays due to the differential rotation, meridional flow and supergranular diffusion. He solved the transport equation analytically for both the equatorial and polar regions and studied the time-asymptotic behavior of large-scale fields.

In order to study the evolution of large-scale fields, the emergence of new bipolar flux sources should also be taken into account and for this a source term $S(\theta, \phi, t)$ has to be added to the right-hand side of Eq. (5.6). Thus the constant emergence of new flux elements from the solar interior and the decay of old ones by diffusion constitute the complete picture of the evolution of solar large-scale magnetic fields.

5.5 Solar Flares and Magnetic Fields

Flares are the most intense solar active features. Generally speaking, flares exist for only about ten minutes, and even the largest ones last no more than one to two hours. However, in such short time intervals they emit a great deal of energy. (For large flares the total amount of energy may attain 10^{32}–10^{33} erg.) Besides visible light, flares can emit intense UV radiation, X- and γ-rays, radio waves, streams of energetic particles and even cosmic rays. Because these radiations may give rise to interruptions of radio communication, polar lights and magnetic storms and cause fatal damage to interplanetary navigation, in recent decades much attention has been paid to the study of solar flares.

How are flares formed? Why can very large amounts of energy accumulate and suddenly be released in flares?

Although these problems are far from being completely solved, much progress has been achieved in the physics of flares in recent decades. It has been firmly established that the magnetic field plays the most important role in the formation, energetics and development of flares.

5.5.1 Flare Energy Source and Storage

It is generally agreed that the flare energy comes from stressed (i.e., non-potential) magnetic fields. This can be explained by arguments showing that there is no other energy source of sufficient capacity. Simple computations tell us that both the thermal and gravitational potential energies are too small to satisfy the need of flare outburst (Tandberg-Hanssen and Emslie, 1988). Besides, the temperature of the preflare atmosphere is too low to induce nuclear fusion. Then the unique candidate seems to be the electromagnetic energy. It can be readily learned that a magnetic field of strength $B = 10^3$ G and volume $V = 10^{27}$ cm^3 contains magnetic energy $B^2V/8\pi = 4 \times 10^{31}$ erg, which is enough for a flare of moderate size.

The next question is the *energy storage* in the magnetic field. It is important to note that energy can be stored only in a sheared or stressed field, while a relaxed, potential field cannot provide any extractable energy. This is because the potential field is in the state of minimum energy. If the flare energy is stored *in situ*, i.e. at the site of flare outburst, some specific magnetic field configuration is required. Because the initial energy release is commonly believed to take place in the corona (de Jager, 1985; Kahler *et al.*, 1986) where the gas pressure is 2–3 orders of magnitude smaller than the magnetic pressure, so specific shapes are needed to store up the energy and keep the magnetic field in stability. It is also possible that energy is accumulated at a location distant from the flare site. If this site is the top of a *magnetic loop*, the energy may be stored at its foot-points located in the photosphere and transported upward through the loop (Spicer, 1982; Henoux, 1983).

The flare energy can also be conserved as the magnetic energy of the *neutral sheet* created in the vicinity of a magnetic field singular line by the focusing of disturbances (Syrovatsky, 1978; Podgorny, 1989). In a simple case the singular line is a circular zero line between two magnetic dipoles of the same orientation placed in line one after another. When the distance between dipoles changes, the variation of fields with circular zero lines leads to the current sheet development. In general a singular line has a series of properties, including the following one. On the singular line there is no magnetic field component perpendicular to it, so the induced electric field along it created by variation of the magnetic flux cannot be compensated by the field $(\mathbf{v} \times \mathbf{B})/c$.

5.5.2 Sudden Release of Flare Energy

Now that the flare energy is stored in a magnetic field, its sudden release is expected to be due to the abrupt change of the field configuration. As shown by observations, flares often occur on or near a *magnetic neutral line* (also called *polarity inversion line*). Such lines can be formed by various magnetic configurations. For example, in Sweet's (1958) model of two approaching local magnetic districts D_1 and D_2 (Figure 5.15) the oppositely directed field lines come together and flatten, like the tires of a heavenly loaded car. They form a neutral current sheet around the magnetic neutral line, whose intersection with the paper plane is denoted by "s" in the figure. Owing to the absence of magnetic pressure, the gas pressure alone cannot keep the magnetic walls apart. At some critical moment they suddenly merge together and annihilate each other. (In other words, the *magnetic reconnection* takes place.) Then the released magnetic energy is transformed into the kinetic energy of the plasma. Under the action of some additional mechanism

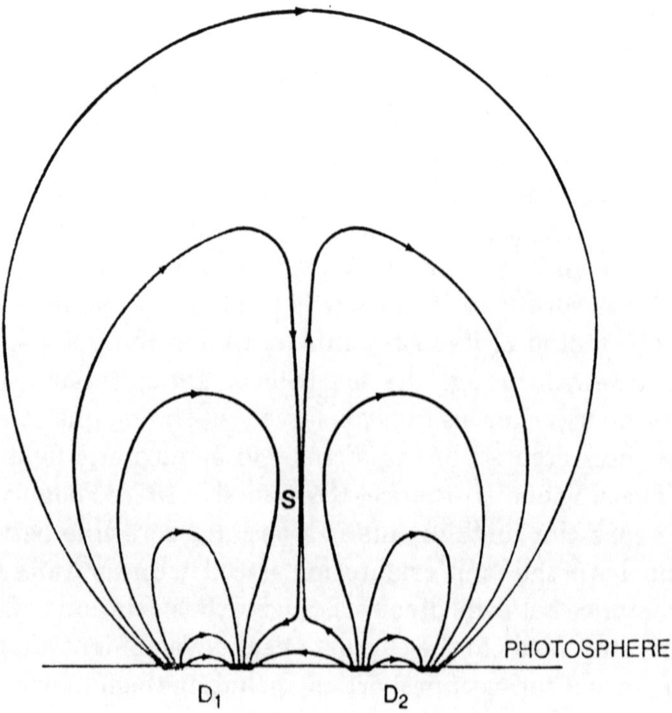

Fig. 5.15. Neutral current sheet in Sweet's flare model.

(Švestka, 1976) the rate of energy dissipation can be greatly enhanced and then a flare occurs in the neutral sheet.

Similar to the above model of Sweet, Gold and Hoyle (1960) proposed another flare model, which is also based on the merging of two magnetic flux tubes. The difference is that these flux tubes are parallel and attract each other. On the surface of their merging a neutral sheet is formed. In Sturrock's (1968) model a helmet streamer produces a neutral sheet terminating at a neutral point, which is situated at the site of bifurcation of an inverted Y-shaped region (Figure 5.16a). Magnetic reconnection takes place around the neutral point. The magnetic lines of force beneath it are closed, and the X-rays and Hα emission are produced in the enclosed region. The field lines above the neutral point are open and high-energy particles are ejected away along them (Fig. 5.16b).

Now it is worthwhile mentioning that the magnetic reconnection plays an important role not only in the energy release of flares, but also in the *cancellation of magnetic features*, *coronal heating* and *dynamo*

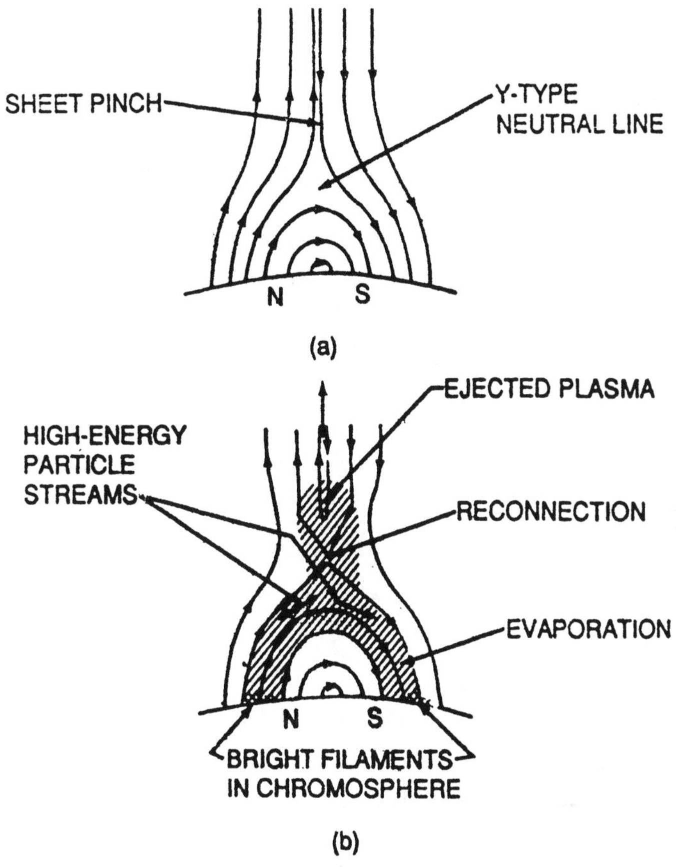

Fig. 5.16. Sturrock's flare model.

action. For recent progress of the magnetic reconnection theory the reader is referred to Priest (1990).

5.5.3 Changes of Local Magnetic Fields Before and After Flares

Almost half a century ago it was noted that flares are associated with variations of sunspot areas and magnetic flux of active regions (Kiepenheuer, 1953a). Several years later Severny (1958a) used the Crimean longitudinal magnetograms to study the *relation between flares and the neighboring magnetic fields*. He discovered that the bright kernels of Hα flares in the initial stage are located close to the neutral line with $B_{\parallel} = 0$. An example is given in Figure 5.17, in which the solid lines represent $B_{\parallel}$ isogauss contours, the broken curve is the magnetic neu-

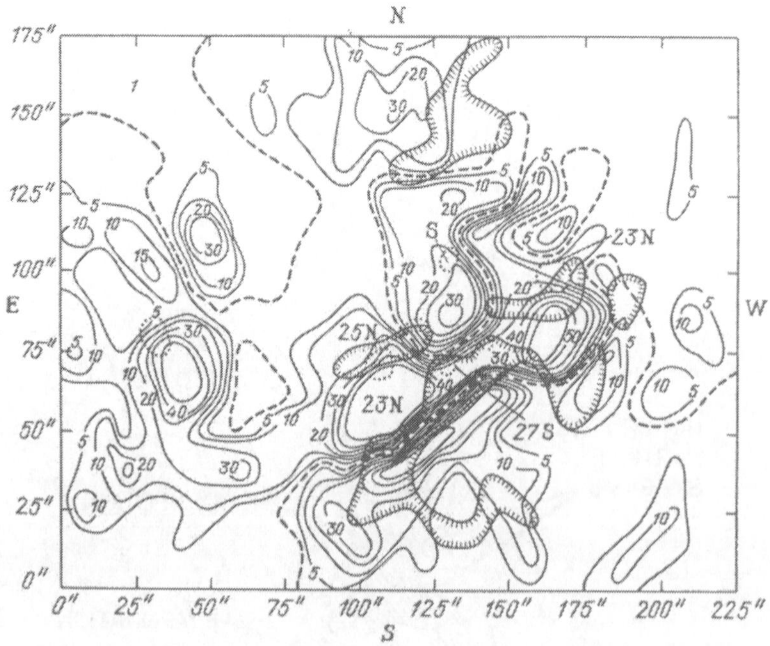

Fig. 5.17. August 18, 1959 flare of importance 3 and neighboring magnetic fields.

tral line and the dotted lines are boundaries of sunspot umbrae. The flare kernels are marked by closed curves with short straight lines on the inner side. Moreover, he (Severny, 1960a) found for 20 flares that they occurred in active regions with complex magnetic fields and large gradients of field strength ($\nabla B_{\parallel} > 0.1\ \mathrm{G\,km^{-1}}$). The local fields around nine flares were recorded completely and it was noted that their flaring process led to the *simplification of field structure* as well as to the *decrease of their gradients* ($\nabla B_{\parallel}$ fell to 0.01–0.02 $\mathrm{G\,km^{-1}}$). The same results were obtained for 51 large flares (Gopasyuk *et al.*, 1963).

The work of Severny and his associates aroused great interest and attention of many astronomers. Some of them (e.g., Bumba, 1958; Bruzek, 1960; Antalova, 1961) confirmed their conclusions with new observations. Nevertheless, more people got negative results, and some instances follow. From an analysis of the Kitt Peak longitudinal magnetograms, Livingston (1973) claimed that the local magnetic field remained unchanged after the occurrence of the 3B flare of August 4, 1972. Rust (1976a) found that, except for some intense flares with large proton events, the changes of magnetic fields were the same whether flares appeared or not. Harvey *et al.* (1970) remarked that local magnetic fields change with a time scale of hours, and that they might be attributed to the global evolution of active regions and not directly concerned with the occurrence of flares. It is still more impressive to note the following fact. Howard and Babcock (1960) could not detect any remarkable change of magnetic field related to the large flare of July 16, 1959, but Severny claimed that after this flare the neighboring $B_{\parallel}$ diminished by three times! The large flares in May 21–23, 1967 also provide examples which contradict Severny's conclusions (Malville and Tandberg-Hanssen, 1969). There were thus many contradictions and debates. In order to clarify the situation, more accurate observations are required.

Recently Lin Yuanzhang *et al.* (1993) studied the Huairou magnetograms of a small flare of October 7, 1987. Their observation has the following merits:

(1) Both the spatial resolution (1.5 arcsec) and temporal resolution (1/12.5 sec) are fairly high.

(2) The longitudinal and transverse fields were measured almost simultaneously and their accuracies are, respectively, 20 and 150 G.

These authors discovered that before and after the flare the *transverse* fields were evidently different, while the *longitudinal* fields had no detectable changes. To be precise, after the flare the directions of strong transverse fields near the upper left part of the flare kernel D and the upper right part of kernel E changed from the approximate E-W orientation to the N-S one. Besides, the transverse fields in the vicinity of kernels A and B also exhibited some variations. (All these facts are shown in Figure 5.18.)

From the above we can say that after many years' controversy and bewilderment it has now become clear that the outburst of flares is accompanied by intrinsic changes of local magnetic fields. The essential point is that *these variations belong basically to the transverse field.* On this the observational evidence provided by Hagyard (1990) and described in Section 3.4.1 is convincing. Quite recently Haimin Wang (1993) produced supplementary data. Besides to Hagyard's prerequisites — i.e. (1) large angular shear, and (2) strong transverse field. A third condition has also to be satisfied. This is a long sheared magnetic neutral line. Therefore, for the study of the outburst of flares and for their prediction the *vector magnetometry* can be considered to be indispensable.

What has been narrated above is the variation of local magnetic fields *after* flares. It is still necessary to say that certain changes of the photospheric magnetic fields have been found *before* flares. As observed by Canfield *et al.* (1974) and justified by Canfield and Fisher (1976), Martin *et al.* (1983) and Rust (1976b), prior to the appearance of flares new magnetic fluxes emerge from below to the photosphere and their convergence with old ones may lead to the occurrence of flares.

5.5.4 Flares and Cancelling Magnetic Fields

In the preceding subsection we discussed the alterations of magnetic field configuration and strength associated with flares. It naturally follows that the magnetic flux may also vary significantly. Using the observational data of six observatories, Mayfield and Chapman (1981) discovered that the fluxes of active regions, where flares occurred, did have large changes. For instance, just after the outburst of flare of August 4, 1972 the flux amounted to 7.2×10^{22} Mx and two hours later it diminished to 5.6×10^{22} Mx. However, the situation may be more complex. Martres *et al.* (1968a, b) found that in an active region where a number of flares appeared, the magnetic flux on one side of the polarity inversion line increased, while the flux decreased on the other side. Although many flares are associated with new magnetic fluxes, there are also flares which happen in the absence of emerging flux (Martin *et al.*, 1984). However, in spite of all these complexities, the following regularity generally holds: Magnetic flux is observed to decrease gradually in regions of both polarities as they migrate together and then a high magnetic field gradient comes into existence until the outburst of the flare.

This naturally leads to the idea that the process of *magnetic field cancellation* may be common to flares.

Livi *et al.* (1989) studied the relation between flares and cancelling magnetic features. They used the BBSO longitudinal magnetograms and Hα filtergrams and found that flares may occur in the following three locations:

(1) in the vicinity of emerging magnetic flux regions, but at places where cancellation has been observed or inferred;

(2) at sites where the magnetic flux is increasing on one side of a polarity inversion line and concurrently decreasing on the other side (Martres *et al.*, 1968a, b; Ribes, 1969);

(3) at positions where cancellation is the unique observed change for at least several hours before the flare (Martin *et al.*, 1985).

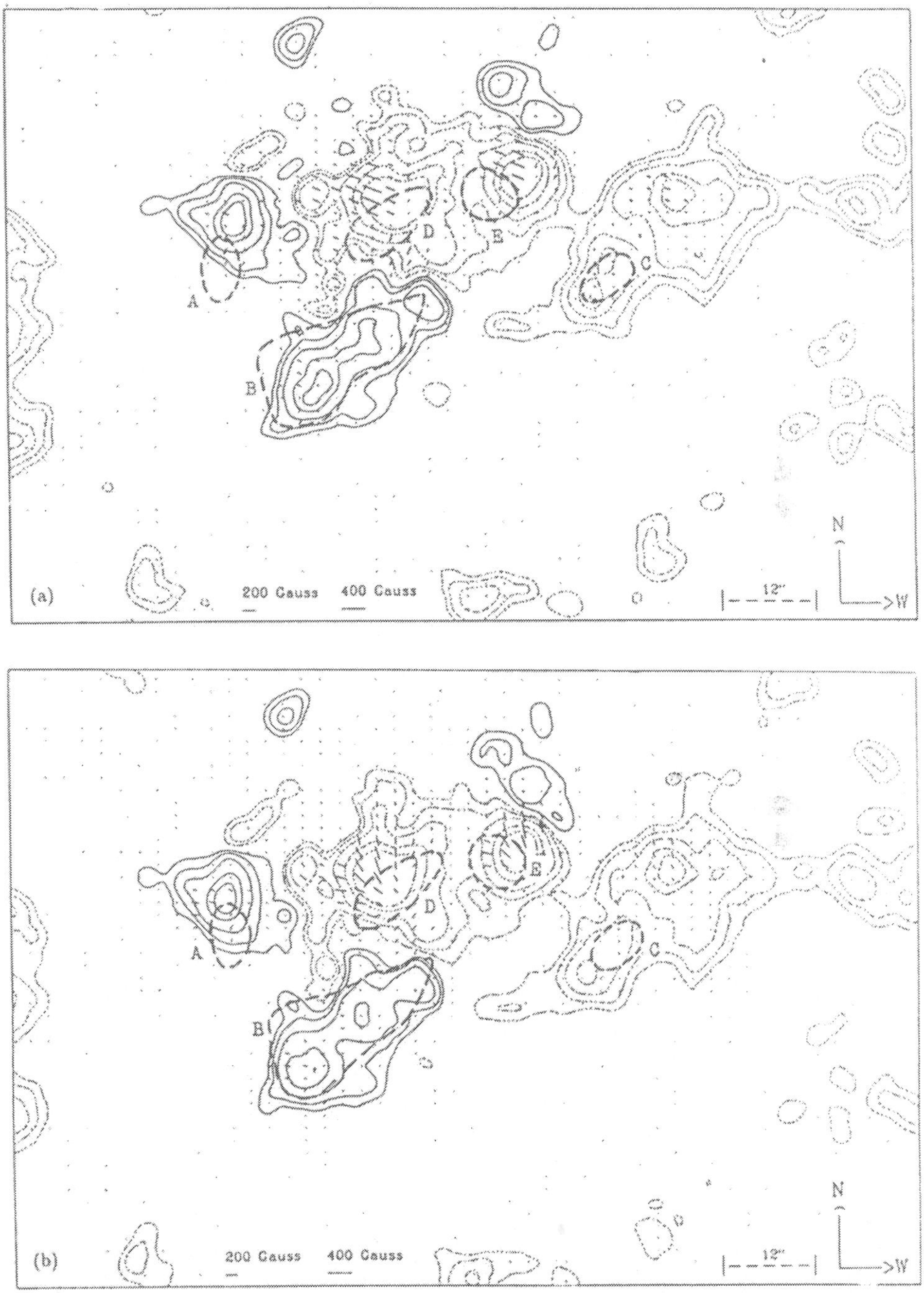

Fig. 5.18. Magnetic field changes connected with a flare of October 7, 1987. (a) magnetogram before flare: $B_{\parallel}$ (00:35 UT), $B_{\perp}$ (00:19 UT). (b) magnetogram after flare: $B_{\parallel}$ (01:14 UT), $B_{\perp}$ (01:24 UT).

Since in all these circumstances, cancellation is the sole common factor, Livi *et al.* proposed that it may be considered to be a necessary precondition for the occurrence of flares.

5.6 Magnetic Fields of Solar Prominences

Many years ago it was foreseen that prominences possess magnetism. This idea was based on the following considerations. First, the trajectory of motion of active prominences and the shape of loop prominences are like magnetic lines of force. Second, the prominence is a cluster of plasma with comparatively low temperature (10^4), but it can be formed and exist for a long time in the hot corona (10^6 K). This may be due to the effect of *heat insulation of magnetic field lines* which wrap up the prominence. Besides, the density of prominences (about 10^{10}–10^{11} hydrogen atoms per cm^3) is much greater than that of the corona. It is possible that prominences are suspended in the coronal space by the support of magnetic lines of force.

All these possibilities were merely speculations in the past, and they have been justified by formal observations and theoretical calculations since the beginning of the sixties.

Zirin was the pioneer in the exploration of prominence magnetic fields. He (Zirin, 1961; Zirin and Severny, 1961) used the Crimean photoelectric magnetograph and preferred to Hβ instead of the commonly adopted working lines, such as Fe I $\lambda\lambda$5250, 6173 and 6302, which are too faint in the prominence spectra to be measured reliably. He found that the longitudinal field strength of quiescent prominences (QPs) is about 50 G, while that of active prominences (APs) is much larger and may attain 200 G. Also, according to his observations, the magnetic polarity of QPs is not concerned with the magnetic polarity of the photospheric district below them, but APs always have the same polarity as that of the neighboring solar surface. A natural explanation is that APs together with their magnetic fields rise from the underlying photosphere and/or chromosphere.

Thanks to the efforts of many other solar physicists, much further progress has been achieved in the study of prominence magnetic fields in recent years. Considering that Tandberg-Hanssen (1974), Leroy (1979) and Kim (1990) have made excellent reviews on this subject, here we need only to briefly mention the following work.

(1) *Prominence magnetographs.* An instrument specially designed for prominence magnetometry (Lee *et al.*, 1965) was operated in 1964–1971. It was attached to a 40 cm coronograph and so the scattered light

and instrumental polarization were for the most part eliminated. A spectral scanning magnetograph was built by Nikolsky *et al.* (1982) and installed at the coudé focus of the Kislovodsk 53 cm coronograph. These prominence magnetographs yielded a lot of fine observational data (see, e.g., Harvey and Tandberg-Hanssen, 1968; Kim *et al.*, 1987).

(2) *Field strengths measured in different spectral lines.* In 1968–1969 Tandberg-Hanssen (1970) observed the magnetic fields of 135 QPs. Except some Balmer lines of hydrogen, he adopted D_3 and $\lambda4471$ of helium, D_1 and D_2 of sodium and b_1 of magnesium. The values of $B_{\|}$ obtained by him are in the range 3–8 G, i.e. much smaller than those which Zirin obtained. Moreover, the field strength of a prominence does not depend on the spectral line used for magnetometry. This probably implies that prominences are physically homogeneous. In other words, the atoms of various chemical elements are excited and emit their radiation in the same region, and in the same magnetic field.

(3) *Classification of prominences by relationship with solar magnetic fields.* Prominences may be said to be the most beautiful and spectacular solar active phenomena. They display a wonderful variety in shapes, movements and evolutionary characteristics. Over the years several classifications have been proposed based on the morphological, dynamical and spectral peculiarities. We would like to state that, according to their relations with the underlying large-scale magnetic fields, prominences can be divided into the following two groups (Hirayama, 1985):

(i) *QPs.* They are located far from active regions. When QPs are projected on the solar disk, they trace out the neutral lines of large-scale magnetic fields. Besides, QPs are generally higher than 35–40″ and possess vertical fine structures with sizes less than 1″. Polar crown prominences also belong to this class.

(ii) *Active region prominences* (more often called *active region filaments*, ARF). These prominences occur in active regions and have heights less than 35″. As a rule, an ARF and the active region beneath it have the same magnetic polarity.

For a long time since astronomers have noted the cyclic variation of prominence heights (Cantu *et al.*, 1968; Makarov, 1983; Kim *et al.*, 1988b). This is interpreted by Kim *et al.* (1988a) to be due to the periodic change of the relative contents of QPs and ARFs. Because these two classes or prominences have different relations with solar magnetic fields, she (Kim, 1990) furthermore suggests that the relative

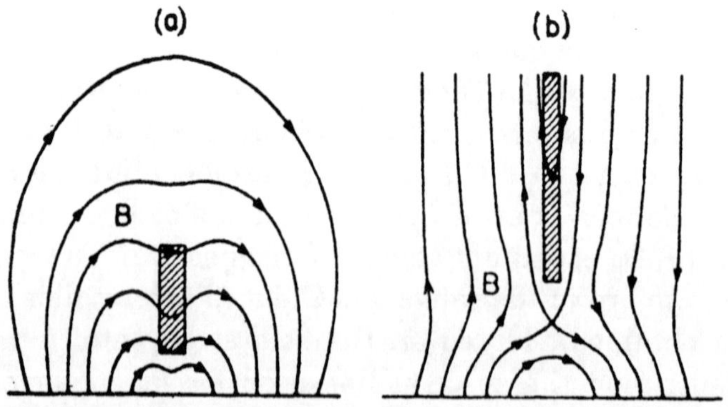

Fig. 5.19. Schematic representstions of (a) the model of Kippenhahn and Schlüter and (b) the model of Kuperus and Raadu.

contents of these two kinds of prominences may be taken to be an index of the solar activity.

(4) *Orientation of the prominence magnetic field.* Both the interpretation of observations (Leroy, 1977) and theoretical considerations show that the field vector is located approximately on the horizontal plane. The angle of deviation determined by Athay *et al.* (1983) has a mean value of 3°. We are then confronted with the so-called *vertical structure paradox* (Leroy, 1988): the vertical fine structures of QPs and a horizontal magnetic vector.

The angle between the field direction and filament long axis (α) was found by Tandberg-Hanssen and Anzer (1970) to be 15°. Some French astronomers got $\alpha = 25°$ (Leroy *et al.*, 1983).

(5) *Magnetic support of prominences.* According to the model of Kippenhahn and Schlüter (1957), the prominence is supported by sagging arches of magnetic lines of force at right angles to the prominence axis. (see Figure 5.19a). However, such a topology, if it really existed, would be unstable. This is because it is not the potential configuration in the lowest energy state. In Kuperus and Raadu's (1974) model the prominence forms in a neutral current sheet suspended above the solar surface. According to this model, convex loops crossing the neutral line should underlie the current sheet. However, in reality the loops cross the neutral line above the filament. So we can say that the problem of support of prominences by magnetic rields has not been completely solved.

(6) *Possible cyclic variations and oscillations of prominence magnetic*

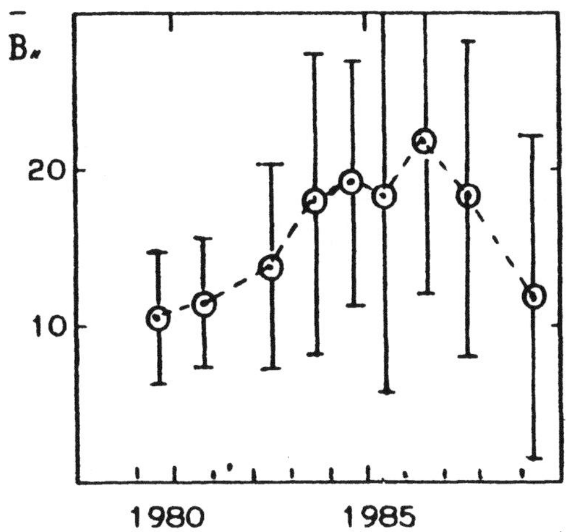

Fig. 5.20. Probable cyclic change of $\overline{B}_{\|}$ of QPs and ARFs as a whole.

fields. Some characteristics of prominences, such as the total number, height and mean latitude of occurrence, vary with the phase of the solar activity cycle. Thus it is possible that the average magnetic field strength of prominences also display periodic changes. According to the observations of some American astronomers (Tandberg-Hanssen, 1974), the mean value of $B_{\|}$ of QPs increased from 5 to 7.3 G in 1965–1969, i.e. in the rising phase of the 20th cycle of activity. Leroy *et al.* (1983) discovered the increase of $B_{\|}$ of polar crown prominences from 6 to 12 G in 1974–1980, viz., in the ascending phase of the 21st cycle, Nevertheless, there is at least one opposite case. Kim (1990) found that in 1980–1985 (the descending phase of the 21st cycle) the average value of $B_{\|}$ of QPs and ARFs taken together, also steadily increased (Figure 5.20). So we are not certain whether there are cyclic variations of prominence magnetic fields and, if they actually exist, how they are related to the solar activity cycle.

The possible existence of the *oscillations* of prominence magnetic fields is also an interesting topic (Jensen, 1983) and some observational evidence of such oscillations was found by Klepikov (1989). As remarked by Kim (1990), this is not certain, because the terrestrial atmosphere might affect the observation.

(7) *Observations by the Hanle effect*. The theory of the Hanle effect was worked out by House (1970, 1971) and applied to prominences by Bommier and Sahal-Bréchot (1978). The prominence emits radiation

basically by resonance scattering of the photospheric light and then a plane polarization parallel to the solar surrace is produced. By measuring the angle between the plane of polarizatlon and the tangent to the solar limb as well as the reduction of the degree of polarization in comparison with the theoretical maximum, one may calculate the strength and direction of the prominence magnetic field. The results of such observations have been summarized by Leroy (1979).

It is meaningful to mention that Bommier (1986) observed the Hanle effect of prominences displayed in both He I D_3 and Hβ. He measured for these two lines four polarization parameters, i.e., two polarization degrees and two polarization directions, and derived from them the following four physical quantities:

(i) the magnetic field strength,

(ii) the angle of inclination of lines of force with respect to the horizontal plane,

(iii) the angle between the field vector projected on the horizontal plane and the prominence long axis, and

(iv) the electron density.

He made such measurements for 14 prominences and found that in seven of them the angle of deviation of lines of force from the horizontal plane is more than 25°. Then he remarked that the magnetic field in QPs deviates more from the horizontal than is currently admitted. Moreover, for all the 14 prominences, he found the mean value of electron density to be equal to 1.3×10^{10} cm^{-3}.

CHAPTER 6

Magnetic Fields of the Solar Atmosphere

One of the key problems of solar physics is the question of the fine structure of the magnetic field.

A.B. Severny, *Some Problems of the Physics of the Sun* (1988).

The Sun is enveloped in a thick atmosphere which can be divided into the following three layers. As the lowest part, the photosphere has a thickness of only about 500 km. Above it is the chromosphere with a thickness of approximately 20,000 km. The outmost part is the corona, which has a scope of several to ten more solar radii. Its shape is not fixed and varies with the phase of the solar activity cycle. The magnetic field pervades the whole solar atmosphere and plays an important role in its structure, dynamics and evolution.

6.1 Photospheric Magnetic Fields

6.1.1 GRANULAR MAGNETIC FIELDS

Our concept about the magnetic field of the photosphere has undergone drastic changes in recent years. Because the visible solar surface, i.e., the photosphere, is crowded with millions of *granules*, for a long time in the past astronomers thought that the magnetic field of the granulation could characterize that of the photosphere. Soon after the invention of the photoelectric magnetograph Kiepenheuer (1953b) tried to measure the strength of the granular magnetic field. His method was crude and he obtained too large a value of 400 ± 10 G. Steshenko (1960) used the apparatus for the measurement of sunspot magnetic fields and adopted the following method. He put a prism between the grating and the imaging mirror so that one half of the diffracted light beam is slightly deflected and forms a spectral strip just below the normal one. These two spectra can be photographed simultaneously and each of them contains one of the two σ-components belonging to the same line. The distance between them yields the magnetic field strength of individual granules. The outstanding merit of this method is that the Doppler displacements of both σ-components caused by local motions should be the same and can be eliminated completely. This is important for the measurement of small $\Delta\lambda_B$'s. The values of $B_{\parallel}$ of individual granules obtained by Steshenko do not exceed 40–60 G, while for granule groups it is 50–60 G. Semel (1962) found that outside active regions there are weak magnetic structures with a mean strength of 24 G. However, it is difficult to affirm a correlation between such structures and the granulation pattern. Howard (1962) scanned the solar surface with the entrance slit of a magnetograph and the slit height was diminished to 2 arcseconds, i.e., close to the diameter of a granule. He obtained the mean value of $B_{\parallel}$ for the undisturbed region equal to 8.2 ± 4.4 G. Using a magnetograph with spatial resolution of 500 km, i.e., about one half of the size of granules, Livingston (1968) found that the upper limit of the strength of the magnetic field correlating with the granulation pattern is 2 G.

From the works mentioned above, the reader may get the impression that various observers obtained quite different results. This can be explained by difficulties encountered in observations. The size of granules is small; they are in incessant motion; their average lifetime is only about 10 minutes; it is hard to get rid of the influence of the atmospheric blurring; etc. However, as pointed out by Bray, Loughhead and Durrant (1984), the main reason lies in the small-scale structure of solar magnetic fields. At the beginning of the preceding chapter it was stated that magnetic elements or flux tubes with an angular size of only 0.1–0.3 arcsec exist not only in active regions, but also in undisturbed photospheric districts. Thus the resolution of $1''$–$2''$ commonly achieved in solar magnetometry may significantly distort the configuration of the observed magnetic structures and greatly diminish the measured field strength of magnetic elements. It is more important to note that the extent of distortion and diminution may vary from instrument to instrument. Therefore, it seems to be of little interest to repeat the above-mentioned observations of Steshenko *et al.*, because they can hardly reflect the intrinsic picture of the fine structures of the photospheric magnetic field.

6.1.2 TWO-COMPONENT MODEL

Taking into consideration the existence of fine structures, many astronomers adopt the two-component model to represent the photospheric magnetic field. In this model the fraction α of the magnetic element is really magnetic, while the remaining part $1-\alpha$ is free from magnetism. The relation between the intrinsic field strength B and the observed one $\langle B\rangle$ is expressed in Eq. (5.1). Strictly speaking, $\langle B\rangle$ is the density of the magnetic flux which in reality comes from the magnetic component but is averaged over the whole element. Now we would like to state that in the framework of this model the observed Stokes parameters can be written as

$$\langle I\rangle = \alpha I_m + (1-\alpha)I_{mn}, \quad \langle Q\rangle = \alpha Q_m, \quad \langle U\rangle = \alpha U_m, \quad \langle V\rangle = \alpha V_m. \tag{6.1}$$

Here I_m, Q_m, U_m and V_m are the parameters of the radiation coming from the magnetic component, while I_{mn} is the unique contribution of the nonmagnetic component. As a first approximation, Stenflo and Harvey (1985) assumed that all the unresolved magnetic elements or flux tubes possess the same field strength, say 1 kG. Then the filling factor and the mean field intensity are simply proportional to each other, viz., $\langle B\rangle \simeq \alpha$ kG. Moreover, it follows from Eqs. (3.18) and (2.1) that $V \propto B$. So we have $\alpha \propto V$. This implies that the parameter V may play a vital role in the determination of the filling factor and then in the elaboration of the two-component model.

Apart from Eq. (6.1), the two-component model can also be expressed in other ways. For instance, Solanki (1990) used this model to express the fine structure of sunspot magnetic fields. As described in Section 4.3, umbral dots and the intermediate umbra may differ greatly in magnetic field strength, and this is so for the bright and dark penumbral filaments. Let "b" and "d" denote the bright and dark counterparts, respectively, then the observed Stokes parameters can be expressed as

$$\begin{aligned}\langle I\rangle &= \alpha I_{\rm b} + (1-\alpha)I_{\rm d},\\ \langle Q\rangle &= \alpha Q_{\rm b} + (1-\alpha)Q_{\rm d},\\ \langle U\rangle &= \alpha U_{\rm b} + (1-\alpha)U_{\rm d},\\ \langle V\rangle &= \alpha V_{\rm b} + (1-\alpha)V_{\rm d}.\end{aligned} \tag{6.2}$$

Another example is provided by the work of Lozitskij and Tsap (1990). As clarified in Section 5.1, the solar background magnetic field consists of flux tubes with fields of approximately 1 kG and the intermediate intrinsically weak field. So these authors assumed that the photospheric field is composed of kG flux tubes and a weak background field. With such a two-component model they achieved a good agreement between theory and the observation made with the Crimean double-channel magnetograph.

Before the end of this section we want to stress that the two-component model is not sufficiently true. As stated above, all the flux tubes are supposed to possess the same field strength and, what is more, identical size and other geometrical characteristics. Of course, this is a rough approximation. Besides, magnetic flux tubes expand rapidly with height and so *magnetic canopies*[1] are formed. Therefore, even at the solar disk center a single ray may pass through more than one component and the 2-D arrangement of magnetic elements has to be considered. All these require that a *multi-component model* should be established for the further investigation of the photospheric magnetic field.

6.1.3 Physics of Photospheric Flux Tubes

According to the two-component model described in the preceding subsection, magnetic lines of force are assumed to be concentrated in the kG flux tubes. Hence the main contents of the study of the photospheric magnetic field can be said to be the determination of various physical quantities and their distribution in the photospheric flux tubes. This can be realized with the diagnostic methods of the *Stokes polarimetry*. In the past twenty years a large amount of work has been done in this field by Stenflo, Solanki *et al.* The existing methods may be roughly divided into the following two categories: the *few-lines* and *many-lines* techniques. Stenflo's line-ratio method, already introduced in Section 3.5.3, may be taken to be the classical representative of the first kind. As for the many-lines methods, the reader can be referred to Solanki and Stenflo (1984, 1985) and Solanki (1986). Solanki's *et al.* (1987) diagnostic methods are based on the V and Q profiles of two magneto-sensitive spectral lines. (These authors used the line pair Fe I $\lambda\lambda$5250 and 5247 with which the reader has already been acquainted.) The physical quantities that can be determined include the three components of the magnetic vector (B, γ and χ), some thermodynamic parameters (temperature, microscopic and macroscopic velocities, etc.) as well as the filling factor. All of them must in certain specific ways be

[1]See Section 6.2.3.

connected with the Q and V profiles of the lines mentioned. The algebraic solution of Unno-Beckers' equations of transfer of the Stokes parameters (4.15) is as follows:

$$\begin{aligned} I(\mu) &= \frac{\beta_0\mu}{1+\beta_0\mu}\left(1-\frac{1+\eta_I}{(1+\eta_I)^2-\eta_Q^2-\eta_U^2-\eta_V^2}\right), \\ Q(\mu) &= \frac{\beta_0\mu}{1+\beta_0\mu}\frac{\eta_Q}{(1+\eta_I)^2-\eta_Q^2-\eta_U^2-\eta_V^2}, \\ U(\mu) &= \frac{\beta_0\mu}{1+\beta_0\mu}\frac{\eta_U}{(1+\eta_I)^2-\eta_Q^2-\eta_U^2-\eta_V^2}, \\ V(\mu) &= \frac{\beta_0\mu}{1+\beta_0\mu}\frac{\eta_V}{(1+\eta_I)^2-\eta_Q^2-\eta_U^2-\eta_V^2}. \end{aligned} \tag{6.3}$$

Here $\mu = cos\theta$, θ and β_0 were defined in Section 2.2.3 and the meanings of the η's were given by Eq. (4.18). In the WFA the above expressions become

$$\begin{aligned} I &= \frac{\beta_0\mu}{1+\beta_0\mu}\left(1-\frac{1}{1+\eta_0}\right), \\ Q &= \frac{\beta_0\mu}{1+\beta_0\mu}\frac{\eta_Q}{(1+\eta_0)^2}, \\ U &= \frac{\beta_0\mu}{1+\beta_0\mu}\frac{\eta_U}{(1+\eta_0)^2}, \\ V &= \frac{\beta_0\mu}{1+\beta_0\mu}\frac{\eta_V}{(1+\eta_0)^2}. \end{aligned} \tag{6.4}$$

By use of the relation between η_Q and η_0 derived by Mathys and Stenflo (1987), i.e.

$$\eta_Q \approx -\frac{1}{4}\sin^2\gamma\cos 2\chi\Delta\lambda_B^2\frac{\mathrm{d}^2\eta_0(\lambda)}{\mathrm{d}\lambda^2}, \tag{6.5}$$

we have

$$Q \approx -\frac{1}{4}\frac{\beta_0\mu}{1+\beta_0\mu}\frac{\sin^2\gamma\cos 2\chi\Delta\lambda_B^2}{(1-\eta_0(\lambda))^2}\frac{\mathrm{d}^2\eta_0(\lambda)}{\mathrm{d}\lambda^2}, \tag{6.6}$$

where $\Delta\lambda_B$ is related to B by Eq. (2.1). The above expression represents the connection of Q with the three components of the magnetic vector. As for the parameter V, Solanki *et al.* obtained

$$V = -\cos\gamma\Delta\lambda_B\frac{\mathrm{d}I}{\mathrm{d}\lambda}. \tag{6.7}$$

With Eqs. (6.6), (6.7) and some other similar formulae, these authors obtained the relations between certain characteristics of the Q and V profiles and some physical quantities of the photospheric flux tubes. For example, the value of γ can be determined by the asymmetry of the profile of Q. To be precise, the degree of asymmetry is expressed by

$$\delta Q = \frac{A_{\sigma_b} + A_{\sigma_r} - A_\pi}{A_{\sigma_b} + A_{\sigma_r} + A_\pi} = \frac{\int Q \, \mathrm{d}\lambda}{\int |Q| \, \mathrm{d}\lambda}, \tag{6.8}$$

where A_{σ_b}, A_{σ_r} and A_π are, respectively, the absolute areas of the σ-component on the blue wing, the σ-component on the red wing and the π-component at the line center. Figure 6.1 illustrates the results of theoretical calculations for $\lambda 5250$ with the Harvard-Smithsonian Reference Atmosphere (HSRA) and with $\mu = 1$. The solid, dashed and dot-dashed curves correspond to $B = 500$ G, 1000 G and 1500 G, respectively. The left panel is the case in which magneto-optical effects are considered, and the right one is without these effects.

It is interesting to note that the asymmetry of the observed Q profile is generally smaller than that of the corresponding theoretical profile. According to the explanation of Solanki *et al.* (1987), this is because the photospheric flux tubes have a temperature higher than that of the HSRA model. Denoting the difference by ΔT, these authors computed the theoretical dependence of δQ on ΔT (Figure 6.2). Thus from the observed asymmetry of the Q profile one may obtain the temperature in flux tubes.

As for the determination of some other quantities, the reader may consult Solanki *et al.* (1987) and the papers cited therein. Moreover, the internal structure of the magnetic element, or the horizontal and vertical distributions of the magnetic field strength inside the flux tube, can also be determined with some appropriate diagnostic methods (Solanki *et al.*, 1988). All these works show that the Stokes polarimetry is a powerful tool for the investigation of solar, including photospheric, magnetic fields.

6.2 Chromospheric Magnetic Fields

The *chromosphere* is an important and distinctive layer in the solar atmosphere. It is the site of flares, flocculi and some other active features and is also a zone of steep changes of physical conditions. From the lower to the higher level of the chromosphere the temperature rises suddenly from about 5×10^3 to 10^6 degrees. The inverse temperature increase, heating mechanism, inhomogeneity, departure from local ther-

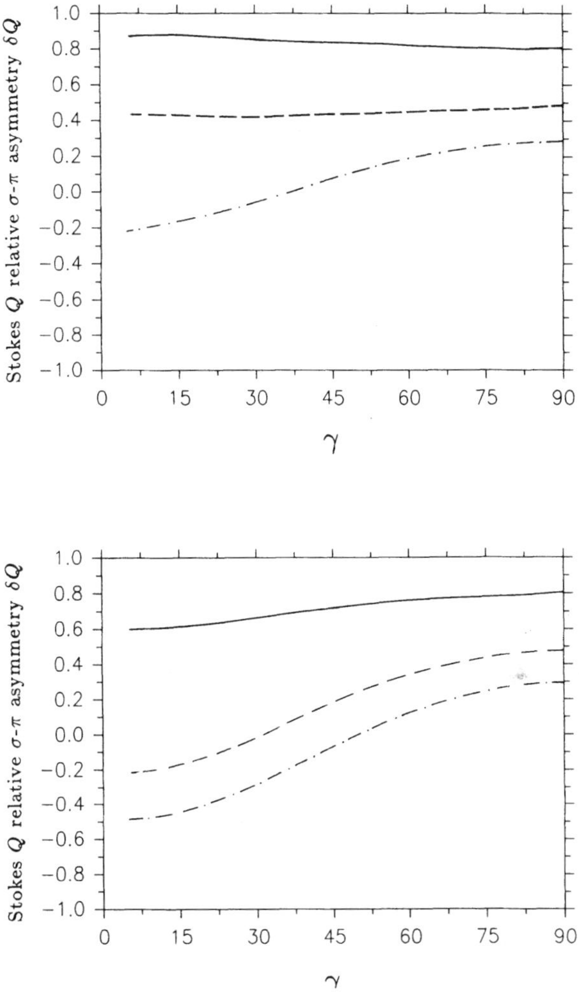

Fig. 6.1. Theoretical relationship between δQ and γ.

modynamic equilibrium, etc., are all important topics in the physics of the chromosphere. It has now to be emphasized that in the study of these problems the magnetic field plays an important role. This is because the density of the chromosphere is so low ($\leq 10^{12}$ g cm^{-3}) that criterion (1.3) is satisfied even for a weak field of a few gauss. This means that the dynamics and morphology of the chromospheric plasma are mainly governed by the magnetic field.

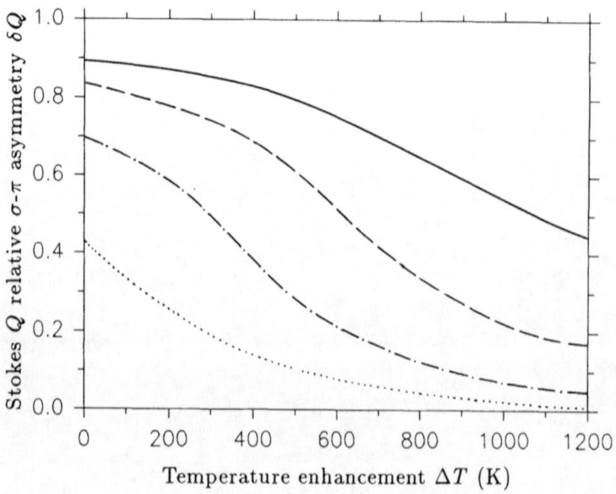

Fig. 6.2. δQ vs. ΔT, calculated with B = 1000 G and $\theta = \gamma = 75°$. The solid, dashed, dot-dashed and dotted curves correspond to $\log gf = -3.938, -4.438, -4.938$ and -5.438, respectively.

(1) *Preliminary consideration.* Early in the beginning of the fifties, some astronomers (e.g., Bugoslavskaja, 1950; van de Hulst, 1953) anticipated the existence of the chromospheric magnetic field. Their judgement was based on the fact that the *spicules* in polar regions have approximately the same directions as coronal rays, which appear under the action of the general magnetic field of the Sun. Severny and Bumba (1958) explicitly proposed the concept that *photospheric magnetic field lines penetrate into the chromosphere*. They had the following two observational evidences. First, the magnetograms scanned with the Crimean photoelectric magnetograph in the three lines Hβ and Fe I $\lambda\lambda$4886 and 5250 are quite similar. For Hβ the observation was made at 0.07 Å from the line center. According to de Jager (1957), the radiation at this wavelength comes from a height of 2000–2500 km in the chromosphere. But the two iron lines are weak and are certainly formed in the photosphere. Another proof is provided by the zigzag patterns of Zeeman splitting in the Hα line center and K_2 Ca II in spot regions. This implies that there are also magnetic fields in the chromosphere

above spots, and these fields are evidently the continuation of sunspot fields in the photosphere.

(2) *Actual measurements.* The observation of chromospheric magnetic fields is much more difficult than that of the photospheric ones. On the technical side, the central parts of such strong lines as Hα, Hβ and H, K Ca II are quite weak, so the magnetographic signals are faint and difficult to measure. On the theoretical side, the existing theory of spectral line formation in the magnetic field has been established basically for the simple Zeeman triplets, to which the strong spectral lines just mentioned do not belong. Therefore, the observational material obtained in them can hardly be used for quantitative studies.

Moreover, even for qualitative work, not all the intense lines can be reliably used. As remarked by Severny (1966) owing to its blending with some neighboring spectral lines, Hβ is inappropriate. The possible errors due to the use of this line were probed by Ioshpa and Obridko (1966). Furthermore, H_3 and K_3 are also unsuitable, because they are too sensitive to the chromospheric fine structures. Severny suggested that Hα is the best line for observations of chromospheric magnetic fields.

As pointed out by Bray and Loughhead (1974), the measurements of the chromospheric magnetic field are quite few. Severny and Bumba (1968) found that the field strengths of the quiet and active chromospheric regions are, respectively, 100–150 G and 300–500 G. In recent years it has become possible to carry out radio observations. For example, Grebinskij (1989) proposed that it is feasible to derive the magnetic field strength (B) and filling factor (α) with the observed degree of circular polarization (ρ_m) of the magneto-bremsstrahlung and the critical wavelength (λ_m) below which there is a sudden decrease of ρ_m. Bogod and Gelfreich (1980) used the Ratan-600 radio telescope to observe the radio emission of flocculi and obtained $\rho_m = 0.016 \pm 0.005$ and $\lambda_m = 3.2$ cm. By analyzing their data, Grebinskij obtained $B = 1670$ G and $\alpha = 0.08$ for the chromospheric flux tubes.

(3) *Magnetic canopies.* As noted by Giovanelli (1980), the magnetograms taken in chromospheric lines display diffuse fringes with reverse magnetic polarity on the limbward side and also on the diffuse centerward extensions with normal polarity (see Figure 1 of his paper). This phenomenon can be explained by the existence of *magnetic canopies* (Figure 6.3). Such structures are formed by magnetic lines of force spreading from compact, nearly vertical flux tubes at boundaries between supergranules into the chromosphere and the corona. Jones and Giovanelli (1982) calculated the magnetographic response of

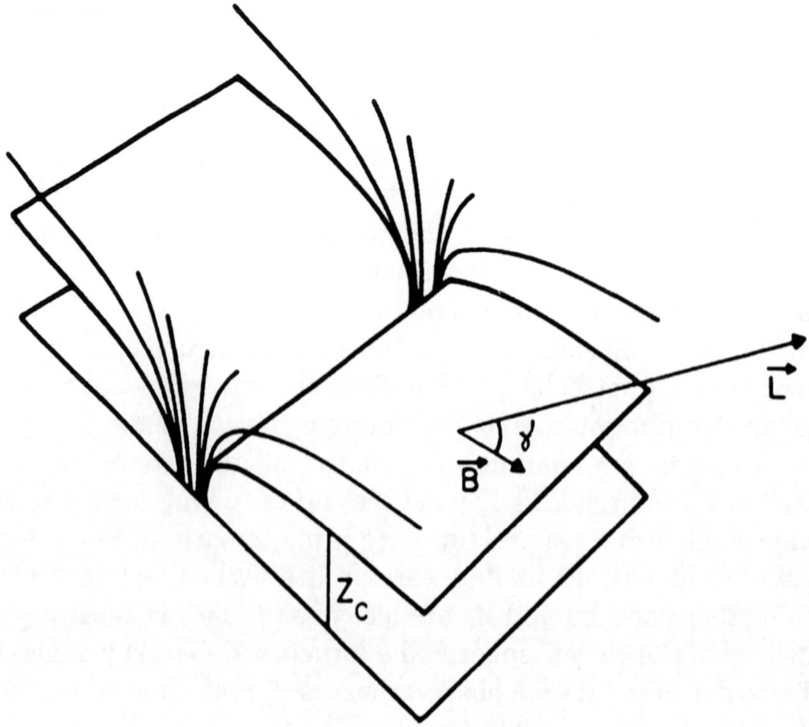

Fig. 6.3. Geometry of magnetic canopies.

a canopy-type field, and then they derived the canopy height Z_c. Furthermore, Jones (1985) used the canopy topology to study the structure and energetics of the chromosphere.

(4) *Magnetic expulsion and formation of spicules.* Spicules are the bright, vertical and narrow ($\sim$ 1000 km in diameter) jets of gas, which ascend from the lower chromosphere into the inner corona, then fall back and fade away. They are closely associated with the chromospheric network, which is located at boundaries of supergranules. Many astronomers have proposed various mechanisms to explain the formation of spicules (for reviews see Beckers, 1968b, 1972; Bray and Loughhead, 1974; Campos, 1984), and those suggested by Pikel'ner (1969, 1971) and Uchida (1969) are directly related to the chromospheric magnetic field. These mechanisms are based on the annihilation and reconnection of magnetic lines of force near a neutral line. At this place lines of force are squeezed together by the opposing horizontal motions of gas in neighboring supergranules (see the left part of Figure 6.4). The compressed gas trapped between them cools and flows downwards under the action of gravity. At the same time a column of matter is carried upwards by reconnected lines of force, and this leads to the formation

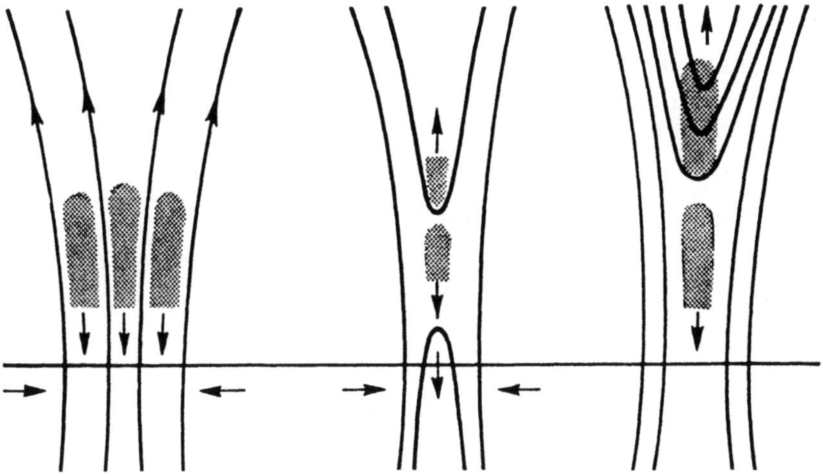

Fig. 6.4. Mechanism of spicules formation.

of a spicule (central and right parts of the Figure). This is the so-called "melon-seed effect". In Pikel'ner's (1971) theory the half-width of the spicule column (R) is given by

$$\frac{R}{\xi} \simeq \frac{B_0}{\sqrt{2\pi p_0}}, \tag{6.9}$$

where ξ is the height scale of the spicule, p_0 and B_0 are its gas pressure and magnetic field strength along its axis. From the observed R, T and p_0 one obtains $B_0 = 6$–8 G for a typical spicule. The intensity of a field far from the axis is $B = 10$–13 G. Besides, the velocity of rise of the spicule is determined not by the magnetic force, but by the rate of field annihilation. This leads to $v = 20\ \mathrm{km\,s^{-1}}$, which agrees well with observation.

(5) *Chromospheric monochromatic images and photospheric magnetograms.* As stated above, due to its low density the chromospheric plasma is well controlled by the magnetic field. Besides, photospheric fields can extend into the chromosphere. When combined together, these two facts may explain the observed similarity between chromospheric networks and photospheric magnetic structures. What has to be stressed here is the high degree of their similarity. As revealed by Leighton (1965), the fine details of calcium flocculi and magnetic field configuration display a very close one-to-one correspondence, and this is so even for a weak field of 2.5 G. The coefficient of correlation found

by him amounts to 0.999! Among the works of many other solar physicists (for a detailed description see Bray and Loughhead, 1974) that of Frazier (1970) is especially worth mentioning. He used the Kitt Peak multi-channel magnetograph and found the coefficient of correlation between K line network and Fe I $\lambda5250$ magnetic structures to be 0.72. It is meaningful that the brightness always increases linearly with the field strength, no matter what the magnetic polarity. In view of this close correlation, Zirin (1972) established some rules by which one may use Hα monochromatic images to infer the polarity and strength of photospheric magnetic fields. Moreover, as remarked by Zirin (1988), this is especially useful for inferring the transverse field, which is difficult to detect with common magnetographs.

6.3 Coronal Magnetic Fields

Due to its extremely low density (about 10^{-15} g cm^{-3}) the coronal plasma is fully controlled by the magnetic force. So magnetic fields are of vital importance to the physics of the solar corona. The orientation of coronal rays or streamers that are mostly situated in the polar regions of the Sun, the change of shape of the corona with the solar cycle, the cooling and suspension of prominences in the corona, etc., all these facts remind us of the existence of coronal magnetic fields. However, it is quite difficult to measure them directly. This is mainly because the coronal emission is exceedingly faint: about one millionth of the photospheric radiation. Until the present, direct observations are scarce and our knowledge about the coronal magnetic field comes mainly from the *theoretical extrapolation* of photospheric fields.

6.3.1 Inferences and Measurements

Before actual observations some estimates of the strength of the coronal magnetic field were made by the following indirect inferences. The streaming material in coronal clouds often trace out curved trajectories. In order that the magnetic force may control the trajectory, it should be larger than the centripetal force, i.e.,

$$\frac{B^2}{8\pi} > \frac{n_H m_H v^2}{r}, \tag{6.10}$$

where v is the velocity of an individual knot and r is the radius of curvature. If we take $r = 10^5$ km, $v = 200$ km s^{-1} and $n_{\rm H} = 10^{11}$ cm^{-3}, then we have $B > 0.1$ G. Other methods of estimating the magnetic

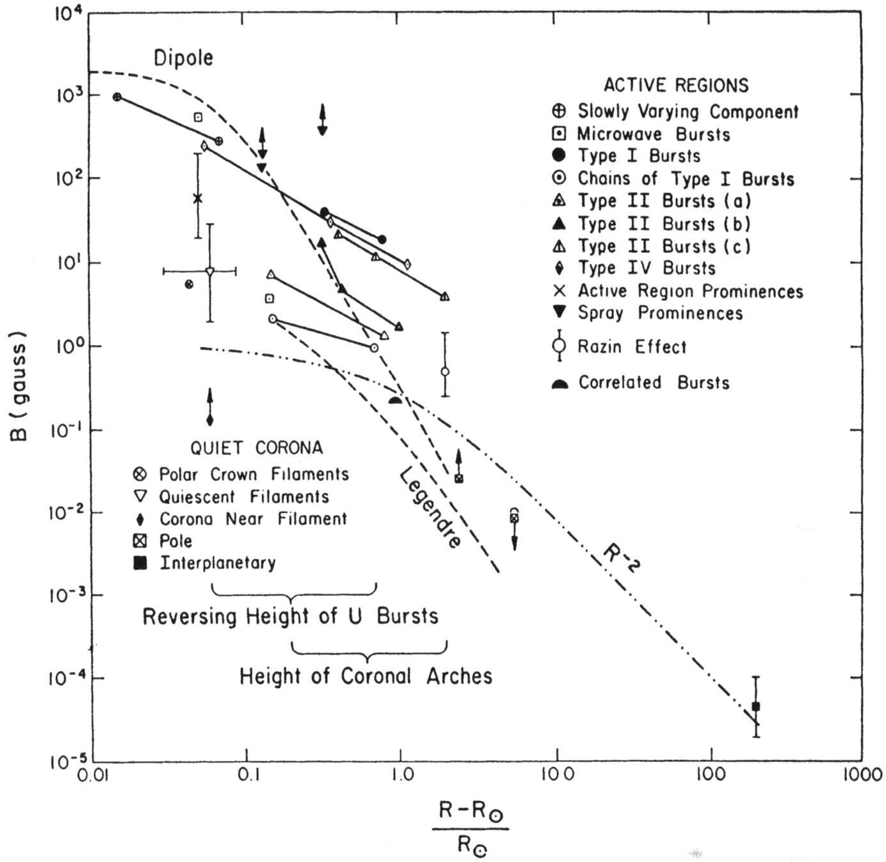

Fig. 6.5. Distribution of the magnetic field strength in the corona.

field strength of the corona are by the damping of prominence oscillations and the confinement of polar plumes as discrete structures. (For a detailed description see Athay, 1976.) These indirect methods can be applied to the quiet corona. As for the active coronal regions, formal measurements may be made with characteristics of the radio bursts of various types, the Razin effect, the change of polarization of the slowly varying component with frequency, etc. (Newkirk, 1967; Kundu, 1990). The field strength in coronal active features may attain 10^2 G. The results of many kinds of measurements as well as theoretical calculations, which will be introduced in the following subsection, are summarized in Figure 6.5. (The detailed explanation can be found in Newkirk, 1971.)

Recently a breakthrough has been achieved with the success of the Coronal Magnetic Structures Observing Campaign (CoMSTOC) (Schmelz *et al.*, 1988; Schmelz, 1990). In this project simultaneous soft X-ray images obtained with SMM's X-ray Polychromator and microwave observations with the Very Large Array make it possible to separate the gyroresonant and thermal bremsstrahlung contributions to the coronal microwave emission. The temperature and emission measure determined from X-ray observations can reveal the importance of bremsstrahlung radiation. When gyroresonance dominates, the local magnetic field intensity is simply $B(\mathrm{G}) = 357\nu(\mathrm{GHz})/n$, where ν is the frequency and n is the harmonic. Moreover, the observations at different frequencies can yield magnetic field strengths of various regions and this allows the exploration of the 3-D structure of the coronal magnetic field.

6.3.2 THEORETICAL CALCULATIONS

The following three methods have been adopted.

(1) *Extrapolation of the photospheric magnetic field to the corona by means of a potential function.* If no electric current flows above the photosphere, the magnetic field on the solar surface can be expressed with a potential function ψ defined as

$$\mathbf{B} = -\nabla\psi. \tag{6.11}$$

For computations on a field not far from the Sun, a small part of the photosphere is sufficient for consideration and it can be treated as a plane. In this case the Cartesian coordinates may be adopted. However, in order to compute the field which is at about or more than $1.3R_\odot$ from the solar center, the observed magnetic field distribution on the whole hemisphere must be taken into account. In this case the potential function may be expressed in spherical coordinates (r, θ, ϕ) such as

$$\psi(r,\theta,\phi) = R_\odot \sum_{n=1}^{\infty}\sum_{m=0}^{n} \left(\frac{R_\odot}{r}\right)^{n+1} (g_n^m \cos m\phi + h_n^m \sin m\phi) P_n^m(\theta), \tag{6.12}$$

where $P_n^m(\theta)$ is the associated Legendre polynomial (for an explanation see Chapman and Bartels, 1940.) The magnetic vector at an arbitrary point in the corona has the following three components:

$$\begin{aligned} B_r &= -\frac{\partial\psi}{\partial r} \\ &= \sum_{n=1}^{\infty}\sum_{m=0}^{n} (n+1) \left(\frac{R_\odot}{r}\right)^{n+2} (g_n^m \cos m\phi + h_n^m \sin m\phi) P_n^m(\theta), \\ B_\theta &= -\frac{1}{r}\frac{\partial\psi}{\partial\theta} \end{aligned}$$

$$= -\sum_{n=1}^{\infty}\sum_{m=0}^{n}\left(\frac{R_\odot}{r}\right)^{n+2}(g_n^m \cos m\phi + h_n^m \sin m\phi)\frac{dP_n^m(\theta)}{d\theta}, \tag{6.13}$$

$$B_\phi = -\frac{1}{r\sin\theta}\frac{\partial\psi}{\partial\phi}$$

$$= \frac{1}{\sin\theta}\sum_{n=1}^{\infty}\sum_{m=0}^{n} m\left(\frac{R_\odot}{r}\right)^{n+2}(g_n^m \cos m\phi - h_n^m \sin m\phi)P_n^m(\theta).$$

As pointed out by Newkirk (1971), this expansion is the equivalent of fitting the photospheric field to a set of multiples situated at the center of the Sun. The coefficients g_n^m and h_n^m can be determined with the observed values of $B_{\parallel}(\theta,\phi)$ at a number of points on the photosphere using the least-squares method.

Now we proceed to calculate the coronal magnetic field. This can be carried out in the following two ways. First, for the potential field, we have

$$\nabla \cdot B = 0,$$
$$\nabla \times B = 0. \tag{6.14}$$

Second, for the *force-free* field which may be considered to be a modified potential field, Eq. (2.34) must be used. Many calculations have been performed for both of these cases (e.g., Levine, 1975; Nakagawa *et al.*, 1978; Wang Zhenzhi and Lin Yuanzhang, 1984; Wu Fei and Wang Zhenzhi, 1984; Ambrož, 1989). Generally speaking, the agreement between the observed coronal configuration and the computed magnetic field is rather satisfactory. Nevertheless, in the simulation of magnetic fields over active regions, these two models may give rise to large discrepancies. This is not difficult to understand, because they differ in physical nature. According to Ampère's law, i.e. $\mathbf{j} = c/(4\pi)\nabla \times \mathbf{B}$, there should be no current at all in the potential field, while in the force-free field we have $\nabla \times \mathbf{j} = \alpha\mathbf{j}$. In fact, the existence of strong electric currents in active regions has been confirmed by observation (Lin Yuan zhang and Wu Fei, 1987), so it is better to adopt the model of force-free fields.

(2) *Interaction of the coronal magnetic field and solar wind.* Beyond the range of 2.5 $R_\odot$ it is necessary to consider the interaction of the solar wind and the coronal magnetic field. In the study of coronal streamers, Pneuman and Kopp (1970) simultaneously solved the continuity, momentum and energy equations. They took into account both the solar wind and the magnetic field. The former may cause heat conduction and convection, while the latter provides thermal insulation. Under the

condition of the balance of magnetic and gas pressures on the lateral side of a streamer, these authors obtained the strength of its magnetic field as 1–10 G and that of the interstreamer region — 0.07 G.

(3) *Extrapolation of the interplanetary magnetic field to the corona.* The result of an R^{-2} extrapolation is shown by the curve in the right and lower part of Figure 6.5.

6.3.3 Transient Changes

The coronal magnetic field does not always remain static, and it often displays rapid variations. The following observational evidence has been discovered for this. When the solar corona occults the Crab Nebula and scatters its radio emission, the nebula seems to be enlarged. It is interesting to find that the increment of its diameter may vary significantly in a few minutes. Besides, when an artificial celestial body is occulted by the corona, its radio signals are transmitted through the coronal magnetic field and the Faraday rotation of the plane of polarization takes place. As described in Section 2.4, one may get the value of $B_{\parallel}$ from the rotation measure. By observations of this kind it has been found that after the outburst of a solar flare there may appear in the corona a "magnetic bottle" (Figure 6.6). This is a temporary magnetic structure. It can stretch to about $10R_{\odot}$ in the coronal space and exists for only 1–2 days.

Finally, it is worthwhile mentioning that the coronal magnetic field is highly inhomogeneous. Its fundamental constituents are coronal loops, which were first detected in the X-ray images of the corona. The microwave counterpart of these ubiquitous structures has been discovered by the high resolution VLA observations (Lang and Willson, 1986).

6.4 Magnetic Field Gradient of the Solar Atmosphere

From the preceding three sections the reader may have noticed that the solar magnetic fields in various atmospheric layers, especially in the photosphere and the chromosphere, are rather similar. This implies that magnetic lines of force may penetrate through almost the whole atmosphere. Nevertheless, the magnetic field changes with height after all. So we are confronted with the problem of its gradient. Since the thirties many astronomers have been engaged in the study of this quantity and much progress has been achieved.

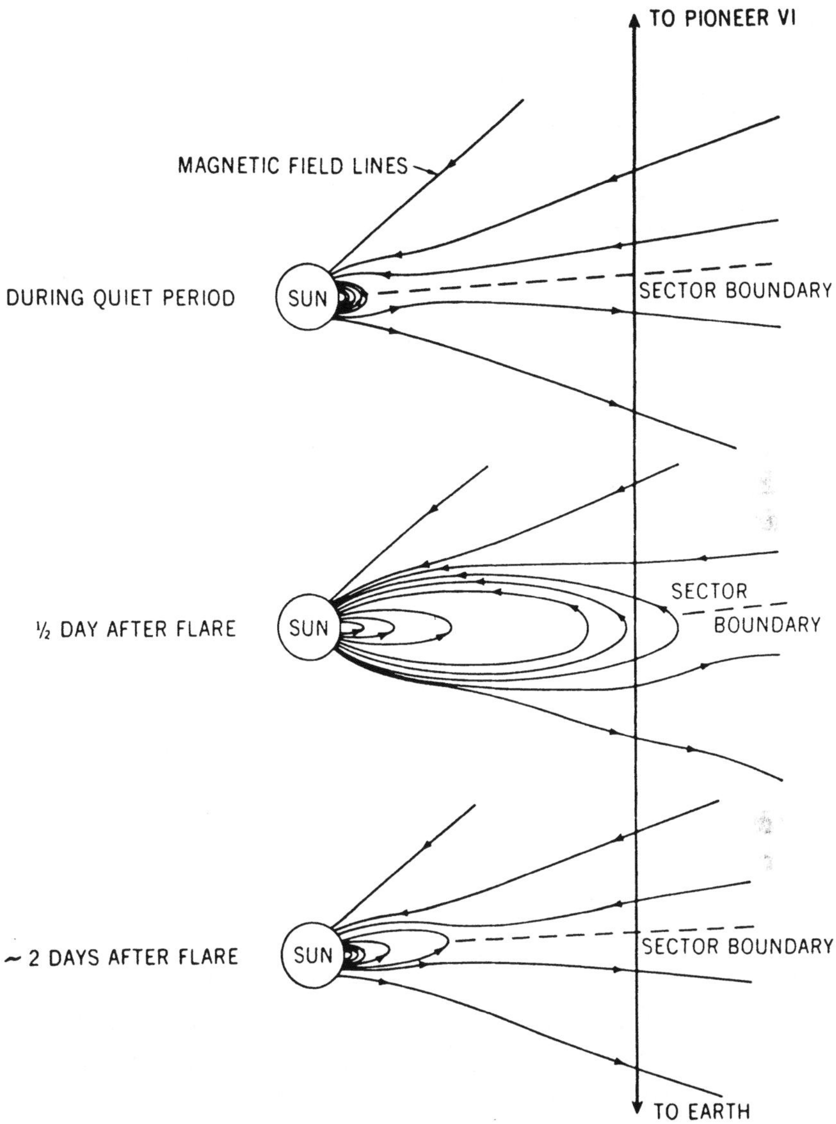

Fig. 6.6. Appearance and evolution of a "magnetic bottle".

6.4.1 Methods of Measurement

(1) *Field gradient given by different lines.* King (1934) discovered that for one and the same sunspot the magnetic field strengths measured with spectral lines of various intensities may differ greatly. It is well known that the main parts of strong lines are formed in the upper photosphere and the chromosphere, while the radiation of weak lines

comes from the photosphere, especially its middle and lower levels. This allowed him to estimate the magnetic field gradient as 3.0 G km^{-1}.

As stated in Section 6.2, Severny and Bumba (1958) proposed the concept of penetration of the photospheric magnetic field into the chromosphere. They measured the strength of the chromospheric field through the Zeeman effect of Hα and some other intense lines. (Their cores are formed at heights of several thousand kilometers above the base of the photosphere.) By comparison with the field intensity observed with photospheric lines, these two authors found a gradient of 1 G km^{-1}.

Henze *et al.* (1982) compared the results of magnetometry made with Fe I λ5250 and C IV λ1548. The former is a well-known photospheric line, while the latter is formed in the chromosphere-corona transition zone and observed with the SMM satellite. The gradient thus obtained lies in the range 0.4–0.6 G km^{-1}.

(2) *Optical thickness of line cores.* Bumba (1960a) calculated the central optical thickness (τ_0) of a series of spectral lines. Besides, he measured with these lines the magnetic field strength (B) of one and the same sunspot. Thus he obtained the relation between B and τ_0. Using this relation as well as Michard's (1953) spot model, he derived the gradient of the spot magnetic field to be 0.3 G km^{-1}.

(3) *Change of spot field strength with angular distance (θ) from the center of the solar disk.* The larger θ, the higher is the level to which the measured B belongs. From the relation between these two quantities, Houtgast and van Sluiters (1948) obtained gradients of 0.5–2.5 G km^{-1}.

(4) *Asymmetry of σ-components.* If there were no field gradient, the profiles of all the π- and σ-components should be symmetric. However, due to the existence of a gradient, the upper level of a spot atmosphere has a weaker field and so it produces a smaller Zeeman splitting. By contrast, the lower level gives rise to a larger splitting. When the contributions from all layers are superposed together, each σ-component possesses an asymmetric profile. So it is possible to derive the magnitude of the gradient from the amount of asymmetry. Hubenet (1954) was the first to apply this method to actual observation. But this was done before the advent of the theory of the inverse Zeeman effect, and so his result is not accurate enough. Baranovskij and Stepanov (1959) and Baranovskij (1960) measured the asymmetry of Na I D_1 and for some spots got gradients of 1.0–1.8 G km^{-1}. We (Ye Shi-hui *et al.*, 1979) used our numerical solution of Unno's equations of transfer of the Stokes parameters and studied the relations between the field gradient and the amounts of asymmetry defined below. We took the middle between the

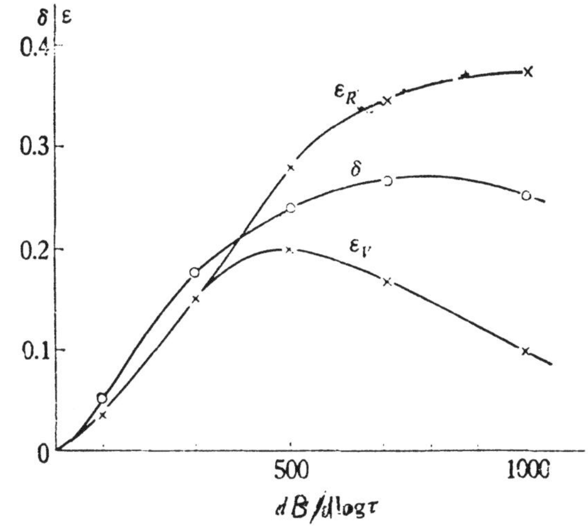

Fig. 6.7. Relations between magnetic field gradient and amounts of line asymmetry and line wing rotations. (δ and ε are in units of Å.)

two points on a σ-profile with half of the central depth and measured its distance (δ) from the line center for the case of no gradient. We call δ the *measure of line asymmetry*. Besides, for a point on the profile also with a depth equal to one half of the central depth, its distance (ε) from the corresponding point in the profile of null gradient is defined to be the *measure of the rotation of the line wing*. For the red and violet wings, respectively, we have ε_R and ε_V.

According to our calculations, the relations between δ, ε's and the gradient $\mathrm{d}B/\mathrm{d}\log\tau$ are shown in Figure 6.7. By fitting our observation of a large spot to the curves in this figure and transforming the optical depth to the geometrical depth, we obtained a gradient of $1\,\mathrm{G\,km^{-1}}$.

(5) *Calculations with transverse field strength.* By use of the observed $B_\perp$ and the condition $\nabla \cdot \mathbf{B} = 0$, it is possible to compute the field gradient (Severny, 1965; Leroy, 1962; Kotov, 1970). The essence of this method can be expressed with the following equality:

$$\frac{\partial B_z}{\partial z} = -\left(\frac{\partial B_x}{\partial x} + \frac{\partial B_y}{\partial y}\right). \tag{6.15}$$

With observed values of $\partial B_x/\partial x$ and $\partial B_y/\partial y$, the required depth gradient $\partial B_z/\partial z$ may be calculated. Kawakami (1983) fitted the observed distribution of the transverse field strength and elevation angle (Θ)

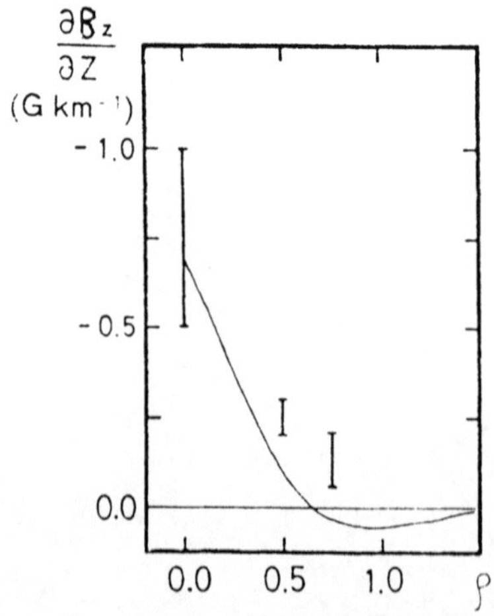

Fig. 6.8. Dependence of $\partial B_z/\partial z$ on ρ.

across the spot radius (R) with the following mathematical expressions:

$$\frac{B_\perp}{B_0} = \frac{\exp(-\rho^{2.5})}{1+\rho^2}\,, \tag{6.16}$$

$$\Theta = \frac{\pi}{2}\left(1-\frac{\rho}{b}\right)\left[1+\left(\frac{\rho}{b}\right)^2\right]\exp\left(-\frac{\rho}{b}\right), \quad b = 1,3 \tag{6.17}$$

where B_0 is the transverse field strength at the spot center and ρ, as defined in Section 4.1, is the normalized distance from the center. If the azimuthal component B_χ is negligible, the condition $\nabla \cdot \mathbf{B} = 0$ leads to the following equation:

$$\frac{\partial B_z}{\partial z} = -\frac{B_0}{R}\left[\frac{1-\rho^2}{\rho(1+\rho^2)^2} - \frac{2.5\rho^{1.5}}{1+\rho^2}\right]\exp(-\rho^{2.5})\cos\Theta -$$
$$-\frac{\pi B_0}{2Rb}\left[2-3\frac{\rho}{b}+4\left(\frac{\rho}{b}\right)^2-\left(\frac{\rho}{b}\right)^3\right]\frac{\exp(-\rho^{2.5}-\rho/b)}{1+\rho^2}\sin\Theta\,. \tag{6.18}$$

Moreover, from his theoretical calculations, Kawakami obtained the variation of the gradient $\partial B_z/\partial z$ across the spot radius. This is shown in Figure 6.8. (The bars represent the results of other workers which are described in Schröter, 1971.) It is interesting to note that the gradient may become positive in the outer part of the spot.

(6) *Radio observations*. The measurements of Hewish (1962) showed that the strength of the coronal magnetic field at 2–3 × 10^4 km above the photosphere is about 500 G and at 5 × 10^4 km it drops to 360 G.

A gradient can then be derived. However, if the complexity of the structure of coronal magnetic fields is taken into consideration, this method must be handled with care.

6.4.2 Some Problems Associated With the Magnetic Field Gradient

We want now to emphasize that the depth gradient of solar magnetic fields may be quite intricate, and this can be illustrated by the following facts.

(i) Koval and Stepanyan (1972) discovered that the field gradient of a spot is related to the flare activity in the neighborhood. Before the outburst of a flare the field in the upper layer of the photosphere in the spot region is stronger than that at the lower level. However, after the flare, the gradient returns to the normal state, i.e. the field strength diminishes with height. This phenomenon was also observed by Babin *et al.* (1974).

(ii) Gelfreich *et al.* (1981) found that the change of field strength with height may not be monotonous. At the height of approximately 600 km above the photosphere the strength increases, but at still higher levels it decreases.

(iii) Bogod *et al.* (1982) reported that the magnetic field measured in Hα is weaker not only than the photospheric field, but also than the field at the base of the corona above a spot. As is generally recognized, the magnetic field observed in the main part of Hα belongs to the chromosphere whose field intensity should be intermediate between those of the photosphere and the corona.

6.4.3 Theoretical Limit of the Sunspot Magnetic Field Gradient

The magnitude of the gradient strongly depends on the spot model used for the interpretation of observational data. A striking example is provided by Guseinov (1974a). For Mattig's model he obtained a gradient of 5–6 $\mathrm{G\,km^{-1}}$, while for Michard's model it is only 0.4–0.5 $\mathrm{G\,km^{-1}}$. Then it naturally follows that for each spot model there may be an upper limit of the gradient. Osherovich (1980, 1984) attacked this problem. Using Schlüter and Temesvary's (1958) model, he derived the following expression for the maximum gradient:

$$\left|\frac{dB_z(0,0)}{dz}\right|_{\max} = \frac{B_z(0,0)}{R_p} \times 1.63 \times 2\sqrt{2}\,, \tag{6.19}$$

where $B_z(0,0)$ is the vertical component of field strength at the spot center and at the photospheric level, and R_p is the radius of the penumbra. Using the return-flux model, which he himself established, the following inequality was obtained:

$$\left|\frac{dB_z(0,0)}{dz}\right| \leq \frac{B_z(0,0)}{R_p}\alpha_p\sqrt{\frac{2(2X-1)}{X-1}}\,, \tag{6.20}$$

where α_p is a parameter and X is the measure of the return flux. The case of $X = 0$ corresponds the Schlüter–Temesvary model in which there is no return flux at all. For a spot with zero line on the outer boundary of the penumbra, i.e., for $B_z(0, R_p) = 0$, α_p is equal to $\sqrt{1.18}$.

CHAPTER 7

The General Magnetic Field of the Sun

The discovery of powerful magnetic fields in sun-spots indicated the possibility of a general observational attack on solar magnetic phenomena.

G.E. Hale, Preliminary results of an attempt to detect the general magnetic field of the Sun, *Astrophys. J.*, **38** (1919), 27.

Does the Sun possess an overall magnetic field like that of the Earth? This is a question of important significance. First, the existence of a *general magnetic field* may affect the structure and evolution of the Sun. Second, the physical interpretation of the solar activity cycle[1] requires that the entire solar globe, especially the high-latitude regions, possesses magnetism. Third, as will be described in the next chapter, the solar wind brings the magnetic field of the solar atmosphere into interplanetary space. Therefore, the solar global magnetism may have a profound influence on the physical status of the heliosphere. Taking these facts into consideration, astronomers have for a long time paid a great deal of attention to the study of the solar general magnetic field, and much progress has been achieved. In recent years observations with high spatial and temporary resolutions have led to the discovery of a lot of subtle and complex phenomena, which are difficult to interpret. It is sad to relate that even the very existence of the solar general magnetic field seems to be questionable. Hence the research in this field has to be carried on further.

[1]A detailed account will be given in Section 11.3.

7.1 Strength of the General Magnetic Field

(1) *Hale's et al. early observations.* Hale was the pioneer in the investigation of not only the sunspot magnetism, but also the solar general magnetic field. Beginning in 1912, he and other Mt. Wilson astronomers (Hale, 1913; Hale *et al.*, 1918; Seares, 1913) tried to observe the general field. They assumed that the Sun has a dipolar magnetic field which is similar to that of the Earth. This implies that the magnetic poles are located not far from the geometric poles and that the field strengths at both poles are equal in absolute values. They derived the following expression for the field intensity on the solar surface at latitude φ:

$$B_\varphi = B_0\sqrt{1 + 3\sin^2\varphi}\,, \tag{7.1}$$

where B_0 is the strength of the general field on the equator. Except at the poles, the lines of force of the general field are not in the radial direction, but form an angle δ with the tangent to the solar surface (Figure 7.1). The relation between δ and φ is $\tan\delta = 2\tan\varphi$. It follows from Eq. (7.1) that the field strength in the polar regions (B_p) is twice that at the equator. Besides, the component along the line of sight ($B_\parallel$) is maximum at $\varphi = \pm 45°$. The relation between $B_\parallel$ and φ is shown in Figure 7.2. According to Hale and his associates, observations of some spectral lines do indeed display such a relationship, but this is not true for all lines. Summing up a large amount of observational data, these authors came to the following conclusions:

(i) The absolute value of B_p lies in the range 25–50 G.

(ii) The S and N polarities are located near the geometric north and south poles, respectively, and this was similar to the geomagnetic field.

(iii) The magnetic and rotational axes of the Sun form an angle of 6°.

(iv) The field strength measured with various spectral lines are different. More concretely speaking, strong lines yield weaker fields and *vice versa*. According to the interpretation of Hale *et al.*, this is because strong (weak) lines are formed at higher (lower) level where the magnetic field is weaker (stronger).

Hale's *et al.* conclusions have been found afterwards to be doubtful, and the main problem lies in the value of the field intensity. Both Langez (1936) and Stenflo (1971) repeated measurements of Hale's *et*

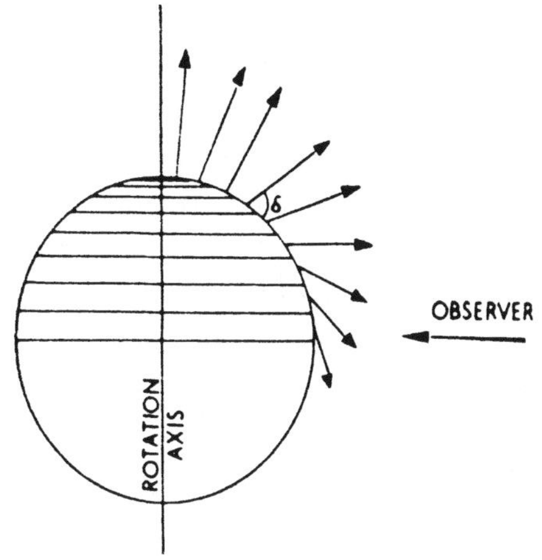

Fig. 7.1. Hale's *et al.* hypothetical general magnetic field of the Sun.

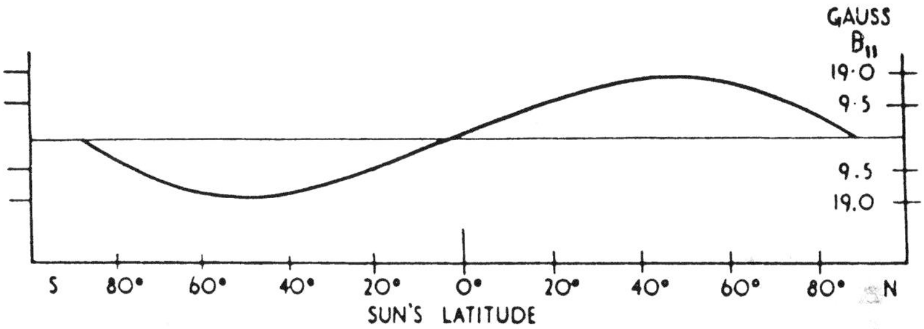

Fig. 7.2. $B_{\parallel}$ vs. φ found by Hale *et al.*

al. original zeemangrams and obtained, respectively, $B_p = 4$ G and $B_p < 5$ G. Hence it has become clear that due to technical errors Hale and his colleagues very much overestimated the strength of the solar general magnetic field.

(2) *Recent work.* Since the thirties, astronomers have tried to observe the general magnetic field with large spectrographs as well as the Fabry-Perot etalon or other devices that might improve the spectral resolution, and they got more accurate results. For instance, von Klüber (1954) obtained $B_p = 3.0 \pm 1.7$ G and Thiessen (1952) found 2.4 ± 0.5 G. (A complete account of such measurements can be found in von Klüber's (1955) review). In recent years, due to the improvement of spatial resolution, our concept about the general magnetic field of the Sun has changed drastically. This field has been discovered to be *highly inhomogeneous* and it consists of many tiny magnetic elements, which

may differ greatly in field strength. Moreover, neighboring elements can possess opposite polarities. When they are observed simultaneously within one and the same small area that can just be resolved, their fluxes are to a great extent cancelled out. The measured intensity of the mixed polarity field then becomes much reduced. As a concrete example, Howard and LaBonnte (1981) used the Mt. Wilson magnetograph, with an aperture of 12.5 arcseconds, and for the polar regions found an absolute mean value $\langle |B| \rangle = 1$ G. However, Murray (1991) used the Kitt Peak magnetograms, scanned with an aperture of 1 arcsecond, and found corresponding values 4–7 times larger. The following fact found by Tang and Haimin Wang (1990) with the BBSO's VMG, is also worthy of mention. Although the high latitude regions have a mean $B_{\parallel}$ of merely 1–2 G, the strength of individual resolved elements can attain 100 G! It is still more meaningful to state that the mean absolute strength $\langle |B| \rangle$ is much larger than the mean value $\langle B \rangle$. For example, the $\langle |B| \rangle$ at the south pole in 1987 was 4 G, while $\langle B \rangle$ in the same region and at almost the same time was +2.75 G. These facts clearly indicate that the solar polar magnetic fields are very inhomogeneous and consist of mixed polarities. Therefore, in order to make accurate measurements of their strength, as high as possible resolution is desirable.

7.2 Fine Structures of the General Magnetic Field

As stated above, the observed intensity of the solar general magnetic field is closely concerned with the spatial resolution. This question was systematically probed by Severny (1988). He found that together with the improvement of the resolution from 50″ to 2″, the average strength of individual magnetic elements increases from 2 G to 30–40 G. Besides, the number of resolved elements is enlarged in 1.5–2 times. The early observations of Hale *et al.* had a rather poor resolution, and so he could not discover the fine structure of solar magnetic fields. In recent years the observations at the Mt. Wilson, Kitt Peak, Crimean, Big Bear, Beijing and some other observatories have revealed that the polar fields, like the background field in other parts of the solar disk, consist of countless tiny elements with different strengths and polarities. In other words, *the Sun does not possess a regular and clear-cut bipolar field*, which was expected by Hale *et al.* This may be illustrated by the full-disk magnetogram in Figure 7.3. The amount of deviations from a horizontal line represents the field strength, and the upward and downward fluctuations belong to different polarities.

More convincing evidence may be the observation of Stenflo (1971)

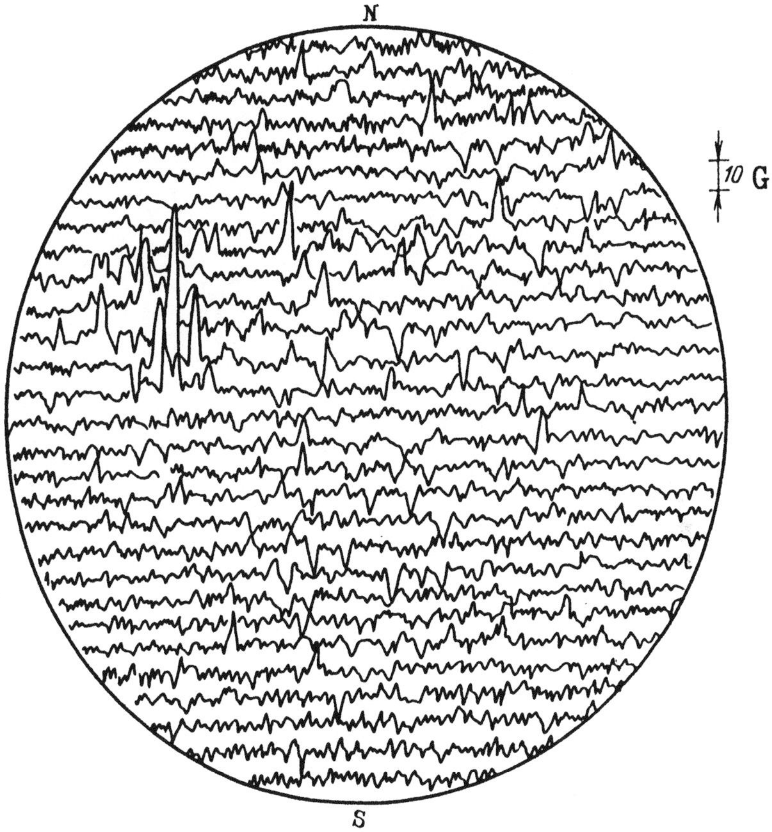

Fig. 7.3. Magnetographic recordings of fine structures on the whole solar disk.

at Mt. Wilson (Figure 7.4). He discovered that the polar field is not homogeneous with respect to longitude. That is to say, there is no rotational symmetry at all. Moreover, there seems to be a sectoral structure with alternate N and S polarities. This reminds us that if the polar field is not observed for at least one full solar rotation, one may get erroneous results.

Not only the polar regions, but also the other parts of the solar general magnetic field are characterized by fine structures. Indirect evidence for this is provided by Severny's *et al.* (1974b) observation of the fluctuations of solar radio emission during the transit of the planet Mercury across the solar disk. This was carried out on May 9, 1970 with the Crimean 22 m radio telescope at wavelength of 8 mm. Figure 7.5 presents a fragment of the original recording of the intensity of radio emission, represented by the antenna temperature (T_A). Repeated observations without Mercury transit did not show such fluctuations.

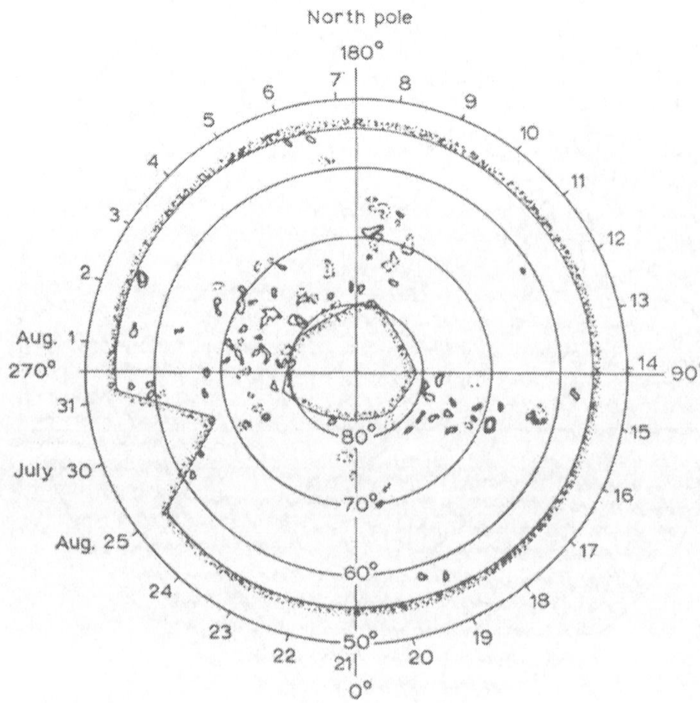

Fig. 7.4. Uneven distribution of polar fields in longitude.

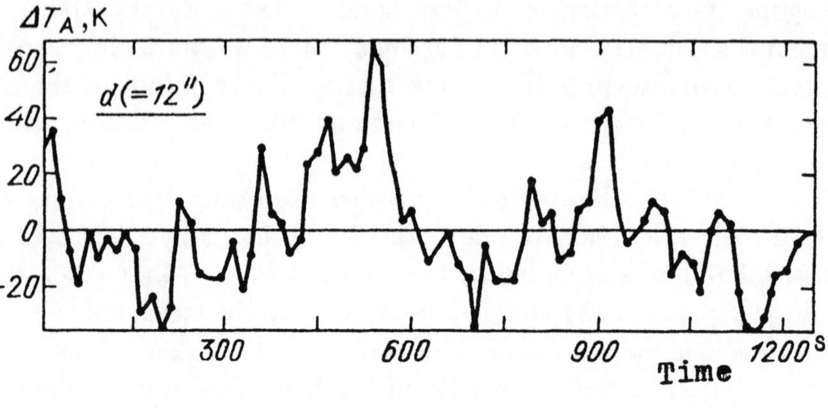

Fig. 7.5. Variations of radio intensity caused by occultation of magnetic fine structures by the Mercury.

It has now become clear that the solar general magnetic field is composed of fine structures and so it is similar to the background field introduced in Section 5.1. It is then tempting to discuss the following question: Are the magnetic fields in the polar regions and in the quiet districts near the equatorial limbs analogous to each other? After a careful comparison of the BBSO videomagnetograms belonging respectively to either of these two kinds of fields, Tang and Haimin Wang (1991) claimed that no difference can be found. From their work we reproduce the curves showing the changes of the mean flux density with the CMD (θ) or the latitude (φ) for the west limb, north pole and south pole respectively. (See the upper, middle and lower panels of Figure 7.6.) The broken line represents the best fit for the east limb data. It is superposed on each of the three curves in the figure for comparison. All these curves, as well as the broken line, agree well with one another in both magnitude and slope. Besides, the videomagnetograms themselves in the polar and limb regions are quite similar. (This is shown by the photographs in Figure 1 of the paper just cited.) These facts tell us that the two kinds of fields mentioned above may be thought to belong to the solar background field and that they are identical in nature. The unique difference is that the former is situated in the extreme north and south of the solar disk, while the latter is close to the east and west limbs. Of course, this has no concern with the physical nature.

As the end of this section, we like to quote the following statement of Severny (1988): "What we call the general magnetic field consists of a multitude of small magnetic elements with different polarities and strengths and the quantitative results have only a statistical significance."

7.3 Polarity Reversals of the General Magnetic Field

The general magnetic fields of the Sun and the Earth differ in some aspects. The main distinction lies in the behavior of their polarities. For the Earth, the magnetic polarity remains practically permanent and reversals occur through very long historical periods. However, the solar field interchanges its polarities every 22 years. Babcock (1959) was the first to discover such a phenomenon, and this caused a sensation in the world solar community. The concrete process of reversal is as follows. The polarity of the geometrically south pole changed gradually from S to N between March and July in 1957, but the field at the other geometrical pole reversed its polarity abruptly in November of 1958. It is curious to find that the reversal did not happen simultaneously at

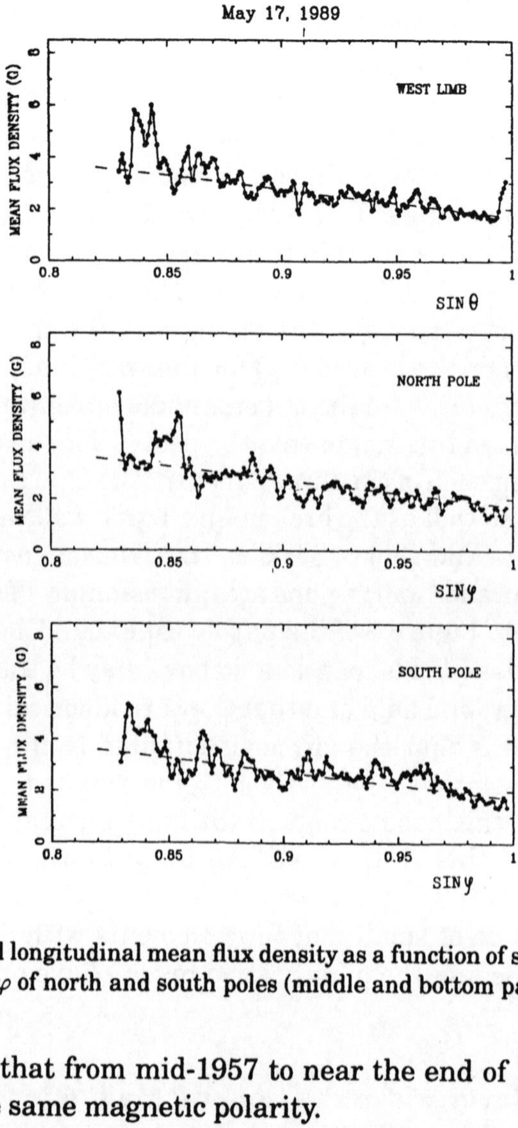

Fig. 7.6. Observed longitudinal mean flux density as a function of sin θ (top panel) and as functions of sin φ of north and south poles (middle and bottom panels).

both poles and that from mid-1957 to near the end of 1958 both polar regions had the same magnetic polarity.

After the discovery of the polarity reversal, Severny (1971) straightened out the observational data of the solar general magnetic field accumulated in the fifty more years since 1912 and found that the reversals did occur regularly. This is illustrated in Figure 7.7 in which the ordinate represents the difference between the N- and S-polarity field intensities and M and m indicate, respectively, the years of maximum and minimum solar activity. It is ready to see that the reversal takes place around the maximum year and that, excluding Hale's *et al.* observations, the general magnetic field attains its peak intensity commonly in the year of minimum solar activity.

Ever since Babcock's discovery during the 19th cycle of solar activ-

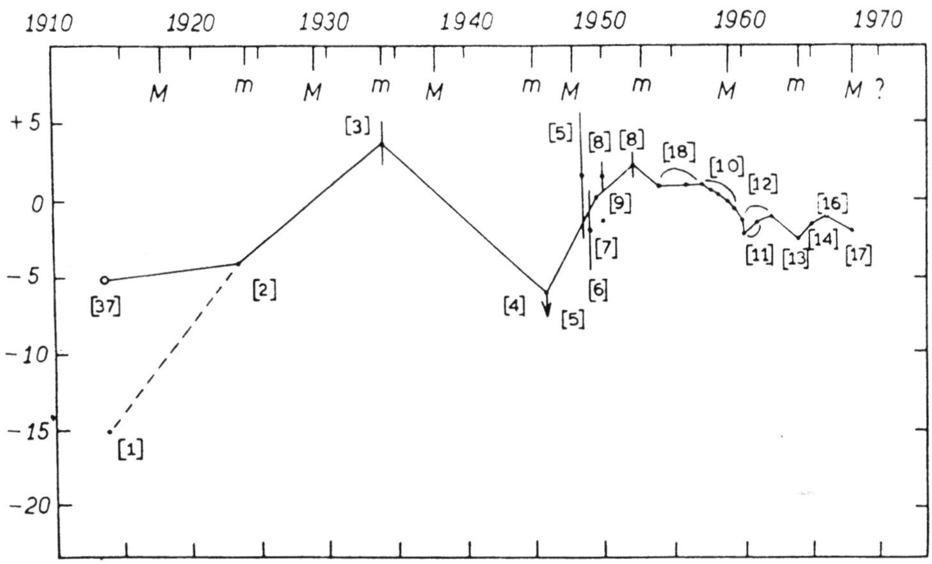

Fig. 7.7. Changes of the polarity and strength of the general magnetic field with time. (Numbers in square brackets pertain to references cited in Severny, 1971.)

ity, astronomers have paid much attention to pursuing further polarity reversals. In the next, 20th solar cycle the southern pole once again reversed first, in mid-1969, and the reversal at the northern pole did not appear until August of 1971 (Howard, 1972, 1974). Moreover, the reversals at both poles were not clear-cut. Their fields fluctuated both before and after the reversal, and this makes it difficult to decide on the exact time of reversal with observations of a short duration. For cycle 21, the north polar magnetic field reversed first in mid-1980 (Howard and LaBonnte, 1981). So it seems to be a rule that the polarity reversal does not take place simultaneously at both poles.

According to the BBSO observations (Tang and Haimin Wang, 1991), the polarity reversal in the 22nd cycle did not happen up to the end of 1989. Nevertheless, near this time there were signs of polarity turnover similar to the preceding cycles, i.e., the decreasing strength of the field with old polarity and fluctuations of the net flux between old and new polarities.

Finally, we want to point out that Babcock, Severny, Howard *et al.* studied polarity reversals solely with magnetic observations. More precisely speaking, they used mainly the routine longitudinal magnetograms of the whole solar surface. In recent years, however, astronomers have begun to utilize also some other observational data, such as those of chromospheric filaments and coronal holes. This is helpful to the determination of the approximate time of reversals of subsequent solar cycles (Webb *et al.*, 1984) as well as cycles prior to the advent of magnetographic observations (Makarov *et al.*, 1983).

7.4 Rapid Changes of the General Magnetic Field

In the preceding section the reader has seen that the solar general magnetic field exhibits long-term variations which may be concerned with the 22-y cycle of sunspot fields described in Section 4.1. In this section we go further to narrate that the general field still has fast and irregular changes as well as five-minute oscillations.

(1) *Short-term variations.* These were first discovered by Babcock and Babcock (1955). As pointed out by Severny (1971), among the numerous anomalies found by them the most remarkable may be the following one. Beginning from July 29, 1954 no magnetic field could be detected in the geometrically south pole and the normal situation was recovered after 13 days. However, during this period the field strength in the north pole remained almost unchanged. These authors claimed that this had been due to the change of the heliocentric latitude of the Earth and for this sake the south polar region had been hidden behind the visible surface of the Sun. Nevertheless, a large number of later observations have negated this idea. In fact, the disappearance of one polarity of the general magnetic field has become a common matter. Besides, it is also often observed that magnetic fields of the same polarity exist simultaneously in both poles of the Sun. Severny (1967) analyzed the Crimean and Mt. Wilson data in 1967 and discovered the rapid variations of the polar fields in a time scale of about one day. (See Figure 7.8 in which the ordinate represents both the field strength and the flux density. The data in the middle panel belong to the Crimean Observatory, and those at both the top and bottom to the Mt. Wilson Observatory.)

It is interesting to note that fields of both polarities rise and fall almost in phase. Besides, together with the change of strength there was inversion of polarity. All these facts tell us that the general magnetic field of the Sun is very unstable. At present it is difficult to explain the physical nature of these phenomena.

(2) *Periodic oscillations.* In addition to the irregular variations described above, there are periodic pulsations. Just like the velocity and brightness fields, the magnetic field also displays 5-minute oscillations.

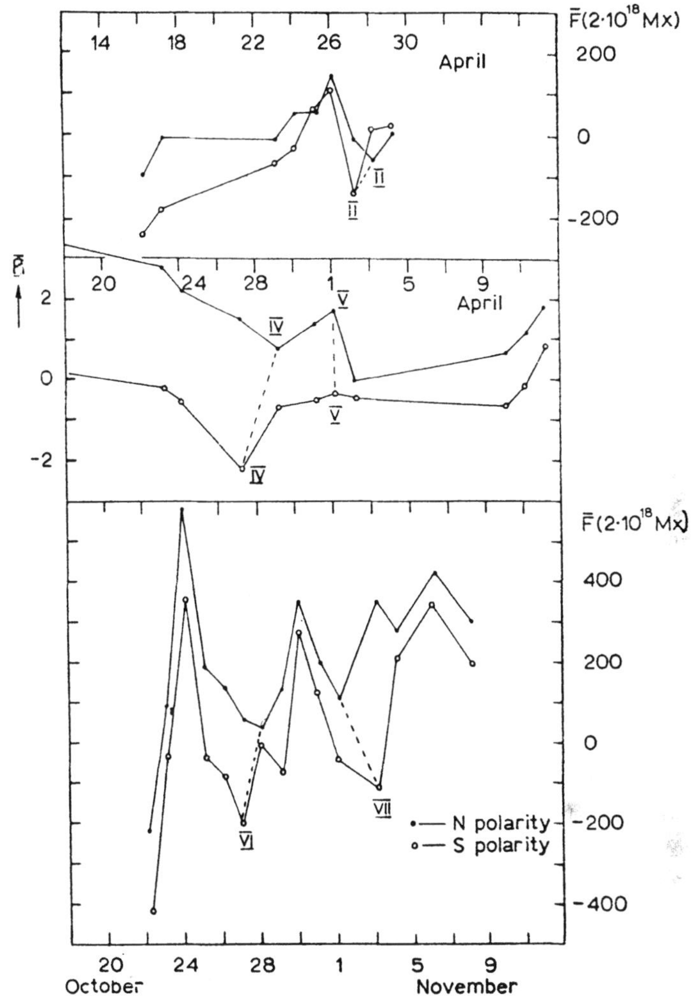

Fig. 7.8. Short-term fluctuations of the mean polar fields of the Sun.

This was observed by Tanenbaum *et al.* (1971). For each time in several hours they carried out magnetographic observation of a definite point on the Sun. (In order to eliminate the influence of solar rotation, they purposely shifted the point of observation with a constant speed towards the west.) A part of their recording is shown in Figure 7.9, and this is the result of a 3-hour long observation. In spite of the "noise", the periodicity of fluctuations is evident. From such observations the following conclusions may be drawn:

(i) The solar background magnetic field exhibits incessant pulsations.

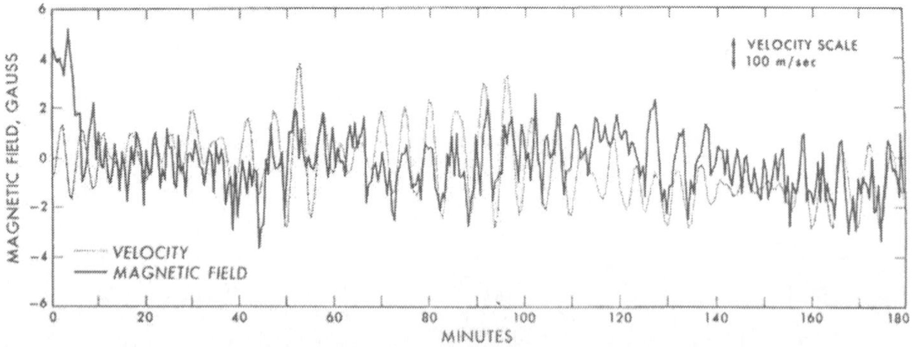

Fig. 7.9. 5-minute oscillations of magnetic and velocity fields.

(ii) The period is the same as that of the velocity field, i.e., about 5 minutes.

(iii) The two kinds of oscillations are approximately in phase.

(iv) Both the photospheric line Fe I $\lambda 5250$ and the chromospheric line Mg I $\lambda 5173$ exhibit such pulsations.

(v) The amplitude of variation of the field strength is about 1–2 G.

(vi) In regions of both weak and strong magnetic fields, with intensities of about 5 G and 80 G respectively, there are similar oscillations.

In order to explain these facts, Tanenbaum *et al.* proposed that plasma motions may affect the intensity of the longitudinal magnetic field. To be precise, this concept is based on the following two ideas. First, vertical oscillations can give rise to periodic changes of the horizontal magnetic field. In the upper panel of Figure 7.10 a lump of plasma oscillates in the vertical direction. Then the magnetic lines of force "frozen" in the plasma also undergo an alternately upward and downward motion. If we observe the plasma element enclosed in the small square indicated in the figure, the lines of force in it are no longer horizontal. The field strength observed in any direction then displays periodic changes. Second, horizontal plasma oscillations may cause periodic changes of a vertical field (see the lower panel of Figure 7.10). This is because the oscillation sometimes makes the plasma in a small element denser, and sometimes thinner. Then the number of magnetic lines of force passing through the small square always varies. This

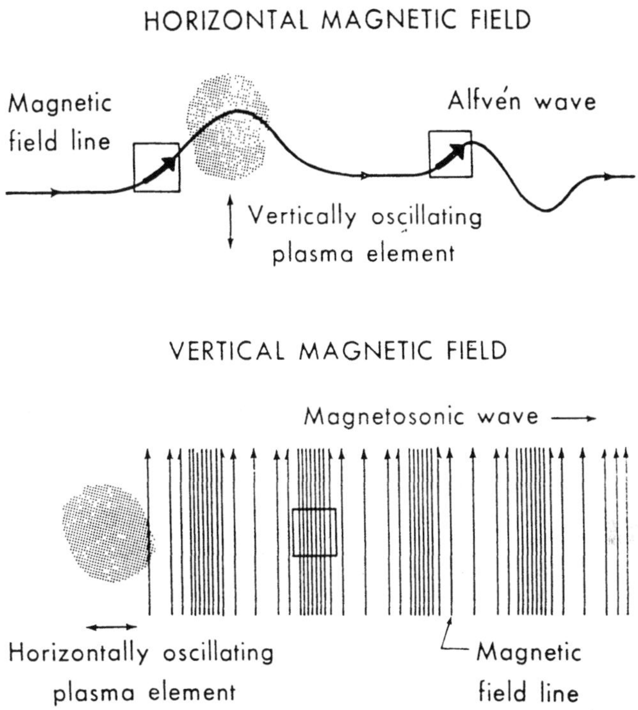

Fig. 7.10. Influence of plasma oscillations on the longitudinal field strength.

implies that the field strength is incessantly changing. These two cases are in principle similar, but they have the following important difference. In the first circumstance, the mean field intensity is equal to zero; but, in the second one, it is not null. The general case may be a combination of these two situations.

7.5 The Global Magnetic Field of the Sun

In this section we are going to discuss this interesting question: How does the solar magnetic field behave, if the Sun is treated as a single point like a star? In the investigation of this problem much progress has been achieved in recent years.

7.5.1 Net Magnetic Flux of the Solar Disk

Using magnetograms with high spatial resolution ($2''.3 \times 9''$), Severny (1967) computed the magnetic fluxes at various places of the solar disk and obtained the total amounts of fluxes of both polarities, F_N and F_S.

Then the *net flux of the whole disk* ($\Delta F \equiv F_N - F_S$) can be obtained. He discovered that within one day ΔF may change significantly and that the quantity $\Delta F/4\pi R_\odot^2$, which should be equal to the magnetic field strength averaged for the entire disk, can vary from -0.6 G to $+0.5$ G. His work had the following defect. The scanning of a magnetogram required a few hours and in this process the solar magnetic field may vary a great deal.

Bumba *et al.* (1967) used a similar method and found that the net magnetic flux behaves like sunspot relative numbers. It is a pity that the spatial resolution of their magnetograms was low ($23'' \times 23''$) and so many magnetic fine structures could not be resolved. Besides, they did not take into consideration the magnetic fields close to the solar limb (with heliocentric distances larger than 60°). It is necessary to point out that the total amount of magnetic fluxes in these regions may be comparable with that of the central part of the solar disk.

Although the works mentioned above are not accurate enough, they may convey to us the following impression. When taken together, the magnetic fluxes from various portions of the whole solar disk may not completely cancel one another out. Therefore, if the Sun is observed as a star, it still possesses a magnetic field of certain strength and polarity.

7.5.2 Characteristics of the Global Field

In order to realize the magnetometry of the Sun as a star, Severny and his associates (Severny, 1969) sent the parallel solar rays reflected from the coelostat of the Crimean solar tower into the spectrograph-slit. Because a solar image is not formed, the magnetic field strength measured in this way is the weighted mean value of the longitudinal fields of various parts of the solar disk and the brightness of the solar surface plays the role of the weight. The field thus observed is called the *solar global magnetic field*. In the upper panel of Figure 7.11 the solid dots represent the intensity of this field (B_s) from March to June of 1968. The open circles belong to the longitudinal component of the interplanetary magnetic field. The lower panel illustrates the changes of the sunspot net magnetic field in the same period. This quantity, denoted by B_{sp}, can be calculated from the expression

$$B_{sp} = 0.15 \sum_i \frac{S_i B_i}{\pi R_\odot^2} ,$$

where B_i is the field strength of the ith spot, S_i is the total area of its umbra and penumbra and 0.15 is the mean value of the ratio of the umbral and penumbral areas.

The following interesting results can be derived from Figure 7.11.

(1) The global field exhibits regular changes and its polarity is alternately N or S. Roughly speaking, the inversion of magnetic polarity occurs twice in each period of solar rotation. A natural explanation is that the solar surface can be divided into several magnetic regions and the neighboring ones have opposite polarities. The dividing lines of these regions are approximately parallel to the solar longitudes (Figure 7.12). Due to the rotation of the Sun, we observe at one time an N-polarity region of the global field, and at another time an S-polarity one.

(2) It is surprising to note that the solar global field is closely connected with the interplanetary magnetic field. Concretely speaking, the variation of the longitudinal component of the latter with time (after correction for the time of propagation of the solar wind from the Sun to the Earth, i.e. about $4\frac{1}{4}$ days) coincides well with the change of the solar global field. This clearly shows that the magnetic field in the interplanetary space is directly controlled by that on the Sun. A more detailed account of this will be given in Section 8.1.

(3) By comparing the upper and lower panels of Figure 7.11, we can see that the spot magnetic fields taken together are not in phase with the global field, and sometimes even possess opposite phases. This means that the contribution of sunspots to the solar global magnetic field is trivial, and this is understandable. Although spots have the strongest magnetic fields on the solar surface, they after all occupy a very small percentage of the surface area.

(4) The period of inversion of the global field may be slightly different for the N- and S-polarities. In 1968 and 1969, the mean interval between two successive appearances of the S-polarity was 27.6 days, while that for the N-polarity was 26.8 days. This may be explained as follows. In these two years the major N spots were closer to the equator than S spots, and due to the differential rotation the former moved more quickly than the latter.

For a long time, astronomers have the following traditional concept. Except for spots and local active regions, the wide space on the solar surface is pervaded by the general magnetic field, which is bipolar in nature and concentrated mainly in the N- and S-polarity regions located near the geometrical poles. In other words, the general solar

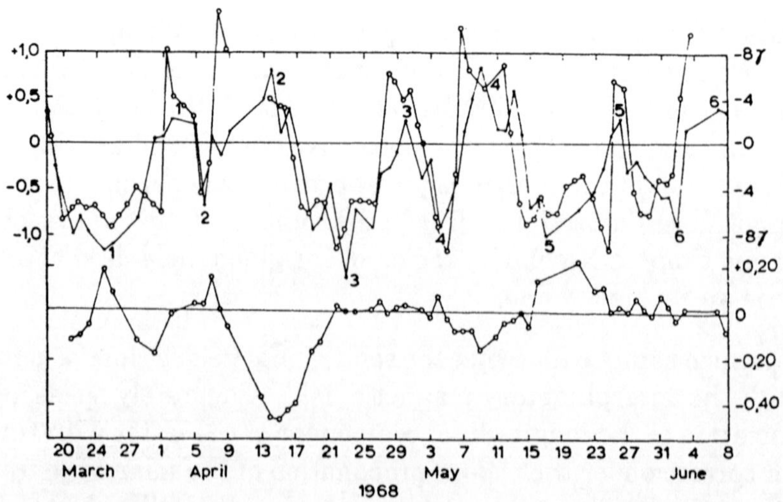

Fig. 7.11. **Fluctuations in solar global and interplanetary magnetic fields as well as in the total flux of sunspots.**

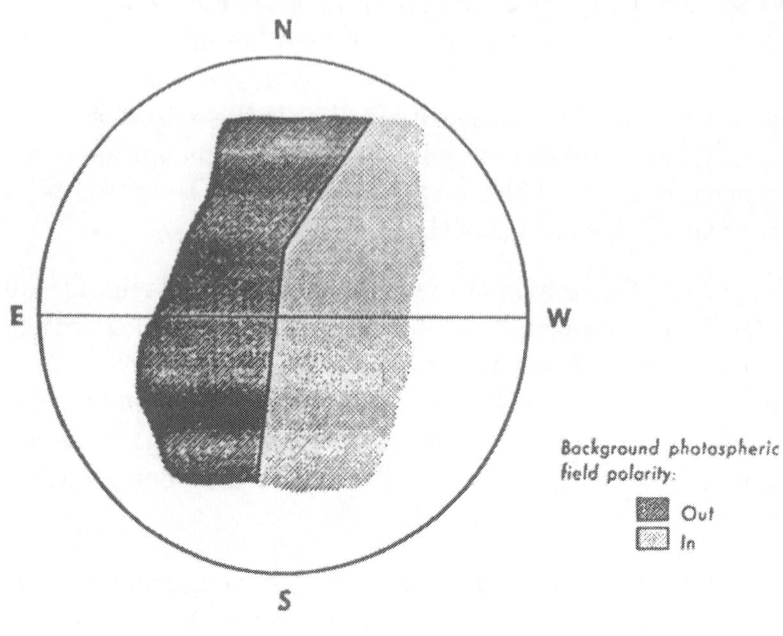

Fig. 7.12. **Magnetic regions on the solar surface.**

magnetic field is characterized by an approximate north-south symmetry. However, what has been stated above shows that the global field is roughly symmetric in the east-west direction. Therefore, the solar magnetic field is the unity of these two seemingly contradictory fields. This may be considered to be a meaningful philosophical proposition. In order to achieve a deep understanding of the solar magnetism, much still remains to be done. This problem will be discussed further in the last chapter.

7.5.3 Rapid Changes of the Global Field

Besides the regular variation of the global field connected with the solar rotation, there are still rapid fluctuations. As can be readily seen in Figure 7.11, both upwards and downwards from the curve representing the solar global field there are deviations which have a time scale of about 1 day and strength amplitude of 1 G. Such changes can hardly be explained by the coming in or going out of a magnetic district on the rotating solar disk. As expounded by Severny (1988), within one day a region with a scale of $l = vt = 2 \times 10^5 \times 10^5 = 2 \times 10^{10}$ cm can rotate onto or outward from the solar limb. The area of this region occupies only a percentage of $(l/d_\odot)^2 = (2 \times 10^{10}/1.4 \times 10^{11})^2 = 2 \times 10^{-2}$ of the disk. This is too small to affect the global field, unless the field intensity of the magnetic region is about 100 G or even stronger. However, this is quite unlikely.

More accurate observations in recent years, e.g. those made in 1975–1978 and 1987 at the Crimean, Mt. Wilson and Sayan Observatories (Demidov *et al.*, 1990), reveal the presence of more or less persistent oscillations with periods near 47, 60, 85 and 160 minutes and with amplitudes of about 10^{-2} G. Among these periods the longest one seems to be the most plausible. The power spectrum of the Crimean measurements made with Fe I λ5250 in 14 days in 1975–1978 (in each day the observation lasted at least 7 hours) is shown in Figure 7.13. Although there are several peaks, the one belonging to the 159-min period is the highest. A similar result was obtained with the Sayan data of 1987. It is still more interesting to compare the oscillations of the global magnetic field with those of the velocity and brightness fields, and this is illustrated by Figure 7.14. The upper panel presents the mean curve of variations of the global field strength with the period of $160^{\mathrm{m}}.0101$. The curves in the middle and lower panels show the averaged changes of the differential velocity and brightness respectively, and these curves are plotted with the same period. The abscissa represents the phase

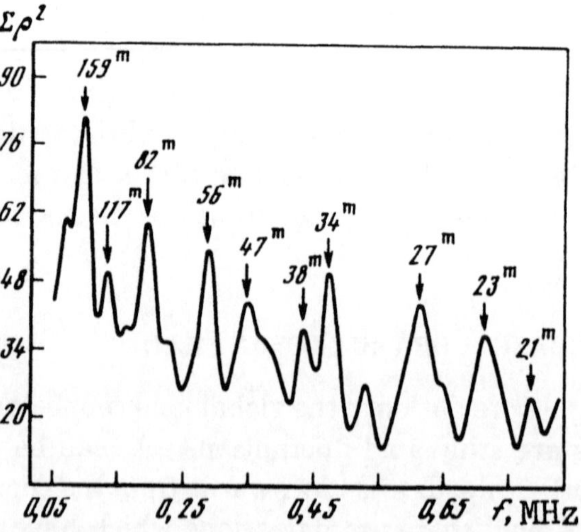

Fig. 7.13. **Power spectrum of fluctuations of the solar global magnetic field.**

and the origin is chosen to be the moment 00^h00m UT of January 1, 1974. Now we would like to mention that the observational data belong to various time intervals (indicated in the figure) and to different sites. (Data of the global magnetic field were secured at the Sayan Observatory, while those of the velocity and brightness fields were got in the Crimea.) Nevertheless, the three curves agree with one another quite well. This implies that *the global magnetic field also possesses a period of 160 min*, which was discovered for solar oscillations by some Crimean astronomers (Severny *et al.*, 1976; Kotov, 1985). After many years of debate it has been firmly established that solar oscillations with this period really exist and that they belong to the g-mode and come from the solar interior. So it is possible that the global field is in some way concerned with the *solar internal magnetic field*, which is the subject of the next section.

Finally, we want to state that except for the 160-min period the global field may also oscillate with certain short periods. Now that we have said in Section 7.4 that the 5-min pulsations of the solar plasma can give rise to periodic changes of the local magnetic field, it is possible that the global field also display 5-min oscillations. This was recently studied by Grigoryev and Demidov (1991) with Sayan observational data. They obtained the power spectrum shown in Figure 7.15. It tells us that among the several possible short periods the 4.9-min period possesses the highest priority. So it is quite probable that the solar global

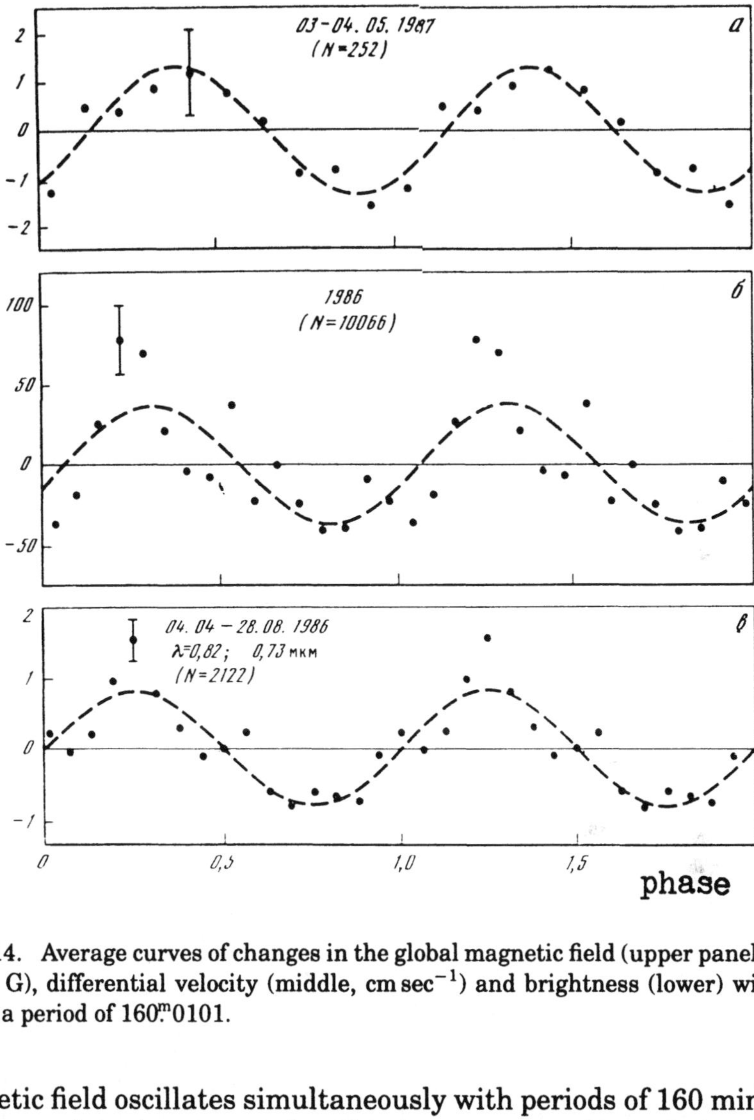

Fig. 7.14. Average curves of changes in the global magnetic field (upper panel, in units of 0.01 G), differential velocity (middle, cm sec^{-1}) and brightness (lower) with phase during a period of 160^{m}0101.

magnetic field oscillates simultaneously with periods of 160 min, 5 min and even other time intervals. This is a very complicated phenomenon and further investigation is needed.

7.6 The Internal Magnetic Field of the Sun

Up to now, all that we have said about the solar magnetic field belongs to the atmosphere, which occupies only a very small portion of the Sun. The bulk of both its mass and volume is contained in the invisible interior. Hence no direct measurement of the internal magnetic field has yet been made. In recent years, using the following indirect methods,

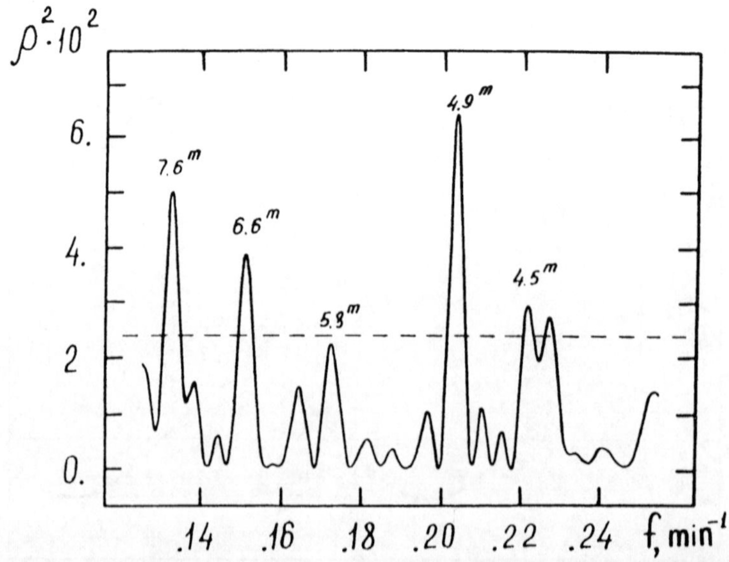

Fig. 7.15. Power spectrum of fluctuations of the global magnetic field with observation of 3 hours on June 9, 1987.

astronomers have begun to draw some inferences about this field.

(1) As a possible way out from the dilemma of the dearth of the *solar neutrino flux*, Chitre *et al.* (1973) and Dicke (1977, 1979) suggested that there is a strong magnetic field of 10^8 G in the core of the Sun.

(2) The Lorentz force caused by the interaction of the magnetic field and the electric current can modify the state of hydrostatic equilibrium. This is expressed by the following equation:

$$\nabla p + \rho \nabla \phi - \frac{1}{c} \mathrm{j} \times \mathbf{B} = 0, \qquad (7.2)$$

where p is the pressure and ϕ is the gravitational potential. Then the existence of an intense magnetic field in the solar interior may lead to the distortion of the solar surface, including its *oblateness*. According to the estimate of Bahcall and Ulrich (1971), a field of about 10^7 G can give rise to the observed oblateness of which the upper limit is 5×10^{-5} (Dicke, 1970).

(3) By solving the dynamo equations Boyer and Levy (1984) derived the strength of the radial component of the internal magnetic field. The approximate symmetry of the two halves of the solar activity cycle requires that this component is merely about 1 G. Pudovkin and Benev-

olenskaya (1984) claimed that at the base of the convective zone it is only 0.5 G. So the internal magnetic field is essentially a toroidal one.

(4) Astronomers have also tried to obtain some information from the observation of solar oscillations. For instance, Dziembowski and Goode (1989, 1990) used Libbrecht's (1989) data and found evidence for a 2 ± 1 megagauss quadrupole toroidal field centered at 0.7 $R_\odot$ which is barely beneath the base of the convective zone. Besides, this field may be associated with the dynamo that drives the Sun's 22-y magnetic cycle.

(5) Parker (1985) studied the meridional circulation induced by the thermal effects of magnetic field inhomogeneities in mixing and depleting ^{7}Li on the solar surface. He found a limit of 4×10^5 G for the field just below the bottom of the convective envelope. This work was followed by Moss (1987). Together with the solution of the MHD and thermal equations of the solar interior, he came to the conclusion that the internal magnetic field strength is unlikely to be much in excess of 10^5 G.

(6) Dudorov *et al.* (1989, 1990) performed a systematic study of the internal magnetic field during the formation and evolution of the Sun. At the initial stage, the Sun was formed by the contraction of the magnetized protosolar cloud, which was permeated by a homogeneous field of strength 4.5×10^{-5} G. This process may be described by the following gas-dynamic equation combined with the induction equation:

$$\frac{\mathrm{d}\mathbf{B}}{\mathrm{d}t} = \nabla \times [(\mathbf{V} - \mathbf{V}_{\mathrm{AD}}) \times \mathbf{B}] + (\mathbf{V} \cdot \nabla)\mathbf{B} - \nabla \times (\nu_m \nabla \times \mathbf{B}), \tag{7.3}$$

where $\mathbf{V} = (V_r, 0, 0)$ is the gas velocity, $\mathbf{B} = (B_r \cos\theta, -B_\theta \sin\theta, 0)$ is the field strength, ν_m is the magnetic viscosity and $\mathbf{V}_{\mathrm{AD}}$ is the velocity of ambipolar diffusion. The solution of this equation shows that after the Sun has just arrived at the main sequence, its magnetic field may be 10^5–10^6 G at the core boundary and 10^4–10^5 G near the base of the convective zone.

Further evolution is determined by the hydromagnetic or resistive instabilities, buoyancy, as well as ohmic and turbulent diffusion. The poloidal (B_p) and toroidal (B_φ) components of the magnetic field are connected with the dynamo number (N_D) and turbulent velocity (V_t) by the relations

$$B_p B_\varphi = \pi\omega H_p V_t, \qquad B_\varphi = \sqrt{N_D}\, B_p\,, \tag{7.4}$$

where ω is the angular velocity and H_p is the pressure scale height. These authors assumed that the protosolar luminosity L is entirely determined by the convective energy transported and in this case $V_t = (L/(MH_p))^{1/3}$. Then they got

$$\begin{aligned} B_p &= \left(\frac{8\pi H_p \rho}{3}\right)^{1/3} \left(\frac{H_p}{R}\right)^{1/2} \left(\frac{\omega}{R}\right)^{3/4} V_t^{3/4}, \\ B_\varphi &= \left(\frac{24\pi H_p \rho}{3}\right)^{1/2} \left(\frac{R}{H_p}\right)^{1/2} \left(\frac{\omega^3}{R}\right)^{1/4} V_t^{1/4}. \end{aligned} \tag{7.5}$$

Adopting the standard solar model with a density near the core $\rho = 10\ \mathrm{g\,cm^{-3}}$, one has $B_p = 10^4$–10^5 G and $B_\varphi = 10^5$–10^6 G. At the base of the convective zone $\rho = 0.1\ \mathrm{g\,cm^{-3}}$, then $B_p = 3 \times 10^3$–3×10^4 G and $B_\varphi = 3 \times 10^4$–$3 \times 10^5$ G. Such a field evolves with time to a state with $B_\varphi = B_p$. As for the further evolution of the internal magnetic field after the Sun's arrival at the main sequence, the reader is referred to the papers by Dudorov *et al.* cited above.

The methods briefly introduced above are based on various principles and they yield different results. At the moment we are still far from a complete and satisfactory recognition of the internal magnetic field of the Sun. Nevertheless, it is certain that this field is very much stronger than that of the atmosphere. Knowledge of the internal magnetic field is closely concerned with many topics of important significance, such as the mechanism of the solar activity cycle, the nature of solar oscillations (in particular, those with long periods) and the cause of the deficiency of the solar neutrino flux. Moreover, as remarked by Sturrock (1986), the solar internal magnetic field is also important for cosmogony. This is because such a field is generally believed to be a relic of the magnetism in the primeval nebula from which the Sun has originated. The results of this investigation may be generalized to many billions of stars in the immense universe.

CHAPTER 8

Magnetic Fields of the Solar System

The interplanetary electromagnetic conditions control the phenomena near the Earth and are responsible for the production of magnetic storms and aurorae. They also produce intensity variations in cosmic radiation, and may play a role for the generation of cosmic radiation.

H. Alfvén and C.-G. Fälthammar, *Cosmical Electrodynamics* (1963).

In the preceding four chapters we have presented a comprehensive and overall description of the solar magnetic field. This chapter will be devoted to the magnetic fields of the other celestial bodies in the solar system as well as of interplanetary space. Just like the overwhelming superiority of its mass, the Sun's magnetic field also occupies a dominant position in the solar system. Due to the action of the solar wind, this field extends outward from the solar atmosphere, permeates the vast space of almost the whole solar system, and affects the magnetic fields of planets and their satellites. Since the beginning of the space era it has become possible to carry out *in situ* exploration of planetary and lunar magnetism. Although work in this field is still in its infancy, some exciting achievements have been obtained in recent years.

8.1 Magnetic Field of Interplanetary Space

8.1.1 EARLY SPECULATIONS

Before the advent of the space age, there were some predictions about the physical status of interplanetary space. It was long ago noted that the cometary tails are not just on the lines connecting the Sun and comets, as suggested by the hypothesis of light pressure, but may lag by a few degrees. Besides, the acceleration of motion of cometary tails is too large to be explained by the light pressure. These facts and some others led Biermann (1951) to suggest that the cometary tails are formed by corpuscular streams from the Sun. This idea is correct, but his value of the outward flow velocity at the solar surface, i.e. 3000 km s^{-1}, is too large.

Chapman (1957) set up the temperature structure of the interplanetary space. He derived the following expression for the thermal flux transported by conductivity:

$$F = -4\pi R_{\odot}^2 K_0 T_0^{5/2} \left(\frac{r}{R_{\odot}}\right)^2 \left(\frac{T}{T_{\odot}}\right)^{5/2} \frac{dT}{dr} = \text{const}, \tag{8.1}$$

where K_0 and T_0 are, respectively, the coefficient of conductivity and temperature at a certain distance from the solar center $r = R_0$. Beyond this R_0, as suggested by him, the energy flux is mainly transported by heat conduction. From the above equation one has

$$\mathrm{d}(T^{7/2}) = \frac{\frac{7}{2} F T_{\odot}^{5/2}}{4\pi K_0 T_0^{5/2}} \mathrm{d}\left(\frac{1}{r}\right) \equiv C\,\mathrm{d}\left(\frac{1}{r}\right), \tag{8.2}$$

where C is a constant. A simple integration leads to

$$T_0^{7/2} - T^{7/2} = C\left(\frac{1}{R_0} - \frac{1}{r}\right). \tag{8.3}$$

Let us assume that when $r = \infty$, $T = 0$. Then we have $C = R_0 T_0^{7/2}$. Substituting this constant back into Eq. (8.3), we obtain

$$T = T_0 \left(\frac{R_0}{r}\right)^{2/7} \propto r^{-2/7}. \tag{8.4}$$

According to this formula, the temperature decreases rather slowly with distance. Taking $T_0 = 10^6$ K, Chapman got $T = 2.19 \times 10^5$ K near the Earth. At first sight it seems surprising that the temperature in the interplanetary space around the Earth is so high.

Furthermore, Parker (1958) suggested that the solar corona is not in hydrodynamic equilibrium and that it expands unceasingly. This gives rise to the *solar wind*. Because of the high conductivity and large scale of the corona, the magnetic field is frozen in its plasma and together with it expands into the interplanetary space. Moreover, under the action of solar rotation, the magnetic lines of force are distorted into Archimedean spirals. All these predictions were soon confirmed by space observations.

8.1.2 The Coronal Expansion

Parker's (1958) model of the expanding solar corona is a stationary and spherically symmetric one. He supposed that, beyond some distance $r = b$, both the solar gravitation and the outward acceleration due to the coronal high temperature may be neglected, so that the outward velocity becomes a constant, v_m. (He took $b = 5\times10^{11}$ cm and $v_m = 1000\,\mathrm{km\,s^{-1}}$.) In spherical coordinates and for $r \geq b$ the velocity components are

$$v_r = v_m, \qquad v_\theta = 0, \qquad v_\phi = \omega(r - b)\sin\theta. \tag{8.5}$$

where ω is the angular velocity of the Sun. The azimuth ϕ_0 of the streamline at $r = b$ is related to r by the following equation:

$$\frac{r}{b} - 1 - \ln\left(\frac{r}{b}\right) = \frac{v_m}{b\omega}(\phi - \phi_0). \tag{8.6}$$

In this model the magnetic field strength has the following components:

$$\begin{aligned} B_r(r,\theta,\phi) &= B(\theta,\phi_0)\left(\frac{b}{r}\right)^2, \\ B_\theta(r,\theta,\phi) &= 0, \\ B_\phi(r,\theta,\phi) &= B(\theta,\phi_0)\frac{\omega b^2}{v_m r^2}(r - b)\sin\theta. \end{aligned} \tag{8.7}$$

For $r \gg b$, Eq. (8.5) reduces to that of the Archimedean spirals, which divide the interplanetary magnetic field into four sectors (Figure 8.1).

In each sectoral region the magnetic polarity is almost everywhere the same, and the neighboring regions have opposite polarities. At 1 A.U. the dividing line forms an angle of about 45° with the Sun-Earth line. Besides, the magnetic field strength is approximately 5×10^{-5} G and the solar wind velocity is $400\,\mathrm{km\,s^{-1}}$.

Although Parker's model is basically correct, the actual conditions of the interplanetary magnetic field are much more complex. For instance, B_θ is not always equal to zero and B_ϕ may differ considerably from that given by Eq. (8.7) (Mariana and Neubauer, 1990). As for the sectoral structure, which will be briefly discussed in the next subsection, the situation is also complicated. As remarked by Wilcox (1968), sometimes (e.g., in the greater part of 1963) the magnetic sectors almost vanish, while at other times (e.g., throughout 1965) the sectoral configuration is quite conspicuous. It has been found that during certain solar rotations there may be only three, or even two, sectors. All these complexities await further investigation.

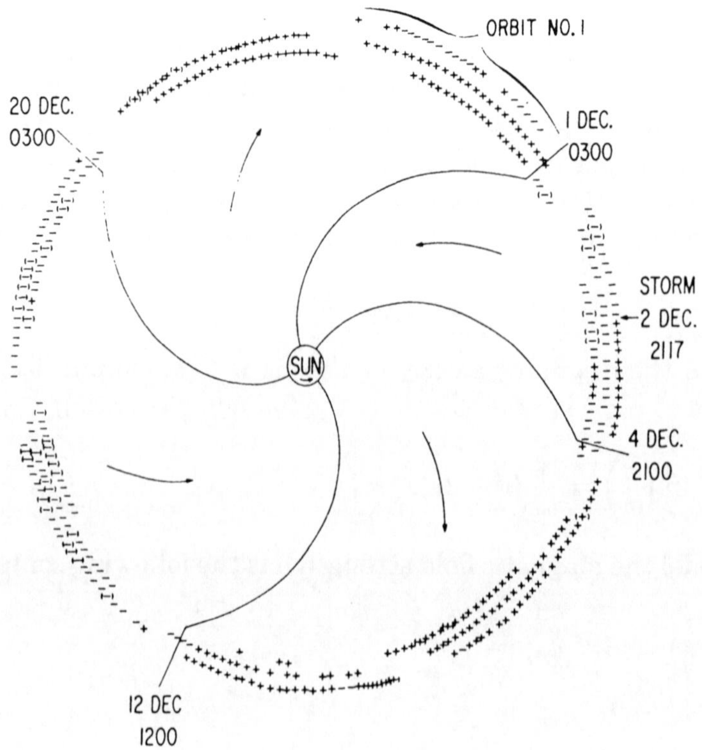

Fig. 8.1. Sector structure of interplanetary magnetic field. (The plus and minus signs indicate, respectively, fields away from and towards the Sun. Observation was made at 1 A.U. in December of 1964.)

8.1.3 The Sectoral Structure

As described in the last subsection, the interplanetary magnetic sectors are formed under the joint action of the extension of the solar global magnetic field and the solar rotation. Because the sectors rotate together with the Sun, at different times our Earth is located at various positions in the sectoral pattern. Therefore, the physical parameters measured outside the Earth in the course of a complete solar rotation can provide some information about the sectoral structure. Some results of such observations made during the solar rotation 1768 in the autumn of 1962 by the spacecraft Mariner 2 are shown in Figure 8.2. They include (a) solar wind velocity, (b) proton density, (c) temperature, (d) interplanetary magnetic field strength, and (e) magnetic polarity. The vertical hatched areas indicate the central meridian passages of *coronal enhancements* observed with the Nancy radio interferometer at 169 MHz (Couturier and Leblanc, 1970). These formations are regions

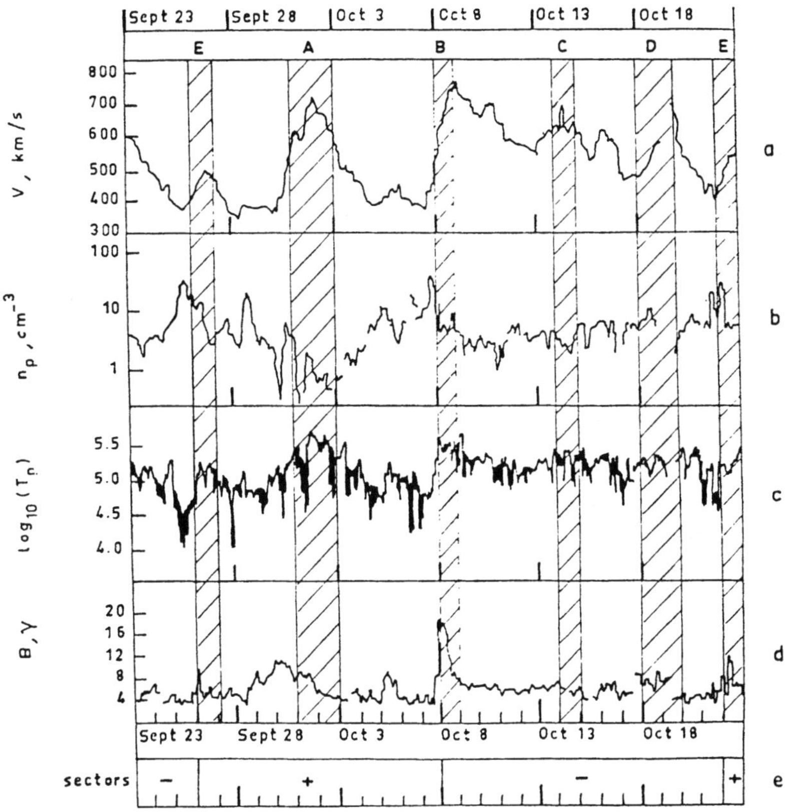

Fig. 8.2. Physical conditions of interplanetary magnetic sectors.

of enhanced electron density and temperature at heights of 0.2–0.5 $R_\odot$ above the photosphere. It is readily seen in the figure that the solar wind velocity attains a maximum during the passage of each coronal enhancement. Besides, each sector boundary corresponds to an enhancement. This also shows how closely the interplanetary magnetic field is related to coronal structures.

It is known from space exploration that a sector boundary sweeps over the Earth in less than 5 minutes. So the thickness of the boundary is estimated to be less than $1\text{–}2 \times 10^5$ km. Furthermore, measurements with a time resolution of a few seconds reveal that sector boundaries contain complex fine structures. For example, as discovered by Behannon and Neubauer (1981), a sector transition may be accompanied by a number of magnetic polarity reversals over a period of a few hours.

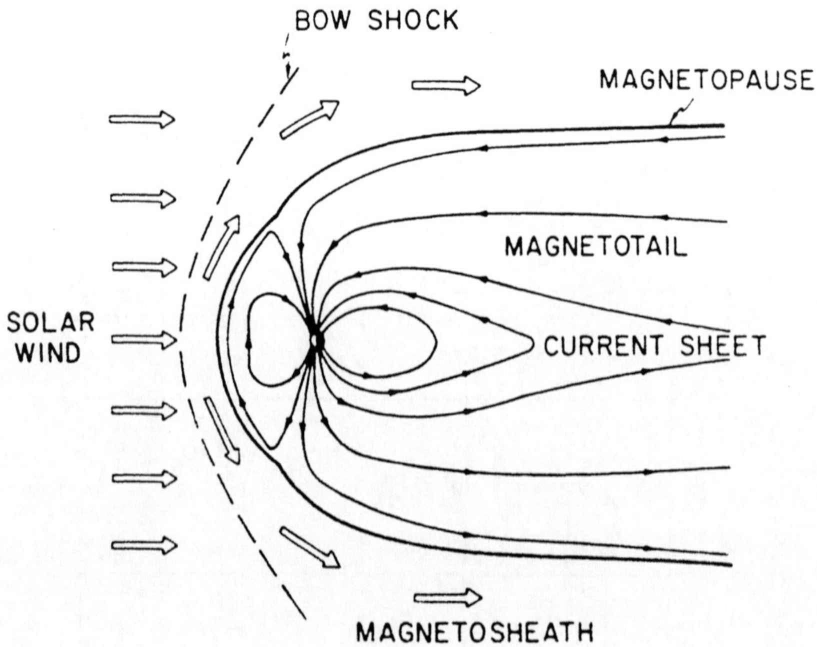

Fig. 8.3. The Earth's magnetosphere.

8.2 Magnetospheres and the Heliosphere

Although the term "magnetosphere" was coined by T. Gold in 1959, the concept itself came into existence much earlier in the work of Chapman and Ferraro (1931). For many years in the past the term magnetosphere was reserved solely for that of the Earth. Now, thanks to the achievements in space exploration, we are aware that maybe all other planets, and even some comets, possess magnetospheres. The magnetosphere of the Sun is called the *heliosphere*.

8.2.1 The Magnetosphere of the Earth

The magnetosphere is a cavity in the terrestrial magnetic field with a long tail extending to a length of hundreds of Earth radii (R_E) in the antisolar direction. The magnetosphere is formed by the interaction between the solar wind and the geomagnetic field (Kamide and Slavin, 1986). The former compresses the latter, while the latter deflects and traps the charged particles of the former. Due to these physical processes the magnetosphere has a rather complex structure (Figure 8.3). Its outer surface is called the *magnetopause*, the location and configuration of which are determined primarily by the following condition of pressure equilibrium. The total (plasma plus magnetic) pressure just outside this surface should be equal to the total pressure just inside. The external pressure can be expressed as

$$P = \xi \rho_0 v_0^2 \cos^2 \psi + P_0. \tag{8.8}$$

Here ρ_0, v_0 and P_0 are, respectively, the mass density, flow speed and thermal pressure of the incident solar wind, ψ is the angle between the local normal to the magnetopause and the upstream solar wind direction, and $\xi = 0.9$ is a constant. P_0 can be neglected in dealing with the front side of the magnetosphere, because in this case ψ is not close to 90°. The internal pressure depends chiefly on the magnetic field and can be taken to be $B^2/8\pi$. Therefore, the condition of pressure equilibrium is

$$\frac{B^2}{8\pi} = \xi \rho_0 v_0^2 \cos^2 \psi. \tag{8.9}$$

In Figure 8.3 the magnetic field lines inside the magnetosphere are represented by the solid lines with arrows and the plasma flow outside the magnetosphere by open arrows. Since the incident solar wind flow is supersonic, a detached bow shock is needed to deflect it. The region of thermalized and deflected plasma between the bow shock and the magnetopause is called the *magnetosheath*. (For a detailed description see, e.g. Vasyliuras, 1983). Due to the frequent variations of the solar wind, the structure of the magnetosphere is not stable. For instance, the average geocentric distance from the center of the Earth to the subsolar point is 10–11 R_E, but occasionally it may be as small as 6.6 R_E or as large as 18 R_E. Because the magnetosphere is within the realm of the geomagnetic dipolar field whose strength diminishes with the square of the distance from the Earth, the intensity of the field on the magnetopause lies between 50γ[1] and 100γ. Outside this layer the magnetic field exhibits irregular fluctuations in the range 20–50 γ. In the interplanetary space the strength falls to a very low level of 5γ.

8.2.2 Planetary Magnetospheres

As far as we know, all the planets already explored by spacecraft possess magnetospheres. Due to the differences of planets in magnetic fields, distances from the Sun, angles between the magnetic and rotational axes etc., the planetary magnetospheres may be quite unlike one

[1] $1\gamma = 10^{-5}$ G.

another. Therefore, they have to be described separately. (A detailed account can be found in Russell, 1987.)

(1) MERCURY. This planet has a magnetic moment more than 2000 times smaller than that of the Earth. Besides, since Mercury is the planet closest to the Sun, the density of the solar wind in its vicinity is much higher than that near the Earth. Due to these two factors, the magnetosphere of Mercury is many times smaller than the terrestrial one. To be precise, the distance to the subsolar point is only 1/20 of that in the Earth's magnetosphere.

(2) VENUS. As will be stated in Section 8.4, no intrinsic magnetic field has yet been detected for this planet. However, it has an ionosphere, which can deflect the solar wind and exclude its magnetic field. Above the ionosphere the magnetic field of the solar wind is compressed and a long magnetic tail is formed. When the pressure of the solar wind is high enough, its magnetic field can penetrate into the ionosphere. In this case the ionosphere becomes strongly magnetized. Thus the magnetosphere of Venus is not formed by its own magnetic field, but by that of the solar wind. This is an important distinction between the magnetosphere of Venus and those of the Earth and maybe most other planets.

(3) MARS. It is still uncertain whether Mars possesses an intrinsic magnetic field. The magnetosphere of this planet is also formed by the interaction of the ionosphere with the solar wind.

(4) JUPITER. As the largest planet in the solar system, Jupiter possesses a magnetic moment about 2×10^4 times larger than that of the Earth. Moreover, its distance from the Sun is more than 5 A.U. Owing to these two facts, Jupiter has a very large magnetosphere. The distance to the subsolar point on its magnetopause is approximately 100 times longer than the corresponding terrestrial quantity. It is interesting to note that four of the Jovian major satellites are located within the gigantic magnetosphere. Besides, energetic particles, mostly produced by the volcanic eruptions on the satellite Io, may be accelerated to extremely high energies in the Jovian magnetosphere.

(5) SATURN. Its magnetic moment is about 600 times stronger than the terrestrial one, so this planet also has a large magnetosphere in which some satellites are contained. Because the magnetic and rotational axes of Saturn almost coincide with each other, the magnetosphere does not undergo such large daily variations as in the terrestrial case.

(6) URANUS. This planet has the following well-known peculiarity. Its rotational axis lies nearly on the ecliptic plane. So it was anticipated

that the magnetic axis might also lie on this plane. In this case one of the polar caps of the magnetosphere should be facing the Sun, and the other cap away from the Sun. Nevertheless, the finding contributed by the spacecraft Voyager 2 in January of 1986 has negated this idea. The magnetic axis of Uranus forms a large angle of 60° with the rotational axis. Hence the magnetosphere may undergo violent and complicated changes in the course of the 17-hour Uranian day.

In passing, it is worth mentioning that the solar wind may interact with comets. When the solar wind penetrates into the cloud surrounding a comet, it picks up cometary ions and then slows down. Besides, the magnetic lines of force are bent to form a long tail. This is similar to the interaction of the solar wind with Venus.

8.2.3 The Heliosphere

As the magnetosphere of the Sun, the heliosphere is the cavity carved out by the solar wind in interstellar space. In the upwind direction of the oncoming interstellar gas, the radius of the cavity is about 100 A.U. The inner corona of the Sun may be taken to be the bottom of the heliosphere. On the outside, the heliosphere has two boundaries. The inner one terminates the supersonic solar wind flow. In other words, passing across this boundary the solar wind changes from a supersonic flow to a subsonic one. Due to the motion of the Sun relative to the interstellar medium, with a speed of about 20 km/sec, the subsonic flow has a long comet-like tail. The pressure equilibrium between the solar wind and the interstellar plasma is attained on the outer boundary. This equilibrium requires that the solar wind ram pressure is equal to the sum of the interstellar magnetic, ram and thermal pressures (Holzer, 1979). Beyond the outer boundary and in front of the heliosphere, the interstellar bow shock possibly exists.

In reality, the structure and physical processes in the heliosphere are quite complicated. On one side, the various plasma waves and dynamical processes induced by solar active features, especially flares, may give rise to transient changes in the heliosphere. For this the famous August 1972 solar-interplanetary events may be taken to be typical examples (Dryer, 1987). As a direct evidence of the influence of flares on solar wind parameters, the spacecraft Pioneer 9 recorded temperatures of millions of degrees. This implies that the plasma compressed by the flare-generated shock waves may be heated to coronal temperatures.

On the other side, the galactic cosmic rays can also affect the phys-

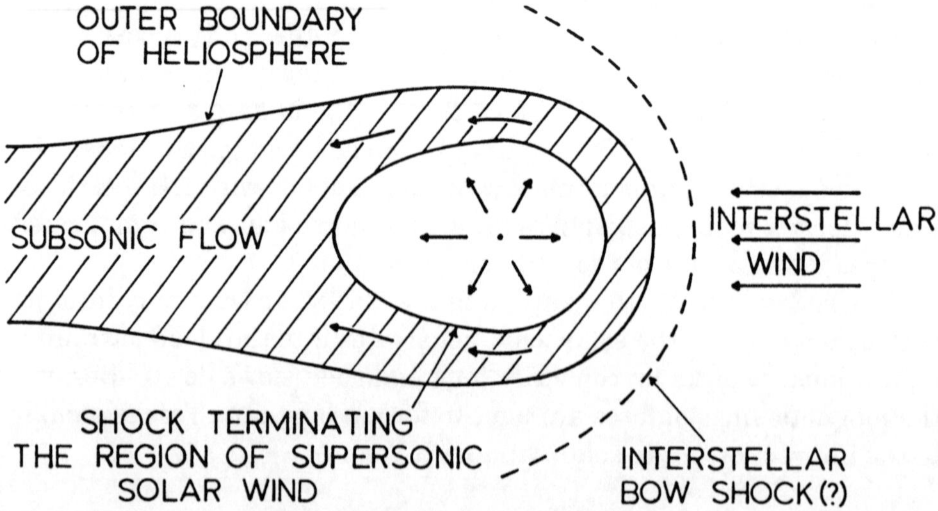

Fig. 8.4. Structure of the heliosphere.

ical state of the heliospheric plasma (Sakurai, 1987). After wandering through the interstellar space for more than 10^7 years since their birth somewhere in the Galaxy, cosmic ray particles may intrude into the heliosphere and interact with the solar wind. The modulation and anisotropy of cosmic rays come into existence due to this interaction. We should mention that these phenomena might be helpful in the study of the heliosphere, especially its outer region. We are aware that until the present no spacecraft has explored deep space beyond the orbit of Uranus.

Taking into consideration the problems on the two sides briefly stated above, we should say that much still remains to be done in heliospheric physics (see, e.g., Schwenn and Marsch, 1990).

8.3 The Lunar Magnetic Field

Although the Moon is the Earth's unique natural satellite and also the nearest heavenly body to us, before the space era we knew nothing about its magnetic field. Even in the first years of space exploration no important discovery on this topic was made. Some sound knowledge has been gained only since the end of the sixties .

8.3.1 SPACE AND *in situ* MEASUREMENTS

The first attempt to measure lunar magnetism was made on September 13, 1959, when the Soviet rocket Lunik 2 hit the lunar surface. Among the instruments on-board there was a magnetometer, but it could not measure fields below 100 γ. Its negative result seemed to mean that on the Moon there were no fields stronger than this value. Some later observations, e.g. those of Lunik 10 (Dolginov *et al.*, 1966) and Explorer 35 told us that the Moon might have no magnetism at all. This is because these spacecraft still could not detect a general magnetic field when they were within several hundred kilometers above the lunar surface. However, thanks to the success of the Apollo missions, the situation has changed drastically. Apollo 11 and 12 brought back rock samples from Mare Tranquilitatis and Oceanus Procellarum, respectively. It was soon discovered that the lunar rock contains alloys of iron with small amounts of nickel and cobalt and that it retains natural magnetism (Doell *et al.*, 1970). Moreover, the *in situ* measurements of astronauts show that the field strengths in some regions can attain 10–$10^2\gamma$. It is meaningful to note that districts with stronger fields are mostly plateaus. In other words, there seems to be a correlation between magnetic and gravitational anomalies. It has also been found that many of the strongest lunar magnetic anomalies correlate in location with swirl-like high- and low-albedo markings. Hood and Williams (1989) proposed that the high-albedo markings represent areas of the lunar surface which have been magnetically shielded from the solar wind ion bombardment and so have not reached optical maturity.

An interesting method of measuring the lunar magnetic field was adopted by Apollo 15 and 16. It is based on the reflection at the lunar surface of electron beams emitted from command modules (Anderson *et al.*, 1975). The electrons move in helical trajectories around the field lines of the interplanetary magnetic field. When they approach the Moon's surface, they are reflected if they encounter magnetic fields there. The fraction of electrons reflected depends on the strength, spatial extent and direction of the lunar magnetic field. The authors claimed that their method is very sensitive. It can detect magnetic features as small as 1 km and fields with strengths of only 0.01 γ. Besides, the direction of the lunar magnetic field may be found to be predominantly upward, downward or horizontal. In principle this method can also be applied to the investigation of the magnetic fields on planetary surfaces.

8.3.2 Origin

Since the existence of the lunar magnetic field has been confirmed, the problem of its *origin* arises. The lunar rock possesses naturally inherited magnetism and this is practically the same for various sorts of rock (Tatsch, 1974). So we may believe that the lunar magnetic field came into existence very early; maybe during the formation of the Moon. But how did it acquire magnetism? A lot of theories have been proposed around this question in recent years (Fuller, 1974; Runcorn, 1987).

(1) *Theories of external origins.* According to these theories, the lunar material was magnetized by some outside field. The possible candidates may be the following ones:

(i) The magnetic field of the solar wind.
This hypothesis has a serious shortcoming, i.e., the interplanetary sectoral field can hardly allow the Moon to maintain a constant field. But it may be assumed that in the early stage of solar evolution the interplanetary field did not have a sectoral structure.

(ii) The geomagnetic field.
In a certain period in the remote past the Earth's magnetic field was much stronger, or the Earth–Moon distance was much shorter than nowadays. Then the geomagnetic field could magnetize the Moon. Nevertheless, in either case the Moon might hardly acquire sufficient magnetism,

(2) *Theories of inherent origin.* At present the most popular theory of the origin of celestial magnetic fields is based on the *dynamo effect.*[2] According to this theory, the Moon possesses a core of liquid metal (mainly iron) whose motion produces a magnetic field. As estimated by Dolginov (1989), the diameter of the core does not exceed 400 km. In order to get a magnetic field in the range 10^{-2}–1 G on the lunar surface, the field intensity on the boundary of the core should be more than 10 G. If such a field is caused by the dynamo mechanism, the necessary rate of energy supply must be no less than 10^{17} erg sec^{-1}. Nevertheless, the gravitational energy released during the formation of the core is only about 10^{15} erg sec^{-1}. This is a serious difficulty for the lunar dynamo. Besides, this theory can hardly explain the inhomogeneity of surface magnetic fields.

Another hypothesis of this category is the *battery theory.* As suggested by Dolginov (1989), the lunar magnetism may be produced by the temperature or compositional nonuniformity of the inner regions. A temperature gradient ∇T gives rise to an electric current $j = -\sigma\eta\nabla T$, where σ is the conductivity and η is the thermoforce. Then the vector potential (**A**) of the magnetic field constructed by j is given by

[2] It will be introduced in detail in Sect. 11.5.

$\nabla^2\mathbf{A} = (4\pi/c)\sigma\eta\nabla T$. The magnetic field strength is $\mathbf{B} = \nabla \times \mathbf{A}$. The expression of the vector potential averaged for a turbulent medium is

$$\nabla^2\mathbf{A} = \frac{4\pi}{c}\langle\sigma\eta\nabla T - \sigma\frac{\mathbf{u}\times\mathbf{B}}{c}\rangle\,, \tag{8.10}$$

where u is the turbulent velocity. According to Vainshtein *et al.* (1980),

$$\langle\mathbf{u}\times\mathbf{B}\rangle = \alpha\mathbf{B} + \beta\nabla^2\mathbf{A}.$$

If the temperature gradient is in the radial direction, the inhomogeneity of $\sigma\eta$ can produce only a toroidal field B_t. The poloidal field B_p arises from flows in the toroidal field. These fields can be computed by the following equations:

$$\mathbf{B}_p = \nabla\times\mathbf{A}_t, \qquad \nabla^2\mathbf{A}_t = -\frac{4\pi\sigma}{c^2}\alpha\mathbf{B}_t; \tag{8.11}$$

$$\mathbf{B}_t = \nabla\times\mathbf{A}_p, \qquad \nabla^2\mathbf{A}_p = \frac{4\pi\sigma}{c}\eta\nabla T. \tag{8.12}$$

If in the early stage the lunar core had $R \simeq 400$ km, $\sigma \simeq 10^{15}$ sec^{-1}, $\alpha = 0.1$ cm sec^{-1} and $B_t = 1$ G, then it follows from Eq. (8.11) that on the lunar surface $B_p = 1$ G.

(3) *Fossil theory*. In its infancy the Moon was a dust cluster, which possessed a weak magnetic field. During the process of condensation, the field gradually became stronger (Urey and Runcorn, 1973).

Although it is difficulty to decide which of the above theories is correct, we can be confident that the magnetic field plays an important role in the formation and evolution of the Moon.

8.4 Planetary Magnetic Fields

The exploration of the magnetic fields of planets other than the Earth was begun after the advent of the space age. Until present man has not landed on other planets and so no *in situ* magnetic measurements have been performed. The planetary magnetism is studied mainly through the magnetospheres and radiation belts of planets, as well as the radio emission of Jupiter.

8.4.1 Methods of Measurement

As stated above, the planetary magnetospheres are mostly formed by the interaction between the solar wind and planetary magnetic fields and each magnetosphere has a definite scope and configuration. The distance between the center of the planet and the boundary of the subordinate magnetosphere is $R = R_0F(\varphi)$, where R_0 is the nearest distance and φ is the angle formed by the Sun–planet and planet–magnetosphere lines. From theoretical calculations (see, e.g., Midgley and Davis, 1963) the following equality can be derived:

$$R_0 = \left(\frac{KM^2}{P}\right)^{1/6}, \tag{8.13}$$

where $P = nmV$ (P, n, m and V are, respectively, the momentum, number density, mass and velocity of solar wind particles), M is the magnetic momentum of the planet, and K is a parameter which depends on the latitude of the planet. For a specific situation of the Sun, the planet and the spacecraft, we may get from Eq. (8.13) the following relation:

$$\frac{M}{M_\oplus} = \left(\frac{R}{R_\oplus}\right)^3 \left(\frac{P}{P_\oplus}\right)^{1/2}, \tag{8.14}$$

where the symbol $\oplus$ denotes the Earth. If we may measure the value of R for a certain planet, the above equation can be used to compute the ratio of magnetic moments of the planet and the Earth. Then the planet's magnetic field strength can be readily obtained.

Besides the above method, the magnetic field of Jupiter can also be studied with its radio emission, which was first discovered in 1955 at the 13.5 m wavelength. Soon afterwards the centimetric and decimetric radiations were also detected. It has been confirmed that the Jovian radio emission contains a nonthermal component, i.e. the synchrotron radiation. This fact indicates that this planet possesses a magnetic field whose strength may be determined with the method described in Section 2.6. It is now generally accepted that the Jovian synchrotron radiation is emitted by high-energy electrons ($E_e \geq 0.5$ MeV) trapped in Jupiter's magnetic field.

8.4.2 A Brief Account of Observational Results

In this subsection we present a brief account of the magnetic fields of several planets as well as the interpretation of some of their interesting characteristics. For more detailed descriptions the reader is referred to Encrenaz *et al.* (1990), Runcorn (1988) and Strom (1987).

(1) JUPITER. As the largest planet in the solar system, Jupiter possesses the strongest magnetic field among all the planets. Thanks to the space exploration of Pioneer 10, 11 and Voyager 1, 2 as well as the ground-based radio observation, the Jovian magnetism is now rather well understood. It is a dipolar field with the axis inclined at about 11° to the rotational axis. Its center is displaced from the geometric center by about one tenth of the Jovian radius in a direction perpendicular to the rotational axis. This asymmetry can explain the observed inequality of field strengths at the poles (14.8 G at the northern pole and 11.8 G at the southern pole). On the equator the field intensity is 4.2 G, i.e. about ten times stronger than the geomagnetic field.

(2) SATURN. It is known from space observations that Saturn's magnetic field is also bipolar but its axis is inclined by only 1° to the rotational axis. Besides, the displacement of the magnetic center from the geometric one is less than 0.01 of the Saturnian radius. Therefore, the magnetic field of this planet may be considered to have an axial symmetry. On the basis of all the available data provided by Pioneer 11, Voyager 1 and Voyager 2, Davis and Smith (1990) set up a model of Saturn's magnetic field.

(3) URANUS. No reliable information had been obtained until Voyager 2 approached this planet in January 1986, when a field of about 0.3 G was detected. Its axis is inclined by 55° to the rotational axis. Due to such a large inclination, the structure of the Uranian magnetosphere is quite complex.

The large inclination and eccentricity of the dipole-like field of Uranus can be interpreted with the following model of Akasofu *et al.* (1991). The offset tilted dipole structure deduced from the spherical harmonics analysis of the observational material of Voyager 2 may be satisfactorily represented with the joint action of an axial dipolar field and an auxiliary one. The former is parallel or antiparallel to the rotational axis and, as the main constituent of Uranus' magnetic field, may be produced by the dynamo mechanism. (Furthermore, as suggested by Ruzmaikin (1991), the dynamo is located in a shell at the bottom of the icy oceans of the planet.) The latter is roughly oriented in the east–west direction and may be formed by the leakage of a toroidal field within the planet. This is like the appearance of pairs of sunspots in the solar photosphere.

(4) NEPTUNE. Its magnetic field was also discovered by the Voyager spacecraft and has approximately the same inclination as that

of Uranus. Its strength is highly asymmetric: 1.2 and 0.06 G in the southern and northern hemispheres, respectively.

The hypotheses of Akasofu *et al.* and Ruzmaikin stated above can also be applied to the magnetic field of Neptune.

(5) MERCURY. As revealed by Mariner 10, which was launched on November 2, 1973, Mercury's magnetic field is similar to the terrestrial magnetic field. The axis of the dipolar field is inclined at $11\overset{\circ}{.}5$ to the rotational axis and the strength on the equator is about 0.3 G. This little planet also has a magnetosphere, which is much smaller than the terrestrial one. No radiation belt has been discovered for Mercury.

Until the present no magnetic field has been reliably detected for Venus, Mars and Pluto. The question of the possible absence of magnetic fields in Venus and Mars is discussed in Strom (1987).

What has been stated in this subsection is about the present status of the planetary magnetic fields. We would like to stress that these fields may have inherited the magnetism of the protoplanetary nebula. Some information can be got about this from the laboratory measurement of the magnetism in meteorites. As remarked by Cisowski and Hood (1991), the parent bodies of the carbonaceous chondrites possessed magnetic fields from 0.1 to several gauss during the first few 10^7 years after the protosolar collapse. The source of magnetism may be the large-scale field located in the proto-planetary nebula and oriented mainly in the ecliptic plane.

CHAPTER 9

Stellar Magnetic Fields

Their magnetism works all night.
Henry Vaughan (1622–1675), *Silex Scintillans* (1650).

The study of the solar magnetic field was initiated in 1908, while that of stellar magnetic fields was only started in 1946. This is because stars are too far from us, and so much fainter and more difficult to observe than the Sun. Thanks to the efforts of astronomers through several generations, tremendous progress in both techniques of observation and theoretical interpretation of the stellar magnetism has been achieved in recent decades. The world of countless stars provides us with very abundant and colorful observational material. The solar general magnetic field has a strength of only 1–2 G, while many stars are characterized by extremely intense magnetic fields. For instance, the strength of the *magnetic star* HD 215441 is 34,400 G, and those of white dwarfs much stronger still: 10^5–10^7 G. It is surprising to learn that the magnetic field of *pulsars*, which are generally interpreted to be *fast rotating neutron stars*, can even attain 10^{12} G! Such intense magnetism may have a profound influence on the spectra, luminosity, chemical composition and physical status as well as the origin, structure and evolution of stellar objects. In this chapter we present a general account of all these problems. For further details the reader may consult the monographs of Cameron (1967) and Glagolevsky and Kopylov (1988) as well as the reviews of Landstreet (1980, 1992), Borra *et al.* (1982), Khokhlova (1985) and Mathys (1989).

9.1 Observations of Stellar Magnetic Fields

Just as for the Sun, the magnetometry of stars is mainly based on the Zeeman effect of their radiation. However, there are a series of significant differences, such as the following. The stars are very much fainter than the Sun. They are seen by us as single bright points instead of full disks. Besides, the rotation can cause further broadening of spectral lines, and this makes the Zeeman splitting and broadening more difficult to detect and measure. Therefore, some specific measures should be taken in observations.

9.1.1 The Pioneering Work of Babcock

With the above constraints in mind, Babcock (1947) adopted the following assumptions: (1) The magnetic axis of the star coincides with the rotational one (i.e., the angle $i = 0$). Moreover, both axes point to the observer. (2) The brightness distribution on the stellar disk obeys the following law of limb darkening:

$$I_\varphi = 0.55 + 0.45 \sin\varphi, \tag{9.1}$$

where φ is the latitude. (3) The equivalent widths of Fraunhofer lines do not vary with the position on the stellar disk. (4) The relation between the magnetic field strength at latitude $\varphi(B_\varphi)$ and that in the polar region (B_p) is

$$B_\varphi = \frac{1}{2} B_p (1 + 3\sin^2\varphi)^{1/2} . \tag{9.2}$$

(This formula is practically the same as that used by G.E. Hale, i.e. Eq. (7.1).) Besides, in order to increase the accuracy of observation, he carried out magnetometry in some sharp spectral lines of mainly early-type stars and used the coudé spectrograph of the Mt. Wilson 100-inch reflector, which was then the largest telescope in the world. Nevertheless, in spite of all these, the Zeeman components are in general not fully separated and so what he measured is the displacement (δ) of the center of gravity of a magneto-sensitive line. This quantity is called the *Zeeman displacement*, which has the following expression:

$$\delta = \frac{aB}{2} \left[\int_0^\pi \sin\varphi \cos\varphi (1 + 3\sin^2\varphi)^{1/2} (0.55 + 0.45 \sin\varphi) \cos\varphi \, d\varphi \right] \times$$

$$\times \left[\int_0^{\pi/2} \sin\varphi \cos\varphi (0.55 + 0.45 \sin\varphi) \, d\varphi \right]^{-1} , \tag{9.3}$$

where $a = 4.67 \times 10^{-13} g \lambda^2$. After integration we have $\delta = 0.31 a B_p$. Then with the measured δ, one can obtain B_p. After performing a large amount of observation and data reduction, Babcock (1958a) published a catalogue of 89 magnetic stars.[1] Their field strengths are all equal to or larger than several hundred gauss. Babcock's classical work was followed by Preston (1969a,b,c; 1971), Wolff and Wolff (1970), *et al.*

9.1.2 MAGNETOGRAPHIC OBSERVATIONS

Severny and his associates (Severny, 1970, 1988; Severny *et al.*, 1974a) were the first to observe the longitudinal magnetic fields of stars with *photoelectric magnetographs*. Initially they used the magnetograph of the Crimean solar tower telescope (Nikulin *et al.*, 1958) and then that attached to the coudé focus of the 2.6 m Shajn reflector. The latter had only one exit slit and the light fluxes in the violet and red wings of a magneto-sensitive line were measured alternately with a certain frequency of modulation. In this way the V-parameter was measured. The working lines were Cr I λ4254.35 ($g = 1.375$) and Fe II λ4233.17 ($g = 1.214$). As remarked by Severny *et al.*, these lines are the most favorable ones for the magnetometry of, respectively, late- and early-type stars. The amount of instrumental polarization, which is especially serious for observation of stars of high declination, can be determined by interpolation of the signals of continua on both sides of the working line. The calibration of magnetographic recordings and the identification of magnetic polarities were made by comparison with the observation of sunspot fields. The accuracy of measurements of field strengths was about 30 G, namely one order of magnitude better than that of Babcock. The observations of Severny *et al.* yielded some interesting results. For instance, they found that the bright supergiant γ Cyg is a *magnetic variable*. Its field strength varies between the limits −200 G and 200 G and the period is $8^{\mathrm{d}}.26$. But some other bright stars, such as α CMa, α Lyr and β Gem, exhibit sporadic changes of magnetic field intensity by amounts of 50–150 G or even less in a few days. These authors suggested that this phenomenon cannot be ascribed to the *oblique rotator model*[2] or convection, but may be due to some specific

[1]A new and much more abundant catalogue can be found in Lang (1992).
[2]This model will be introduced in Section 9.4.

oscillations on stellar surfaces.

The magnetographic observation can yield high-resolution intensity and circular polarization distributions within magneto-sensitive lines, i.e. their I- and V-profiles. Borra (1980) used such profiles across Fe II λ4520.2 to derive the geometric configuration of magnetic fields on 78 Vir by means of computer simulation. Also using this method, Hatzes (1991) constructed magnetic maps on the surface of θ Aur with the changes of absorption line profiles of Si II and Cr II. Stift and Goossens (1991) modelled the surface magnetic field of HD 126515 within the framework of the oblique rotator hypothesis. Another technical improvement is to make magnetographic measurements not in a single line, but in many lines simultaneously. For example, Mathys and Stenflo (1986) studied the strongly magnetic star HD 147010 with the I- and V-profiles of 16 lines of Fe II and 5 lines of Si II.

The observation of the *transverse magnetic field* is a difficult task. This is mainly because it is based on the measurement of linear polarization, which is in general very weak. So works of this kind are few. We should mention the observation of α^2 CVn by Kodaira and Unno (1969) and that of β CrB by Borra and Vaughan (1976). It is interesting to note that both observations provided evidences for the oblique rotator model.

9.1.3 Diagnostic Methods

The observations described above belong chiefly to early-type stars with sharp spectral lines, e.g., Ap and Bp stars. However, late-type stars may also possess magnetic fields. For their detection, a breakthrough in the study of stellar magnetism took place at the beginning of the eighties due to the invention of Robinson's (1980) method of diagnosis. Before introducing this method it is necessary to discuss the question of *detectability* of stellar magnetic fields.

As stated in the preceding two subsections, the observations of Babcock, Severny and others are in principle polarimetric ones. To be precise, most authors measure the Stokes parameter V, and a few others measure Q and U. However, as remarked by Mathys (1989), this is not a rational method of stellar magnetometry. The reason is that the magnetic field of a late-type star very likely contains a large number of small dipolar regions scattered over the stellar surface and so in the measurement of the total amount of polarization the effects of numerous magnetic elements with opposite polarities are cancelled out. This can be clearly seen from the following mathematical expressions. In

Mathys' notation the second order solution of the equation of transfer of the Stokes parameters in the Milne–Eddington atmosphere is

$$\begin{aligned} I(\tau=0,\mu) \approx B_0 \Bigg(1 + \beta_0\mu \Bigg\{ & \frac{1}{1+\eta_l\psi(\lambda-\lambda_0)} - \\ & - \frac{S_2\eta_l\Delta\lambda_Z^2 H^2}{[1+\eta_l\psi(\lambda-\lambda_0)]^2}\frac{\mathrm{d}^2\psi(\lambda-\lambda_0)}{\mathrm{d}\lambda^2} - \\ & - \frac{\eta_l\Delta\lambda_Z^2 H_z^2}{[1+\eta_l\psi(\lambda-\lambda_0)]^2}\Bigg[D_2\frac{\mathrm{d}^2\psi(\lambda-\lambda_0)}{\mathrm{d}\lambda^2} - \\ & - \frac{\bar{g}^2\eta_l}{1+\eta_l\psi(\lambda-\lambda_0)}\left(\frac{\mathrm{d}\psi(\lambda-\lambda_0)}{\mathrm{d}\lambda}\right)^2 \Bigg] \Bigg\} \Bigg), \end{aligned} \tag{9.4}$$

$$Q(\tau=0,\mu) \approx B_0\beta_0\mu \frac{D_2\eta_l\Delta\lambda_Z^2}{[1+\eta_l\psi(\lambda-\lambda_0)]^2}\frac{H_Q^2-H_{Q+\pi/2}^2}{2}\frac{\mathrm{d}^2\psi(\lambda-\lambda_0)}{\mathrm{d}\lambda^2},$$

$$U(\tau=0,\mu) \approx B_0\beta_0\mu \frac{D_2\eta_l\Delta\lambda_Z^2}{[1+\eta_l\psi(\lambda-\lambda_0)]^2} H_Q H_{Q+\pi/2}\frac{\mathrm{d}^2\psi(\lambda-\lambda_0)}{\mathrm{d}\lambda^2},$$

$$V(\tau=0,\mu) \approx B_0\beta_0\mu \frac{\bar{g}\eta_l\Delta\lambda_Z^2 H_z}{[1+\eta_l\psi(\lambda-\lambda_0)]^2}\frac{\mathrm{d}\psi(\lambda-\lambda_0)}{\mathrm{d}\lambda}.$$

The meanings of various symbols can be found in his paper, just cited. Now we want only to emphasize that V, Q and U are, respectively, proportional to H_z, $(H_Q^2 - H_{Q+\pi/2}^2)$ and $H_Q H_{Q+\pi/2}$. Here H_z is equivalent to the symbol $B_\parallel$ used in this book and H_Q and $H_{Q+\pi/2}$ are defined by the following equalities:

$$\begin{aligned} \sin^2\gamma\cos 2\chi &= (H_Q^2 - H_{Q+\pi/2}^2)/H^2, \\ \sin^2\gamma\sin 2\chi &= 2H_Q H_{Q+\pi/2}/H^2. \end{aligned} \tag{9.5}$$

Hence the effects of subdivided magnetic fields, whose azimuths differ by $\pi/2$, may mostly cancel out when they are integrated over the stellar disk. By contrast, the contribution of the magnetic field to the intensity of radiation I is embodied in H^2 (the square of the field modulus) and H_z^2 (the square of the line-of-sight component). It is important to note that both these quantities are always positive. This implies that the contributions of various magnetic elements do not compensate one another in the process of summation. Therefore, in order to study the stellar magnetic field with a large-scale organization, it is advisable to analyze the profile of I instead of those of the other Stokes parameters. Just in this way, i.e., through the subtle magnetic broadening of intensity profiles, Robinson (1980) established his diagnostic technique.

Robinson's *non-polarimetric method* is based on the idea that magnetic field information can be extracted from a comparison between the profiles of a magnetically broadened line and a line unaffected by the magnetic field. These two lines are required to possess similar spectral characteristics. For example, Fe I $\lambda4080.88$ with $g = 3.00$ and Fe I $\lambda4065.39$ with $g = 0$ form such a pair. Now, due to lack of space, we cite from Mathys (1989) merely the following equality, which expresses the effect of the magnetic field on the observed line profiles.

$$r_{\mathcal{F}_I} = \left(1 - \alpha + \alpha\frac{\sin^2\overline{\gamma}}{2}\right) r_{\mathcal{F}_{I,0}1}(\lambda - \lambda_0) + \\ +\alpha\frac{1+\cos^2\overline{\gamma}}{4}[r_{\mathcal{F}_{I,0}1}(\lambda - \lambda_0 - \overline{g}\Delta\lambda_{H_1}) + r_{\mathcal{F}_{I,0}1}(\lambda - \lambda_0 + \overline{g}\Delta\lambda_{H_1})], \tag{9.6}$$

where $r_{\mathcal{F}_I}$ and $r_{\mathcal{F}_{I,0}}$ are, respectively, the depths of the magnetically sensitive and insensitive lines, α is the filling factor and the other symbols have their common meanings. In passing, we should mention the following. For the application of this method, it is necessary to measure $r_{\mathcal{F}_I}$ and $r_{\mathcal{F}_{I,0}}$ very accurately. So the skillful diagnostic method must be matched with advanced technology in stellar spectroscopy, in particular the development of electronic detectors.

Robinson *et al.* (1980) applied the method described above to two late-type dwarfs and determined their magnetic field intensities, as well as the percentages of stellar surfaces occupied by magnetic regions. For one of the stars these two quantities are 2550 ± 390 G and 20–45%, and for the other 1880 ± 350 G and 10%. It is interesting to state that these authors found for solar active regions an average field strength of 1800 ± 550 G and the solar surface percentage of active regions covered by such a field is 10%.

Finally, it is worthwhile pointing out that Robinson's method can avoid the drawbacks of the classical polarimetric technique, i.e. the cancellation of effects of magnetic elements with opposite polarities and the dilution of polarimetric signals by nonmagnetic regions on the stellar surface. Nevertheless, at least nowadays, the diagnostic method is not sensitive enough to weak magnetic fields.

9.2 Characteristics of Magnetic Stars

In Babcock's (1958a) catalogue most magnetic stars belong to the spectral type A and have sharp spectral lines. Because such stars possess some peculiar properties, they are called *Ap stars*. Besides, in comparison with ordinary stars of the same type, their metal lines are especially strong, and so they have also the name of *Am stars*. The chief characteristics of magnetic stars are as follows.[3]

(1) *Variations of magnetic field strength*. The field intensity of all magnetic stars changes with time. Hence we may say that every magnetic star is a *magnetic variable* too. According to the conditions of changes, magnetic stars are divided into four types. The first three can be represented by α^2 CVn, β CrB and γ Equ, and so they are called types α, β and γ. Following this sequence, the fourth class is named type δ. The magnetic fields of α-type stars exhibit periodic variations. The periods are in general several days, but some magnetic stars have quite long periods. For instance, the periods of HD 187474 and HR 7552 are about 7 years. It is interesting to note that the spectral line intensities of these stars (except HD 32633) vary with the same periods as those of magnetic field changes. The field strength of β-type stars shows irregular variations, and the polarity may undergo *inversion*. Stars of type γ also display irregular field strength changes, but their polarities are fixed. β- and γ-type magnetic stars do not show spectral changes and the amplitudes of variation of their field intensities are generally less than those of α-type stars. (The average amplitudes for stars of types α, β and γ are, respectively, 3500–4200 G, 1950 G and 1150 G). As for the δ-type magnetic stars, the spectral changes are irregular and have no evident correlation with the magnetic field variations.

(2) *Spectral line intensity*. As stated above, the intensities of spectral lines of α-type magnetic stars change with the same periods as those of magnetic fields. It has to be added that the variations of *line intensities* are very complex. The behaviors of lines of different chemical elements, and even of various lines of one and the same element, may be quite unlike. For example, when the S-polarity field of α^2 CVn attains its maximum, lines of Eu II are most intense. But the Cr lines have their largest intensities during the maximum of the N field. However, the situation of another α-type star (HD 125248) is just the opposite.

(3) *Crossover effect*. When the magnetic field of some (mainly α-type) stars changes from one polarity to the neutral state and further to the opposite polarity, the widths of splitting patterns of magneto sensitive lines may show corresponding variations. Besides, this change can be different for various lines. This is the so-called crossover effect, which may be explained in the following way. On a magnetic star there are

[3]The chemical composition of magnetic stars and its anomalies will be treated in the next section.

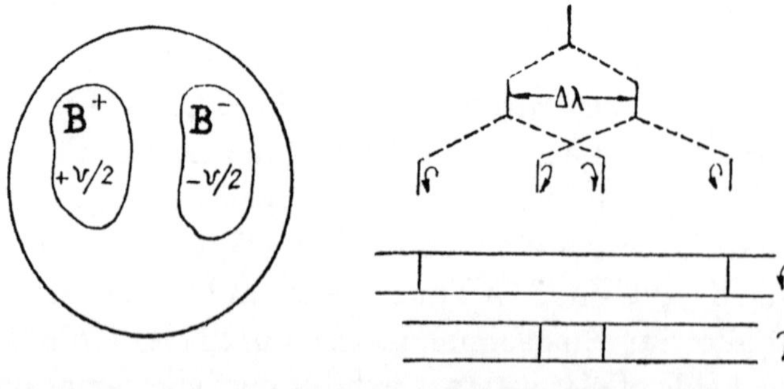

Fig. 9.1. Interpretation of crossover effect.

two regions with opposite polarities as well as very different radial velocities $+v/2$ and $-v/2$ (see the left panel of Figure 9.1). Then on both sides of a spectral line we see two Doppler-shifted lines. Moreover, due to the normal Zeeman effect, each line is split into two circularly polarized components. Therefore, through the polarization analyzer we see alternately two pairs of Zeeman components with different amounts of splitting (right panel of Figure 9.1).

(4) *Radial velocity*. The radial velocities of many magnetic stars also vary unceasingly. The important matter is that these changes are related to magnetic fields. Let us take HD 125248 (α-type) as an example. The radial velocities measured with lines of several elements are similar and are in phase with the periodic variation of the magnetic field (Figure 9.2). This may be interpreted by suggesting that in the atmosphere of this star atoms of various elements flow outward along magnetic lines of force and that the flow velocity varies in phase with the magnetic field changes. However, the radial velocity of another magnetic star HD 153883, also of the α-type, almost remains constant. In other words, its spectral lines do not exhibit such periodic displacements as those of HD 125248.

(5) *Luminosity*. This quantity also constantly changes, although the amplitude is in general merely several hundredths of a magnitude. As an instance, the variations of the luminosity and the magnetic field strength of α^2 CVn are shown in Figure 9.3. It is evident that these two quantities vary correlatively. Speaking more precisely, the minima of luminosity coincide with the maxima of field strength. Similar cases have been found for many other magnetic stars. This implies that magnetic stars are also variables in brightness.

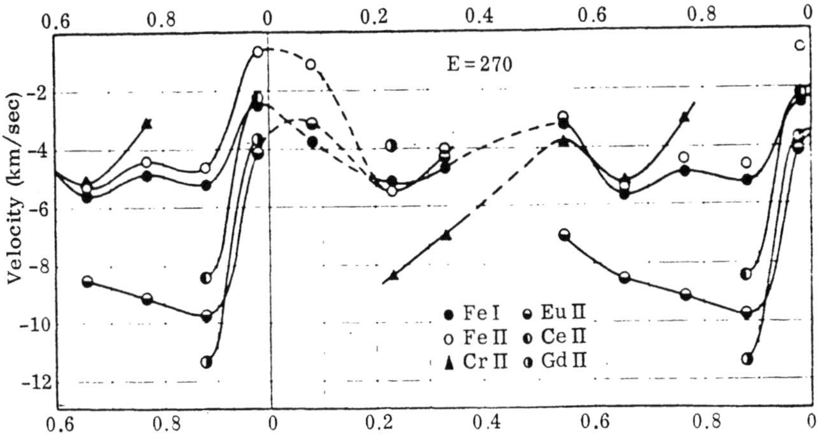

Fig. 9.2. Radial velocity variation of magnetic star HD 125248. (The abscissa represents the phase of change of the magnetic field.)

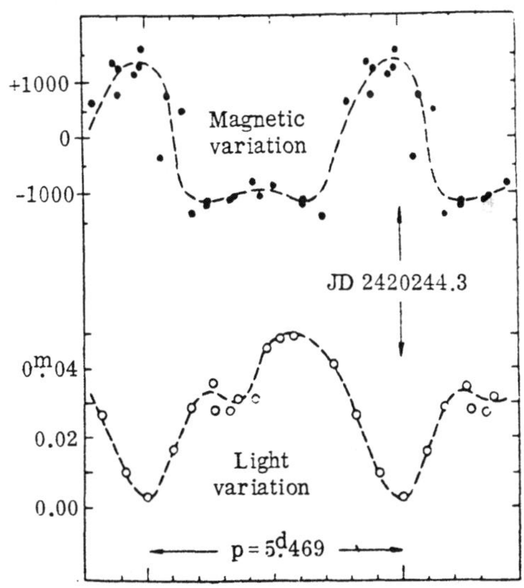

Fig. 9.3. Magnetic and luminosity changes of α^2 CVn.

Now it is worthwhile to state separately something about HD 215441. This is the magnetic star with the most intense magnetic field known up to now. It is an Ap star of magnitude 8.6. The field strength has the maximum of 34,400 G and the minimum is 4140 G. Its magnetism is so intense that the σ- and π-components of many lines are completely split up. The changes of both the magnetic field and radial velocity are irregular. Recently, Landstreet *et al.* (1989) constructed a detailed model of the magnetic field and elemental distribution on the surface of this star. Besides, as estimated by Babcock (1960a), its total amount of magnetic energy has a lower limit of 10^{41} erg, which is equivalent to the total quantity of radiative energy emitted by the Sun in a year. So it is reasonable to believe that the magnetic field plays an important role in the structure and evolution of this star.

In this section we have given a more or less comprehensive description of the characteristics of magnetic stars. What is unfolded before the reader is an abundant, colorful and dazzling picture with numerous complicated phenomena. Much remains to be done for the revelation of their physical nature.

9.3 Chemical Composition of Magnetic Stars

9.3.1 Anomalies in Chemical Composition

The peculiarities of magnetic stars in respect of their *chemical composition* are also striking. This can be shown by the following facts:

(1) Most magnetic stars are poor in helium.

(2) Some light elements, such as aluminium, silicon, phosphorus, sulfur and chlorine, are 10^2–10^3 times more abundant on magnetic stars than on the Sun.

(3) The rare-earth elements (lanthanum, europium, gadolinium, dysprosium, etc.) are in general 300–1000 times richer than the solar content.

(4) It is surprising that the magnetic star HR 465 contains promethium (Pm) (Aller, 1971). This is an unstable element. All its isotopes can be formed only by nucleosynthesis and have quite short half-lives; the longest is merely 10–20 years. This possibly implies that some nuclear reactions are continuously occurring in the atmosphere of this star.

(5) The isotopes of mercury also have something uncommon. For example, on the magnetic star HR 4072 Hg^{204} occupies 97% of the total content of mercury, while Hg^{202}, which is the most abundant isotope on the Earth, constitutes only the remaining 3%.

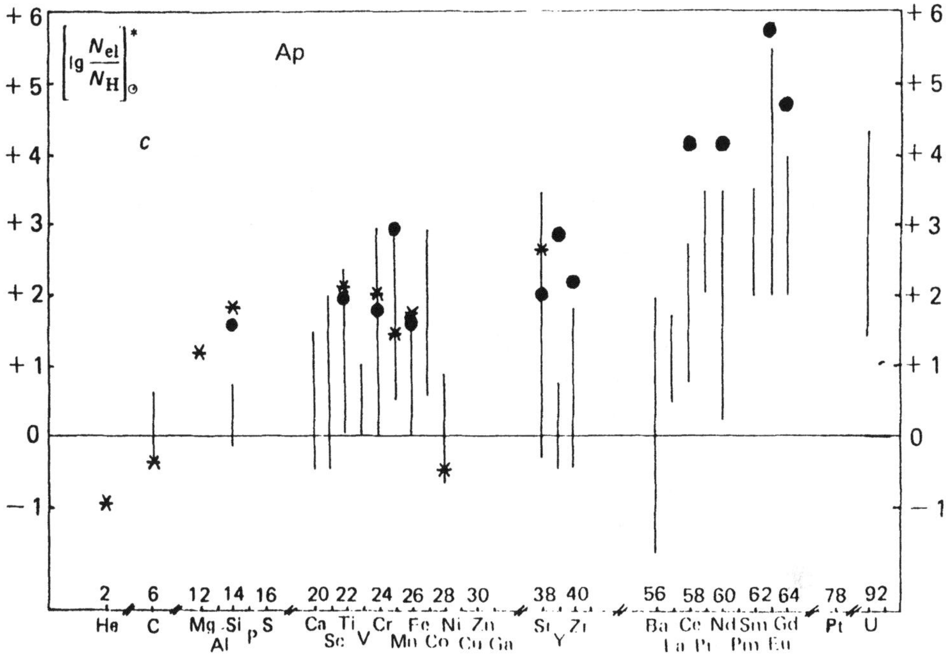

Fig. 9.4. Chemical composition of magnetic stars.

An overall idea of the anomalies may be got by looking at Figure 9.4 in which the abscissa is the atomic number and the ordinate represents the difference between the abundances of elements in units of hydrogen abundance in the atmospheres of Ap stars and the Sun. The vertical bars indicate the range in which the measured abundances fall. The circles and asterisks belong, respectively, to α^2 CVn and HD 34452. From the figure the reader may get the impression that the difference increases with the atomic number.

Before trying to interpret the anomalous chemical composition of magnetic stars, one should ask whether it belongs to the whole stars, or only their atmospheres. For reasons stated in Khokhlova (1985), we believe that the anomalies are merely surface effects and that the magnetic star as a whole has a normal chemical composition. Then the next problem confronting us is the cause of anomalies existing on the stellar surface.

Since 1970 the *diffusion theory* has been developed by many astronomers (e.g. Watson, 1970; Pikel'ner and Khokhlova, 1972). In this theory the anomalies are due to the selective diffusion of atoms and ions under the action of gravity and radiative pressure. The former causes the heavier atoms to sink, while the latter pushes them outward. The magnetic field cannot affect the diffusion of neutral atoms, but it radically alters the diffusion of ions. To be precise, ions may be captured and confined by the magnetic force and they can form spots above stellar surfaces.

According to the *supernova hypothesis* (Guthrie, 1968; Kuchovicz, 1975), the products of explosion of a nearby supernova fall on the magnetic star and this may increase its abundance of heavy elements. It is also possible that the magnetic star is one component of a binary system and the chemically anomalous substance produced by nuclear reactions on the other component may spray to the surface of the magnetic star. Each of these theories has its respective merits and difficulties. (For details see, e.g., Khokhlova, 1985.)

9.3.2 Inhomogeneous Distribution of Chemical Elements on the Surface of Magnetic Stars

As an observed fact, the spectral line profiles of many magnetic stars vary with time. Deutsch (1970) suggested that this is due to rotation of the magnetic star on which the distribution of the relevant element is nonuniform, and he tried to derive the *elemental distribution* on the stellar surface from spectral changes. He used a Fourier expansion of the periodic variations of the equivalent width (W) and the weighted-mean Doppler shift of the spectral line as a whole. His method was modified by Pyper (1969), who separated a line profile into components and expanded the curve of variation of W in a Fourier series for each of the components. One can then determine the position of the regions from which the most part of the spectral line radiation comes. The essence of this method may be expressed as the solution of the following integral equation:

$$R^*(\lambda, \omega t) = \frac{\int_{\cos\theta>0} I_c(M,\theta) R[M,\theta,\Delta\lambda_D(\omega t)] \cos\theta \, \mathrm{d}M}{\int_{\cos\theta>0} I_c(M,\theta) \cos\theta \, \mathrm{d}M} . \tag{9.7}$$

The left side of the equation is the observed depth of the line at phase ωt. The right side consists of integrals over the visible hemisphere of the star, taking into account the contribution of local profiles $R[M,\theta,\lambda + \Delta\lambda_D(\omega t)]$ at wavelength λ, the variation of the continuum intensity I_c. and the effect of limb darkening. The phase ωt enters the integrals implicitly via $\Delta\lambda_D$ and θ. For brevity, M denotes the coordinates of a specific point on the stellar surface. By solving this equation one can use the observed line profile represented by $R^*(\lambda, \omega t)$ to obtain some local physical parameters. The mathematical aspect of the problem and the uniqueness of the solution are discussed by Khokhlova (1985,

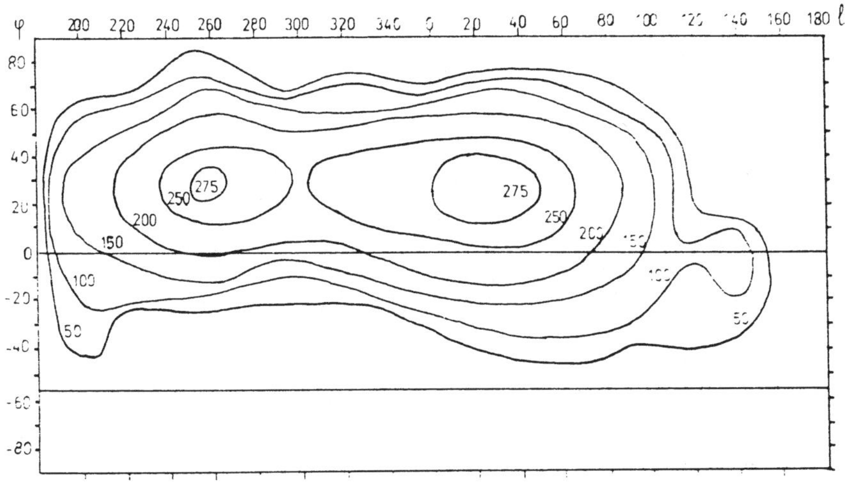

Fig. 9.5. Map of distribution of Sr over surface of χ Ser.

1986). Some interesting results have been obtained by application of this method and an example is given in Figure 9.5, which illustrates the inhomogeneous distribution of strontium on the surface of χ Ser (Goncharskii *et al.*, 1983).

In a similar way Hatzes *et al.* (1990) mapped the surface elemental distribution of Ap stars. They applied the principles of *maximum entropy image reconstruction* to iteratively solve a matrix equation, which connects the observed changes of spectral line profiles and the equivalent width distribution on the stellar surface. By use of the high resolution line profiles obtained at the Lick Observatory, they obtained the equivalent width maps of Si II $\lambda 6437$ on γ^2 Ari. This is shown by Figure 9.6, which consists of eight maps with equidistant phases. It is interesting to note that on the surface of this star there are two approximately circular spots of depleted abundance of silicon and one of them is surrounded by an incomplete annulus of enhanced abundance. Moreover, in the spots the magnetic field lines are vertical and in the annulus they are horizontal. The authors suggested that this can be explained with the theory of selective diffusion of elements.

9.4 Models of Magnetic Stars

In the preceding two sections the reader may have been impressed by the fascinating, mysterious and thought-provoking properties of mag-

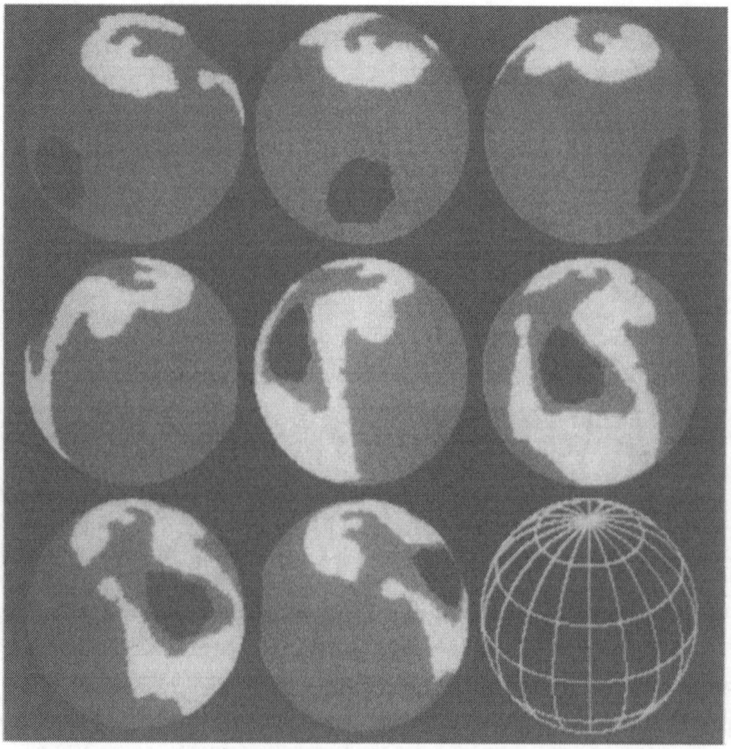

Fig. 9.6. Abundance distribution of silicon on γ^2 Ari. White shows regions of enhanced abundance and black denotes the depleted abundance. The phantom star, lower right, illustrates the location of the rotation pole.

netic stars. In order to properly interpret them, astronomers have set up various models that will be briefly described in the following.

(1) *Magnetic oscillator model.* Schwarzschild (1949) suggested that the magnetic star is a magnetized sphere composed of a perfectly conducting, incompressible and homogeneous fluid and that it pulsates with a period which depends on the radius, density and mean magnetic field intensity of the star. Besides, he took the field strength in the interior to be 10^6 G, that on the surface 10^4 G and the radius $2R_\odot$. A period of about 9 days was then obtained. This model can to some extent explain the periodic variations of the luminosity, spectrum, radial velocity, etc. of the magnetic star. Nevertheless, it cannot interpret the polarity reversal in polar regions. Moreover, as pointed out by Cowling (1952) and others, if the gravity, which is expected to have a larger effect

than the magnetic force, is taken into consideration, then the period of oscillation should become only a few hours instead of several days.

(2) *Solar cycle model.* As the essential point of this model, magnetic stars are like solar active features, but enlarged by several orders of magnitude. Babcock (1958b,1960b) suggested that the magnetic star is similar to the Sun and possesses two different magnetic fields. One of them is the general field that is mainly concentrated in polar regions and the other is the toroidal field, which is located mainly in low latitudes. Moreover, both fields vary with certain periods. The stronger is the field strength, the shorter is the period of magnetic activity. The Sun has a general field of a few gauss and a magnetic cycle of 22 years. The field strength of magnetic stars may be taken to be 10^3 G and the period is merely several days. However, the differences of these quantities between the Sun and magnetic stars are so large that the empirical relation stated above can hardly be persuasive.

(3) *Oblique rotator model.* This is at present the most popular and generally accepted model. It was advanced by Stibbs (1950) and refined by Deutsch (1954, 1958). In this model the axis of a static magnetic field is inclined at angle i to the rotational axis of the magnetic star. When our line of sight forms an arbitrary angle β with the axis of rotation, the observed field exhibits periodic variations as a consequence of rotation. The relation of these angles to the change of the field strength can be derived as follows (Preston, 1967). For a dipolar rotator the intensities of the effective field (B_e) and the polar field (B_p) are related to each other in the following way:

$$B_e \propto B_p(\cos i \cos \beta + \sin i \sin \beta \cos 2\pi\phi), \tag{9.8}$$

where $\phi = t/p$ is the phase. Let the ratio of the extrema of the effective field strength be denoted by $r \equiv B_e(\min)/B_e(\max)$, then we have

$$\tan i = \frac{1-r}{1+r} \cot \beta. \tag{9.9}$$

This gives rise to a group of curves in Figure 9.7. The value of r can be found by observation. Then we may choose an appropriate curve in this figure for a specific magnetic star. If the angle β can be measured by some method, we may obtain the value of i.

Unlike the two models introduced above, the oblique rotator model can give rise to some predictions, and so it may be verified by observation. It is successful at least in the following aspects. First, it can explain the relation between the period of magnetic field variation (p) and the spectral line width (W). As a fact of observation,

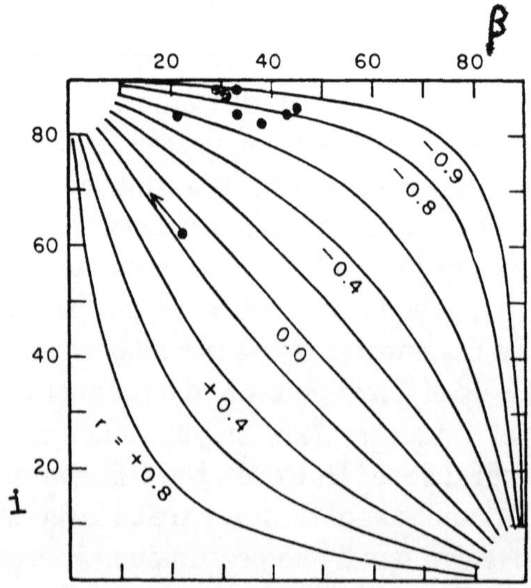

Fig. 9.7. Relations between i and β for various values of r.

these two quantities are inversely proportional to each other. With the oblique rotator model Deutsch derived their mathematical relation to be $W = 3.6 \sin i/p$, and the units of W and p are, respectively, angström and day. Second, this model can predict polarity reversals. Third, it may quantitatively interpret the crossover effect of spectral lines. Moreover, after Deutsch's improvement the variations of equivalent widths and radial velocities can also be explained satisfactorily.

Nevertheless, the oblique rotator model is not an ideal one after all. It is unable to explain why many magnetic stars exhibit irregular changes of magnetic fields and why the periods of luminosity and magnetic field variations are the same. Besides, the periods of magnetic field changes of some stars are very long (as we mentioned, those of HR 7552 and HD 187474 are about 7 years), and it is difficult to conceive that these stars might rotate so slowly.

Finally, if a star possesses an intense general magnetic field, the MHD theory requires that the magnetic and rotational axes of the star should not differ much. (In fact, this is so for the Sun, the Earth and Jupiter, although their general fields are not strong.) If this reasoning is true, it may be a fundamental objection to the oblique rotator model.

(4) *Radial bipolar model.* According to this model, there are many small magnetic regions on the surface of the magnetic star. The numbers of regions of N- and S-polarities are approximately equal and their

lines of force are basically in the radial direction. Of course, the lines of force are continuous, viz., the condition $\nabla \cdot \mathbf{B} = 0$ is satisfied, and the turning of their direction occurs mainly in the outer atmosphere above magnetic regions. (This is the situation for solar local magnetic fields.) This model may have two extreme cases. First, between regions with opposite polarities there is no systematic relative motion, but they are different in the mean brightness. In this case we can detect a residual magnetic field with the spectral line radiation coming from the star as a whole. Second, exactly on the contrary, the two kinds of magnetic regions do not differ in brightness but they have relative motions. Then, due to the joint action of Doppler and Zeeman effects, the crossover effect takes place. Of course, the general situation lies between these extremes. Preston (1967) applied this model to interpret the crossover effect of 78 Vir. From the intensity ratio of the resolved components in left-circularly polarized light coming from regions "1" and "2" of the spectral line j, i.e. I_{j1}/I_{j2}, one may compute the temperature difference $\theta_2 - \theta_1$ ($\theta \equiv 5040/T$) with the formula

$$\lg(I_{j1}/I_{j2}) = (\theta_2 - \theta_1)\chi_j + \text{const}, \tag{9.10}$$

where χ_j is the excitation potential of the lower level of line j. By analysis of the Zeeman spectrograms of 78 Vir Preston found a temperature difference of 300° and a relative motion velocity of 2 km s^{-1}.

(5) *Binary model.* Rensen (1963a, b) suggested that all the Ap stars are *close binaries* and that the magnetic star is the main component of a binary system. The other and generally invisible component should be a small but heavy companion, which causes tides on the main star and so gives rise to its magnetic field. Since the relative positions of the two stars revolving around the common center of gravity always vary, the tide may sometimes be high and at other times low. This can explain the periodic change of the magnetic field. Because the eccentricity of orbit may be large or small, the cycle of magnetic variations can be quite short or very long. This is an outstanding merit of the binary model, because all the other existing models cannot explain why the periods of various magnetic stars differ very much. Moreover, the tidal effect may arouse not only magnetic, but also spectral, brightness and radial velocity variations of the magnetic star. Because all these changes are due to the same cause, it is natural that they may have the same period.

Indeed many stars, especially A-type ones, are binaries. Besides, it has been ascertained that some magnetic stars, e.g. HD 98088, belong to binary systems. Nevertheless, we are not certain that all the magnetic stars are binaries.

In the above we have introduced five models of magnetic stars. Each of them has its own merits and shortcomings. Now we want to stress that two different mechanisms may possibly operate on one and the same star. It is also probable that two models are responsible for one and the same phenomenon. The difficult question of complementarity and uniqueness of various models has been discussed by Preston (1967).

9.5 The Magnetic Field of Pulsars

The preceding sections of this chapter dealt with *nondegenerate* stars. The remaining two sections will be devoted to two kinds of *degenerate* objects, i.e. *neutron stars* and *white dwarfs*. These compact celestial bodies are in many respects quite different from ordinary stars, and one of their striking peculiarities is the exceedingly intense magnetism. Hence they are worth special attention in the study of celestial magnetic fields.

9.5.1 Discovery and Identification of Pulsars

The first *pulsar* (later designated as CP 1919) was discovered by Hewish *et al.* (1968) on November 28, 1967 in Cambridge. It is a radio source which emitted pulsating signals at 81.5 MHz with a stable period of 1.377 sec. Soon after the announcement of this epoch-making discovery, astronomers of many countries plunged into the search for this new kind of celestial objects and at present several hundred pulsars have been found. Their periods lie in the range 1.6 ms to 4.3 s and the durations of individual pulsars are several to several tens of milliseconds. It was surprising to discover that all the pulsars are remarkably good clocks and their periods may be measured even to 13 significant digits! The periods increase very slowly, but seldom decrease. In order to interpret the physical nature of pulsars, some hypotheses have been proposed (see, e.g., Shapiro and Teukolsky, 1983) and now the model advanced by Gold (1968) is generally accepted. According to his theory, pulsars are rapidly rotating neutron stars and such objects were predicted long ago by Baade and Zwicky (1934) and Oppenheimer and Volkoff (1939).

The neutron star has a mass close to that of the Sun, but at the same time a radius of merely about 20 km. So its density is extremely large, approximately 10^8 tons cm^{-3}. In this star the very high pressure forces electrons to combine with atomic nuclei and form neutrons. The surface temperature of neutron stars may attain 10^7 K and the total radiation flux is about one million times larger than that of the Sun.

For the reason to be explained in the following subsection, a star in this superdense and ultrahigh temperature conditions rotates very rapidly. In each revolution the region of enhanced radio emission sends us a signal, and this is the pulses that we observe.

9.5.2 ROTATION AND THE MAGNETIC FIELD

In order that a rapidly rotating stellar object is stable, i.e. its mass does not decrease by a large amount, the necessary condition is that its gravitational acceleration should be larger than the centrifugal one. In the mathematical language this condition is

$$\frac{GM}{r_0^2} > \frac{v_0^2}{r_0} = \Omega^2 r_0, \tag{9.11}$$

or

$$\Omega < \sqrt{\frac{GM}{r_0^3}} = \sqrt{\frac{4\pi G\overline{\rho}}{3}}, \tag{9.12}$$

where M, r_0, Ω, v_0 and $\overline{\rho}$ are, respectively, the mass, radius, angular velocity of rotation, surface velocity and average density of the star. Then the period of rotation is found to be

$$p = \frac{2\pi}{\Omega} > \sqrt{\frac{3\pi}{G\overline{\rho}}}. \tag{9.13}$$

This implies that *the rotational period decreases with the increase of the mean density*. When $\overline{\rho} \leq 10^8\ \mathrm{g\,cm^{-3}}$, $p > 1$ s. The average density of white dwarfs is 10^8–$10^9\ \mathrm{g\,cm^{-3}}$, so they can hardly emit pulses in short periods. In fact, no white dwarf has been identified as the optical counterpart of a pulsar. However, the case for neutron stars is different. In the late stage of stellar evolution the energy source is almost exhausted and the star rapidly contracts. Under the action of intense pressure electrons are squeezed into atomic nuclei, and this gives rise to the birth of a neutron star. Its mean density may be as high as $10^{14}\ \mathrm{g\,cm^{-3}}$. Substituting this value into Eq. (9.13), we get a very short rotational period of about 10^{-2} s, and this is one of the chief characteristics of neutron stars.

Before becoming a neutron star, an ordinary star has to undergo drastic contraction. In this process two violent changes take place. First, according to the law of conservation of angular momentum, the angular velocity of rotation increases with the diminution of radius.

Second, the magnetic field intensifies very much. This is because, for a field frozen in plasma, the total number of lines of force is fixed, i.e., $\pi r_0^2 B = const$, and so *the field strength B is inversely proportional to r_0^2 or directly proportional to $\bar{\rho}^{2/3}$*. If we assume that before contraction $r_0 = 3 \times 10^{10}$ cm, $\bar{\rho} = 1\,\mathrm{g\,cm^{-3}}$ and $B = 1$ G and that after contraction $r = 3 \times 10^6$ cm, $\bar{\rho} = 10^{12}\,\mathrm{g\,cm^{-3}}$, then B becomes 10^8 G. Taking into consideration that even before contraction the stellar magnetic field may have a strength of 10^3–10^4 G and that the average density of the neutron star is as high as $10^{14}\,\mathrm{g\,cm^{-3}}$, it is no wonder that the surface field intensity of pulsars can attain 10^{12}–10^{14} G. In such exceedingly strong magnetic fields a lot of unimaginable phenomena may take place.

9.5.3 SUPERSTRONG MAGNETIC FIELDS

In the regime of such fields the classical theory of electromagnetism is no longer valid and certain *quantum effects* may appear. For this some new theories have been established (e.g., Ginzburg and Usov, 1972; Hillebrandt and Müller, 1976; Ginzburg, 1985). We would now like first to distinguish the realms of applicability of the classical and the new theories.

According to the classical theory, the motion of an electron in a magnetic field **B** is governed by the equation

$$m\frac{d\mathbf{v}}{dt} = \frac{e}{c}\mathbf{v} \times \mathbf{B}. \tag{9.14}$$

Let us take the z-axis to be parallel to **B**, then the solution of this equation is

$$\begin{aligned} x &= x_0 + R_L \sin(\omega t + \alpha), \\ y &= y_0 + R_L \cos(\omega t + \alpha), \\ z &= z_0 + v_z t. \end{aligned} \tag{9.15}$$

It can readily be seen from these expressions that the electron spirals around the magnetic line of force and the projection of its trajectory on the (x, y)-plane is a circle with center at (x_0, y_0) and radius R_L. Besides, the angular velocity is ω and the phase angle is α. As projected on the z-direction, it performs a uniform linear motion with velocity v_z. The *Larmor radius* $R_L = cp_\perp/(eB)$, where $p_\perp$ is the component of momentum of the electron on the (x, y)-plane. It then follows that the stronger is B, the smaller is R_L. When R_L is so small as to be comparable with the *de Broglie radius* (λ_B) of the electron, the condition of quantization becomes effective. In this situation R_L is no longer arbitrary and can take only certain definite values.

By its definition, $\lambda_B = h/mv$, or

$$\lambda_B = \frac{1.22 \times 10^{-7}}{\sqrt{E}} . \tag{9.16}$$

In the last equality the units of λ_B and E are centimeter and electron volt, respectively. When the energy of an electron is 1 eV and the magnetic field has a strength of 1 G, the Larmor radius is 1 cm. However, for a field of 10^7 G, the Larmor radius of an electron of 1 eV is diminished to 10^{-7} cm and this is comparable to the de Broglie radius. So it is clear that for white dwarfs which have $B \geq 10^7$ G, and especially for neutron stars, we have to take into account the *quantization of the magnetic field.*

In what follows we shall state how the matter behaves in the *superstrong magnetic field.* As is known from atomic physics, the difference between energy levels of an atom, e.g. a hydrogen atom, is approximately $E_a \simeq (e^4 m)/(2\hbar^2) = 10$ eV. On the other hand, the energy difference between the levels of a non-relativistic free electron in a magnetic field is $E_B \simeq (e\hbar B)/(mc) = 10^{-8} B$ eV. In ordinary conditions B is not very large and the external magnetic field does not affect the internal structure of the atom. In this case we have the inequality $E_B \ll E_a$ or

$$B \ll \frac{e^3 m^2 c}{\hbar^3} = 3 \times 10^9 \text{ G}. \tag{9.17}$$

If the atom under consideration is not that of hydrogen, but that of a heavy element with atomic number Z, the above criterion should be modified to

$$B \ll \frac{e^3 Z^3 m^2 c}{\hbar^3} = 3 \times 10^9 Z^3 \text{ G}. \tag{9.18}$$

This means that the strength of the magnetic field capable of violating the criterion becomes still larger. In the past nobody doubted its validity for celestial bodies. Nevertheless, the idea has changed much after the discovery of pulsars. In these special objects, the inequality (9.17) is no longer valid for the hydrogen atom. Moreover, the criterion (9.18) for heavy atoms also breaks down. Due to the profound influence of external magnetic fields on the interior of atoms, a lot of abnormal phenomena come into existence. For instance, the electronic orbits around nuclei become very oblate ellipses. Two heavy atoms, e.g. iron atoms, combine into a molecule (Fe_2) and these molecules can form polymolecular substances, which are constituents of the hard surface of neutron stars. Moreover, even the β-decay of neutrons and the production of neutrinos may be influenced by superstrong magnetic fields. Detailed discussion of these interesting topics can be found in, e.g., Ruderman (1971), Cameron and Canuto (1974) and Ginzburg (1985).

The study of superstrong magnetic fields is a new area of physics. Because at present it is still difficult to create magnetic fields with intensities more than 10^8 G in the terrestrial laboratory, the research on pulsars may provide valuable inspirations for physicists.

9.5.4 Magnetic Dipole Model of Pulsars

This model was proposed by Pacini (1967, 1968) and Gunn and Ostriker (1969). They suggested that the neutron star has a dipolar magnetic field whose axis forms an angle i with the rotational axis. The relation between the magnetic moment m and the strength at magnetic pole B_p is

$$|\mathrm{m}| = \frac{B_p R^3}{2}, \tag{9.19}$$

where R is the radius of the star. Due to the change of **m**, the star radiates energy at the following rate

$$\dot{E} = -\frac{2}{3c^3}|\ddot{\mathrm{m}}|^2 . \tag{9.20}$$

In the system of rectangular coordinates with unit vector $\mathrm{e}_\parallel$ parallel to the rotational axis and with $\mathrm{e}_\perp$ and $\mathrm{e}'_\perp$ perpendicular to it, m has the following expression:

$$\mathrm{m} = \frac{1}{2}B_p R^3(\mathrm{e}_\parallel \cos i + \mathrm{e}_\perp \sin i \cos \Omega t + e'_\perp \sin i \sin \Omega t). \tag{9.21}$$

Substituting (9.21) into (9.20), we get

$$\dot{E} = -\frac{B_p^2 R^6 \Omega^4 \sin^2 i}{6c^3}, \tag{9.22}$$

where Ω is the angular velocity of rotation. The radiative energy emitted by the star is supplied by its kinetic energy of rotation $E = \frac{1}{2}I\Omega^2$, where I is the moment of inertia. Then we have

$$\dot{E} = I\Omega\dot{\Omega}. \tag{9.23}$$

Because $\dot{E}$ is negative, $\dot{\Omega}$ should also be negative. This agrees with the fact of observation mentioned above, viz., the rotation of the pulsar gradually slows down. From this we may determine its characteristic age, defined as follows:

$$T = -\left(\frac{\Omega}{\dot{\Omega}}\right)_0 = \frac{6Ic^3}{B_p^2 R^6 \sin^2 i\Omega_0^2} . \qquad (9.24)$$

Then from Eqs. (9.22) and (9.23) we have

$$\Omega = \Omega_i \left(1 + \frac{2\Omega_i^2}{\Omega_0^2}\frac{t}{T}\right)^{-1/2} , \qquad (9.25)$$

where Ω_i is the initial angular velocity. By setting $\Omega = \Omega_0$ into this equality, we can find the present age of the pulsar to be

$$t = \frac{T}{2}\left(1 - \frac{\Omega_0^2}{\Omega_i^2}\right) . \qquad (9.26)$$

For $\Omega_0 \ll \Omega_i$, we have $t = T/2$. The value of T of the Crab pulsar determined in this way by Groth (1975) and Gullahorn *et al.* (1977) in 1972 is 2486 y. This implies that the age of this pulsar is 1243 y. It is well known that this object was produced by the supernova explosion observed in 1054 by Chinese royal astronomers of the Sung dynasty. So the actual age should be taken to be 918 y. The agreement may be said to be fairly good. Some other interesting deductions of this model are described in Shapiro and Teukolsky (1983).

Recently a *unified model* of the magnetic field of neutron stars was proposed by Romani (1990). According to this model the field is generated after the birth of the neutron star during the initial cooling phase, either by thermomagnetic effects in the solid crust or by dynamo processes in the liquid envelope above the solid interior. Besides, this model can trace the further growth and decay of the field. As estimated by Krishnamohan (1987), the time scale of decay of the pulsar magnetic field is about 20 My.

9.6 The Magnetic Field of White Dwarfs

The white dwarfs are stars in the last phase of their evolution. Their nuclear fuel has been exhausted and they contract until their gravity is balanced by the pressure of degenerate electrons. They have exceedingly large densities which, according to the reason explained above, leads to superstrong magnetic fields.

(1) *A brief historical account.* In 1850 F.W. Bessel discovered from his meridional observation that the motion of Sirius (α CMa) is irregular and he suggested that this is caused by an invisible companion. This prediction was justified by A. Clark in 1862. Further studies have revealed that the companion of Sirius, now called *Sirius B*, is by no means an ordinary star. Its volume is small (close to those of planets), its luminosity is low (only about 0.3% of that of the Sun), but its mass is of the same order as the solar one. Hence, as the first white dwarf ever discovered, it is a compact object with an average density $\overline{\rho} \simeq 10^5\ \mathrm{g\,cm^{-3}}$. Since the twenties of this century, Sirius B has made a distinguished appearance in astrophysics. The *gravitational redshift* of its spectral lines has become one of the three classical proofs of A. Einstein's general theory of relativity. S. Chandrasekhar took into consideration the relativistic effects in the degenerate electron equation of state and constructed models of white dwarfs. He demonstrated that such stars cannot possess masses exceeding 1.4 $M_{\odot}$, which is called the *Chandrasekhar limit*.

For a detailed description of these interesting topics the reader is referred to, e.g., Luyten (1971), Shapiro and Teukolsky (1983) and Wegner (1989). In the following our discussion will be restricted to the ultrahigh magnetic fields of white dwarfs.

(2) *The magnetic field and polarization of the continuous spectrum.* As stated in the preceding section, for magnetic fields frozen in plasmas the field strength B is proportional to $\overline{\rho}^{2/3}$. A rough estimate with the known value of $\overline{\rho}$ leads to $B \simeq 10^5$–10^7 G. However, it is difficult to justify this with the Zeeman effect of magneto-sensitive lines. This is because the compact objects, such as white dwarfs, emit continuous spectra with few spectral lines. Therefore, it is desirable to measure intense magnetic fields with certain characteristics of the continuum. This can be realized with Kemp's (1970) argument that the light from a thermal source situated in a sufficiently strong magnetic field possesses a spectrally diffuse circular polarization. This may be explained as follows. The Zeeman components of a spectral line, which is radiated along the magnetic field direction, are circularly polarized. If for some reason, say high temperature, the line is greatly broadened, all the components may be blended together and so a single broad line is formed. When this happens simultaneously for many lines, the continuum acquires a *fractional circular polarization* $q(\nu)$, where ν is the frequency. Kemp derived the following expression from his theoretical calculation:

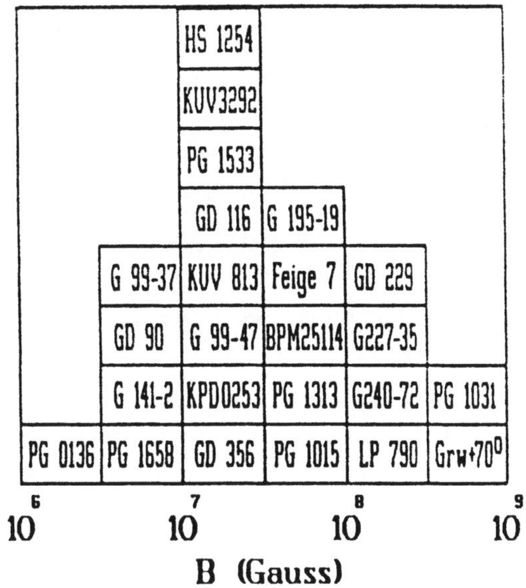

Fig. 9.8. Distribution of white dwarfs with polar field strengths (Schmidt, 1987).

$$q(\nu) = -\frac{eB}{2m\nu}, \tag{9.27}$$

where m is the electron mass. This theory was justified by experiment (Kemp *et al.*, 1970a). When incandescent metals or some other light sources are placed in a magnetic field of $B = 10^4$ G, circular polarization was observed and its amount was $q = 10^{-5}$, which agrees with the value given by the above formula. Besides, it was discovered by Kemp *et al.* (1970b) that there is a measurable circular polarization in the continuum of the white dwarf Grw+70°8247. To be precise, in the spectral range 4000–7000 Å the degree of circular polarization is 3.29%. It was then found with the above formula that this star has a magnetic field of approximately 10^7 G. A comprehensive account of measurements of this kind can be found in Landstreet (1979) and Schmidt (1987, 1989).

(3) *Magnetic field strength and structure.* According to the statistics of Schmidt (1989), there are altogether 26 white dwarfs whose magnetic fields have been measured and they occupy only 2–3% of all the known stars of this kind. The field intensities are in the range 1–1000 MG, and their distribution can be seen in Figure 9.8. As remarked by Schmidt (1987), most white dwarfs are expected to have surface field strengths well below 1 MG.

Generally speaking, the magnetic fields of white dwarfs are dipolar in character, but the magnetic axis does not always coincide with the rotational one. In particular, for the stars Feige 7 and PG 1015−014, these axes are perpendicular to each other (Wickramasinghe, 1987). Through analysis of the polarimetric and spectroscopic data, Ferrario and Wickramasinghe (1989) suggested that on the surface of the magnetic white dwarf EXO 033319−2554.2 there are two cyclotron emission regions with magnetic field intensities 5.6×10^7 G and 2.8×10^7 G, respectively and that these regions are separately by only about 90°. It can be inferred from the large difference in field strengths and the proximity of the two regions that the magnetic field structure is more complex than that of a dipolar field.

Finally, we would like to briefly describe the interesting structure of the surface magnetic field of the white dwarf PG 1031+234. It possesses the most intense polar magnetic field ($\sim$ 500 MG) among all the white dwarfs ever observed up to the present. Its grossly asymmetric variations in the spectrum and polarization are interpreted by Schmidt *et al.* (1986) and Latter *et al.* (1987) as being due to the existence of a "spot" with extremely strong magnetic field of about 1000 MG. With this model these authors performed numerical simulations of the changes of the degrees of linear and circular polarizations (P and V) as well as the angle between the line of sight and magnetic vector (θ) within a period of rotation. The results of simulations are represented by the solid curves in Figure 9.9. Moreover, Wickramasinghe (1987) published a review of such models from a theoretical point of view.

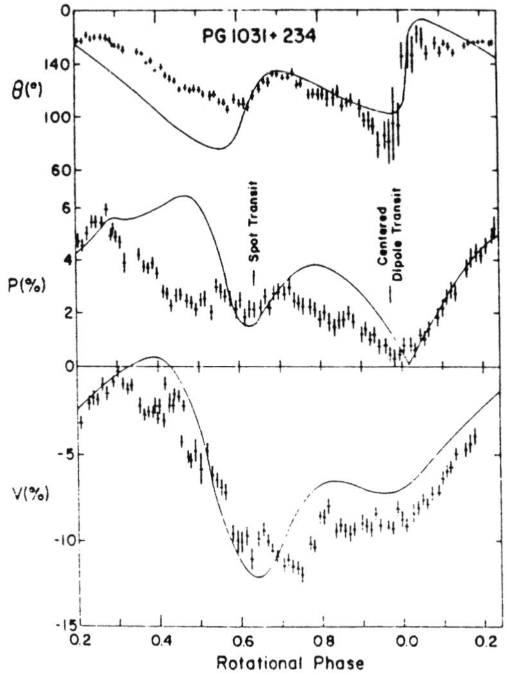

Fig. 9.9. Observed and theoretically calculated variations of magnetic field orientation and polarization of PG 1031+234.

CHAPTER 10

Magnetic Fields of Galaxies and Intergalactic Space

In interstellar space, and in intergalactic space, electromagnetic phenomena may be of decisive importance.

H. Alfvén and C.-G. Fälthammar, *Cosmical Electrodynamics* (1963).

After discussion of the solar and stellar magnetic fields we devote this chapter to the magnetism in the immense universe: in the *Milky Way*, *extragalactic nebulae* and *intergalactic space*. The main difference between magnetic fields of these two categories lies in the following. The Sun and magnetic stars have relatively strong fields in comparatively small volumes, while galaxies (including our Milky Way) and the intergalactic medium possess extremely faint fields in quite vast scopes. Therefore, the methods of measurements as well as the structure and characteristics of these two kinds of magnetic fields are to a great extent different.

10.1 The Interstellar Magnetic Field

Although our Milky Way contains more than one hundred billion (> 10^{11}) stars, its average density is quite trivial: 10^{-23} g cm^{-3}, i.e. even smaller than that of the highest "vacuum" which may be attained in the modern physics laboratory. The overwhelming majority of the volume of our Galaxy is the interstellar space, which has a mean density of 6×10^{-24} g cm^{-3}. In other words, there are only a few atoms in a cubic centimeter. But such a rarefied interstellar matter is by no means insignificant. This is because, due to the vast dimensions, its total mass is almost equal to that of all the stars in the Milky Way. Now we want to stress that the interstellar magnetic field is just like this.

10.1.1 Evidence of Existence and Indirect Measurements

Alfvén (1937) first suggested that there is magnetism in interstellar space. This idea was based on the theoretical interpretation of the *confinement of cosmic rays*. But formal measurement of this magnetic field was not carried out until thirty years later. To sum up, all the following phenomena indicate that Alfvén's prediction is true.

(1) *Cosmic rays*. This radiation is composed of natural streams of high energy particles, which are mainly protons and α-particles with energies (E) in the range of 10^8–10^{20} eV. One of the main characteristics of cosmic rays is their *isotropy*. This means that the intensities of the cosmic radiation from all directions are almost the same. Concretely speaking, for cosmic rays with $E < 10^{17}$ eV, the degree of anisotropy is less than 1%. For $10^{17}\ eV < E < 10^{19}$ eV, it is a little larger but still does not exceed 3%. Only for superhigh energy particles with $E > 10^{19}$ eV, a conspicuous anisotropy can be detected. Why are cosmic rays isotropic? There are two possible answers. First, the sources of the cosmic radiation are uniform in spatial distribution. If this were true, the cosmic rays in the metagalaxy should be isotropic and then their energy density would have to be much larger than that of the stellar radiation. However, in reality the energy density of cosmic rays in the vicinity of the Earth is about 10^{-12} erg cm^{-3}, which is close to that of the stellar radiation in the Milky Way. Second, in interstellar space there is a magnetic field which can affect the direction of motion of charged particles and cause the cosmic radiation to become isotropic. The radius of curvature of the trajectory of a particle with charge Z (and because most cosmic particles are protons, Z is effectively equal to 1) in a magnetic field of strength B is

$$r = \frac{E \sin\theta}{300BZ}, \tag{10.1}$$

where θ is the angle between the direction of motion of the particle and the magnetic lines of force and the unit of E is electron volt. By this formula, even for a weak field with $B = 3 \times 10^{-6}$ G and for particles with superhigh energies, $E = 10^{19}$ eV, the value of r cannot exceed the dimension of our Galaxy ($\sim 10^{23}$ cm). This implies that cosmic rays can be confined in the Galaxy and become isotropic.

(2) *Form and orientation of interstellar nebulae*. In the Milky Way almost all the diffuse nebulae are elongated in the direction parallel to

the Galactic plane. As shown by theoretical calculations, this cannot be due to the differential rotation of our Galaxy. So Shajn (1955) suggested that this is caused by the interstellar magnetic field. Nevertheless, van de Hulst (1967) remarked that some nonmagnetic dynamical effects can also give rise to the phenomenon stated above. Although Shajn's interpretation may not be exclusive, it is still of some interest. In passing, we mentioned in Section 2.7 that the vertical dust lanes in some spiral galaxies were interpreted by Sofue (1987) to be connected with the poloidal magnetic fields of these extragalactic objects.

(3) *Polarization of starlight.* Independently of each other, Hiltner (1949) and Hall (1949) discovered that the light from many stars close to the Galactic plane is linearly polarized. The farther away the stars are, the larger is the degree of polarization. (Its maximum may be 10%.) This phenomenon can be interpreted with the scattering of stellar light by interstellar dust particles, which are aligned by magnetic field lines. However, the collisions of these particles with neighboring atoms may violate the regular orientation. The necessary condition for the maintenance of the magnetic alignment is that the characteristic time of recovery to the original orientation ($\tau_{\rm or}$) is shorter than the relaxation time of collisions ($\tau_{\rm rel}$), i.e. $\delta = \tau_{\rm rel}/\tau_{\rm or} > 1$. The mathematical expression of δ is

$$\delta \simeq 3 \times 10^8 \frac{B^2}{T_D T^{1/2} n a} . \tag{10.2}$$

Here T_D and a are the temperature and size of dust particles, respectively, and T and n are the temperature and number density of the interstellar gas atoms. The value of B can be estimated from this equality. (For further details see Ruzmaikin *et al.*, 1988.)

(4) *Synchrotron radiation and its polarization.* The Galactic radio emission was discovered by K.G. Jansky in 1932 at the wavelength of 14.6 m. Its nonthermal component is the *synchrotron radiation* produced by the motion of relativistic electrons in magnetic fields. The intensity of this radiation is proportional to the transverse component of the Galactic magnetic field strength. The mathematical expression is

$$I \propto K \nu^{-(\gamma-1)/2} B_{\perp}^{(\gamma+1)/2} , \tag{10.3}$$

where ν is the frequency of radiation and K is a coefficient which depends on the number density of relativistic electrons. Besides, the index γ is defined by

$$N_e \, dE = KE^{-\gamma} \, dE \,, \tag{10.4}$$

where N_e is the number of relativistic electrons with energies lying between E and $E + dE$. As found from analysis of the background radio emission of the Milky Way, $\gamma = 2.7$ for $E \simeq 1$–3 GeV and $\gamma = 2.2$ for $E \simeq 1$ MeV (Ruzmaikin *et al.*, 1988). Taking N_e to be approximately equal to the number density of protons in cosmic rays, it is found that $B_{\perp} \simeq 3 \times 10^{-6}$–$10^{-5}$ G.

The polarization of the synchrotron radiation can also be used to study the magnetic field of the source. For a single relativistic electron, this radiation is elliptically polarized. For an ensemble of electrons with isotropic distribution of velocities, the degree of circular polarization (p_v) decreases greatly. When the velocity distribution is anisotropic, p_v becomes larger. The value of p_v varies with the frequency in the following way:

$$p_v = p_1 \nu^{-1/2} \,, \tag{10.5}$$

where the coefficient p_1 is directly concerned with the transverse magnetic field strength ($B_{\perp}$). Their relation may be expressed as

$$B_{\perp} = \left[\frac{p_1/100C}{400(\gamma + 2)} \right]^2 \times 10^9 \text{ G}. \tag{10.6}$$

Here γ is the index defined by Eq. (10.4). For optically thin sources $C = 1$, and for optically thick ones C is a known function of γ. Therefore, a determination of the degree of polarization of the synchrotron emission may lead to an estimate of $B_{\perp}$. This method has been applied to the observation of the magnetic field of extragalactic sources, whose degrees of polarization are very low ($< 0.5\%$) and so are difficult to measure (Weiler and de Pater, 1983). A detailed discussion of the method can be found in Ruzmaikin *et al.* (1988).

(5) *Maser radiation and molecular radio lines*. The process of star formation takes place mainly in the densest regions of the interstellar medium and it is also there that the maser radiation is generated. The polarization of this radiation can be used to estimate the magnetic field strength of the region of its formation (Western and Watson, 1984). Another method is connected with the polarization of radio lines, e.g. those of the CO molecules (Goldreich and Kylafis, 1981, 1982).

10.1.2 Direct Measurements

Although the above phenomena may imply the existence of the interstellar magnetic field and provide us with some preliminary knowledge of its strength and orientation, direct and more accurate measurements are needed. Since the sixties much progress has been achieved by application of the following methods.

(1) *Zeeman effect.* Bolton and Wild (1957) proposed measuring the Zeeman splitting of the absorption line of hydrogen at 21 cm, but the actual observation was carried out only ten years later (Verschuur, 1968). This is because the recording of the very small splitting, which will be indicated below, required an exceedingly long measurement time. As estimated by Verschuur (1979), for the first five reliable records, 5000 hours were spent on observation!

The radio spectral line of hydrogen at 21.11 cm can be split into the following three components:

$$\pi\text{-component} \qquad \nu = \nu_0 = 1420.41 \text{ MHz},$$

$$\sigma\text{-components} \qquad \nu = \nu_0 \pm \frac{1}{2}\Delta\nu_B = 1420.41 \pm \frac{eB}{4\pi mc} \text{ MHz}.$$

The amount of Zeeman splitting is

$$\Delta\nu_B = \frac{eB}{2\pi mc} = 2.8B \text{ MHz}. \tag{10.7}$$

When $B = 10^{-5}$ G, $\Delta\nu_B \simeq 30$ Hz. Needless to say, this splitting is quite trivial; even less than 1% of the half-width of the 21 cm hydrogen line. Moreover, this small splitting may be smeared by Doppler shifts. In order to accurately measure such a tiny quantity, astronomers resorted to a technique, which is similar to that of the photoelectric magnetograph (Davies *et al.*, 1960).

In some dense interstellar clouds hydrogen is situated mainly in the molecular state and so the 21 cm line is too weak to be measured. In this case the OH line with a wavelength of 18 cm can be used (Heiles and Stevens, 1986). In H II regions most of the hydrogen atoms are ionized, and observation can be done on the recombination lines of hydrogen (Heiles, 1976). It is interesting to note that from these measurements the relation between magnetic field strengths and densities of various interstellar clouds has been revealed, and this is shown in Figure 10.1. As we have shown in the last chapter, in the process of isotropic contraction the exponent k in the relation $B \propto \rho^{-k}$ is equal to 2/3. However, a cloud may contract into an oblate object with its major axis lying along magnetic lines of force. In this case of anisotropic contraction k is smaller and lies between 1/3 and 1/2 (Troland and Heiles, 1986).

(2) *Faraday rotation.* The principle of this effect was explained in Section 2.4. In recent years the Faraday rotation of the polarized

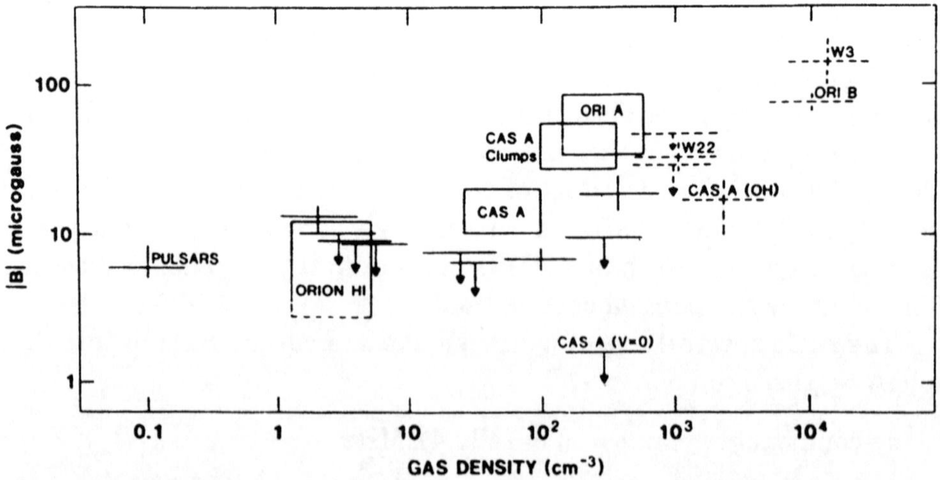

Fig. 10.1. Cloud densities versus magnetic field intensities determined with Zeeman measurements.

radiation of pulsars has been applied to the study of the interstellar magnetic field. The *rotation measure* (RM) defined by Eq. (2.25) may be rewritten to be

$$RM = 0.81\overline{B_{\|}} \int n_e \, dr \,, \tag{10.8}$$

where $\overline{B_{\|}}$ is the mean longitudinal component of the interstellar field strength. The integral $\int n_e \, dr$ is defined as the *dispersion measure* (DM) with the following physical significance. The signals of pulsars may undergo dispersion in the interstellar medium so that a broadband pulse arrives later at lower frequencies than at higher ones, and the time difference is proportional to DM. If both RM and DM have been determined by observation, it is easy to find the value of $\overline{B_{\|}}$. (For a more detailed account see, e.g., Verschuur, 1970.)

In order to apply the method described above, one has to measure the angle of rotation of the plane of polarization and at the same time to use the value of DM determined in some other observation. As an improvement, Smirnova (1991) suggested measuring the time modulation of pulses. This quantity is produced by the joint action of the Faraday rotation and the dispersive retardation of signals in the interstellar medium. Then the measurement of the period of modulation may directly yield the value of $\overline{B_{\|}}$ and the data of DM are no longer needed. She applied this new method to eight pulsars and found $\overline{B_{\|}}$'s in the range of about 1–3 μG.

Fig. 10.2. Axisymmetric (left) and bisymmetric (right) spiral configurations.

Before the end of this section we would like to talk briefly about the origin of the interstellar magnetic field. Two basic hypotheses are the *dynamo mechanism* and the *fossil field* (Beck, 1987). According to the former, magnetic fields are produced by the motion of charged particles, while the latter asserts that the present fields are remnants of the ancient magnetism which existed at the very beginning of the universe. These two theories will be discussed in some detail in the last chapter, and what we now hope to say is merely that they should give rise to different large-scale structures of the interstellar magnetic field. To be precise, the former may produce an axisymmetric or bisymmetric field with magnetic lines of force closed within the Galaxy (see Figure 10.2), while the latter can give rise to a bisymmetric structure connected with the intergalactic field. (Therefore, in order to judge which hypothesis is correct, one has to study the structure of the Galactic magnetic field. This is the subject of the next section.) Furthermore, it should be noted that the two models mentioned above are not completely independent of each other. The former needs a primordial "seed field", while the latter requires the field amplification by a "dynamo". We shall continue to discuss this question in Section 11.5.

10.1.3 Structure of the Galactic Magnetic Field

In the last section we have briefly described various methods of the measurements of the Galactic magnetic field. Now it has to be added that the classical method based on the Zeeman effect can be applied only to dense interstellar clouds with comparatively strong magnetism, such as Tau A, M 13 and Orion A (Verschuur, 1970). For the very weak

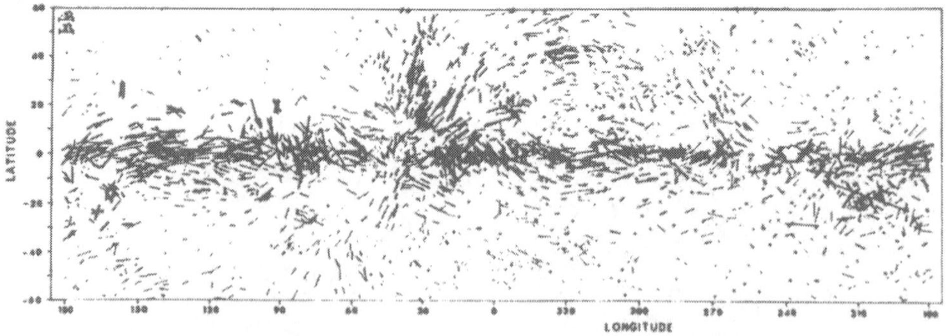

Fig. 10.3. **Map of the sky with orientations of electric vectors of polarized emission from distant reddened stars.**

fields in the vast space of the Milky Way, astronomers prefer to the measurements of *RM*s of pulsars and extragalactic radio sources as well as to the observation of starlight polarization. The main results of work in the last thirty more years can be briefly summarized as follows.

(1) *Overall structure*. A rough impression of the Galactic magnetic field can be gained by looking at Figure 10.3 (Mathewson and Ford, 1970). It illustrates the positions of polarization vectors of about 6000 stars projected on a cross-section perpendicular to the Galactic plane. The direction of the longitudinal field may be inferred from the focusing of polarization vectors. This occurs at longitudes 80° and 260°, which are close to, but smaller than the values 94° and 274° given by *RM* measurements. As explained by Heiles (1976), this is because these two methods are applied to different parts of the Galaxy. To be precise, the *RM* data represent the interarm regions, while the optical ones belong to the spiral arms. Moreover, the difference between *RM*s derived from pulsars and those from extragalactic radio sources may imply that the Galactic magnetic field cannot be represented simply by either a radial or a longitudinal model. In fact, it is quite complex in structure and consists of spurs and loops, as well as chaotic components. (For details see, e.g., Zeldovich *et al.*, 1983; Trimble, 1990.)

(2) *Spiral arms*. The Milky Way is a spiral galaxy with several arms. On the average an arm has a width of 1400 lightyear and a density of about one atom per cm^3. Both its density and magnetic field strength are approximately one order of magnitude larger than those of interarm regions. According to Vallée (1988), the configurations of magnetic field lines in the Perseus, Orion, Sagittarius and Scutum arms are shown by the dotted lines in Figure 10.4 and the solid curves represent the arms. The Sun is located at the origin of coordinates and the center of

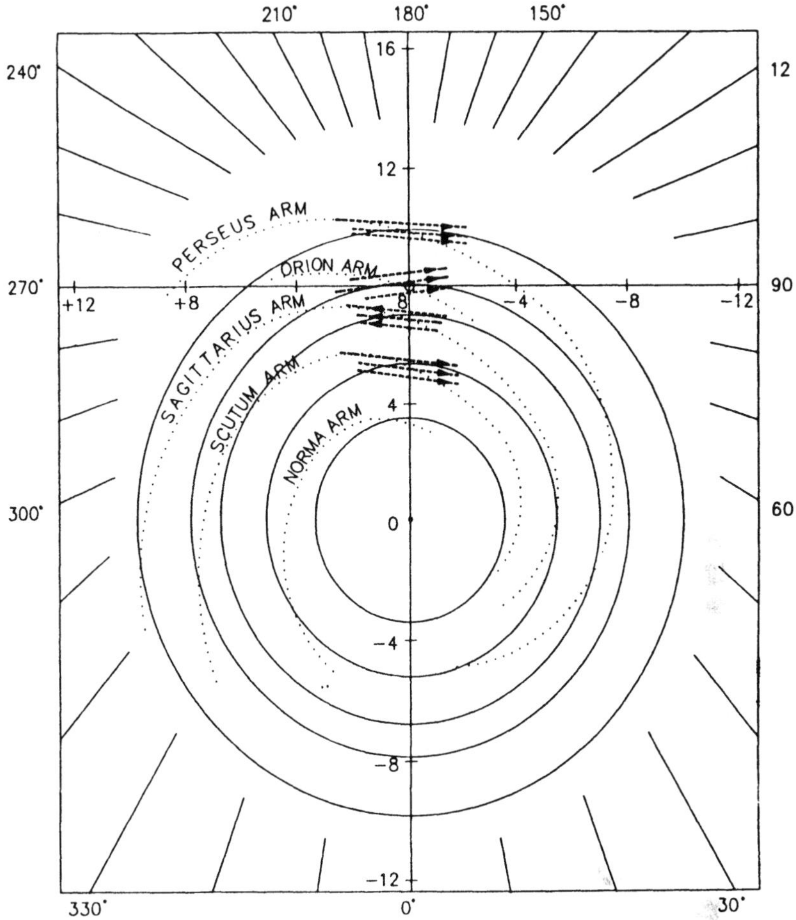

Fig. 10.4. Magnetic fields in spiral arms of the Galaxy.

the Galaxy is at point 0. The magnetic field lines and spiral arms do not coincide exactly. Their *pitch angles*[1] are somewhat different. The strengths and senses of magnetic fields as well as the pitch angles of the four spiral arms just mentioned are listed in Table 10.1.

The *stability of the spiral arm* is a question of great theoretical interest. As proposed by Chandrasekhar and Fermi (1953), the criterion of stability of a gas column permeated by magnetic field lines is

$$\sqrt{\overline{B^2}} \le 2\pi R\sqrt{\rho\rho' G}\,, \tag{10.9}$$

where R is the radius of the column, ρ is the gas density, ρ' is the density of gas and stars counted together, and G is the gravitational constant.

[1]The pitch angle is the angle of deviation of a straight line from the tangent of a circle.

TABLE 10.1
Magnetic field strengths and configuration of spiral arms

arm name	magnetic field strength (μG)	sense of magnetic field	pitch angle: magnetic field	pitch angle: spiral arm
Perseus arm	2	clockwise	$-7^\circ \pm 4^\circ$	$-14^\circ \pm 3^\circ$
Orion arm	3	clockwise	$+16^\circ \pm 4^\circ$	$-18^\circ \pm 3^\circ$
Sagittarius arm	6	counter-clockwise	$-6^\circ \pm 3^\circ$	$-18^\circ \pm 3^\circ$
Scutum arm	5	clockwise		$-19^\circ \pm 3^\circ$
median value	4		$-6^\circ \pm 3^\circ$	$-18^\circ \pm 3^\circ$

If we take $\rho = 2 \times 10^{-24}$ g cm^{-3}, $\rho' = 3\rho$ and $R = 250$ pc (1 parsec = 3×10^{18} cm), then $\sqrt{\overline{B^2}} \leq 4\ \mu$G. Because the Galactic spiral arms do possess magnetic fields of such a strength, they can be considered to be stable.

(3) *The local arm.* The Sun is situated in the Orion arm whose magnetic field is known better than those of the other arms. Recently Rand and Kulkarni (1990) analyzed the *RM*s of 116 pulsars nearer than 3 kpc and found that the local field has an intensity of $1.6 \pm 0.2\ \mu$G and the direction toward the Galactic longitude of $96^\circ \pm 4^\circ$. Besides, this field has a reversal at a distance of 600 ± 80 pc toward the inner Galaxy. Abramenkov and Krymkin (1990) observed the nonthermal radio emission in the Galactic disk and from the density of this emission, which according to Eq. (10.3) is proportional to $B_\perp^{(\gamma+1)/2}$, they derived the parameters of the magnetic field in the local arm over a distance of less than 1 kpc in the direction of the Galactic anticenter. Besides, they found that the local field consists of two ingredients, i.e., a regular (direct) component and an isotropic one. The ratio of their strengths is 0.62 ± 0.22. The field intensity in the interarm region is approximately one tenth of that in the local arm.

(4) *The Galactic center.* The central region of the Galaxy has a size of less than 1 kpc. It may be said to be a miniature Galaxy (Oort, 1977), because it also possesses a rotating gas disk. It has been found from radio observations that within about 200 pc from the Galactic center the magnetic field is a poloidal one and has a strength of approximately 1 mG, i.e., in 3–4 orders of magnitude larger than those in the other parts of the Milky Way. The existence of such a comparatively intense field may have the following implications: (1) There is a strong current system around the Galactic nucleus. (2) The field can significantly affect the dynamic processes of the molecular clouds in the Galactic central

region. However, as shown by far-IR polarization data, the magnetic field within 5 pc in the Galactic center is no longer poloidal (Morris, 1990). Besides, the field around the center may be responsible for the filamentary structures observed by Anantharamaiah and Pedlar (1990) with the Very Large Array at 333.1 MHz and with angular resolution of 33″ × 17″. (An example is given in Figure 10.5.)

Although much progress in the study of the Galactic center has been achieved in recent decades, there are still many puzzles and contradictions, which were recently reviewed by Townes (1989). Now we want only to mention that the degree of importance of the magnetic field for the dynamics of the Galactic central region is uncertain. It may be believed that magnetic fields play a decisive role in the interstellar plasma motions. However, according to the rough estimate of Aitken *et al.* (1986), the magnetic energy density in the above-mentioned region is merely $\frac{1}{3}$ to $\frac{1}{4}$ of the kinetic energy of the gas.

(5) *The Galactic corona.* The Galactic nonthermal radio emission comes not only from the Galactic disk, but also from high latitude regions. This is because around the disk there exists the *Galactic corona* or, as it is called by some authors, the *halo*, and this district also emits synchrotron radiation. From its intensity the magnetic field strength of the Galactic corona is found to be about twice as weak as that of the Galactic disk. The coronal magnetic field should be somewhat chaotic, unless there would be no isotropy of cosmic rays. The radius of the corona is estimated to be approximately 15,000 pc, but it does not have a definite boundary and in the outside it merges with the vast intergalactic space.

On the basis of the *RM*s of 185 pulsars and 802 extragalactic radio sources, Andreassian and Makarov (1990) established a two-component model of the Galactic magnetic field. The *plane component* consists of the magnetic fields in spiral arms, and the *halo component* stretches far from the Galactic disk and completely surrounds the plane component. In passing, it is worthwhile noting that the halo fields in the southern and northern hemispheres are oppositely directed.

Finally, we want to state that in spite of many complexities the overall structure of the Galactic magnetic field may be roughly characterized by the following simple descriptions (Wielebinski, 1990). First, in the disk of the Galaxy the field is an azimuthal one. As determined by Vallée (1988), the pitch angles of the magnetic field lines in spiral arms are small, possibly less than 6°. Second, the field in the center of the Galaxy is perpendicular to the Galactic plane. In other words, it may be a part of an extended poloidal field. It is likely that the magnetic

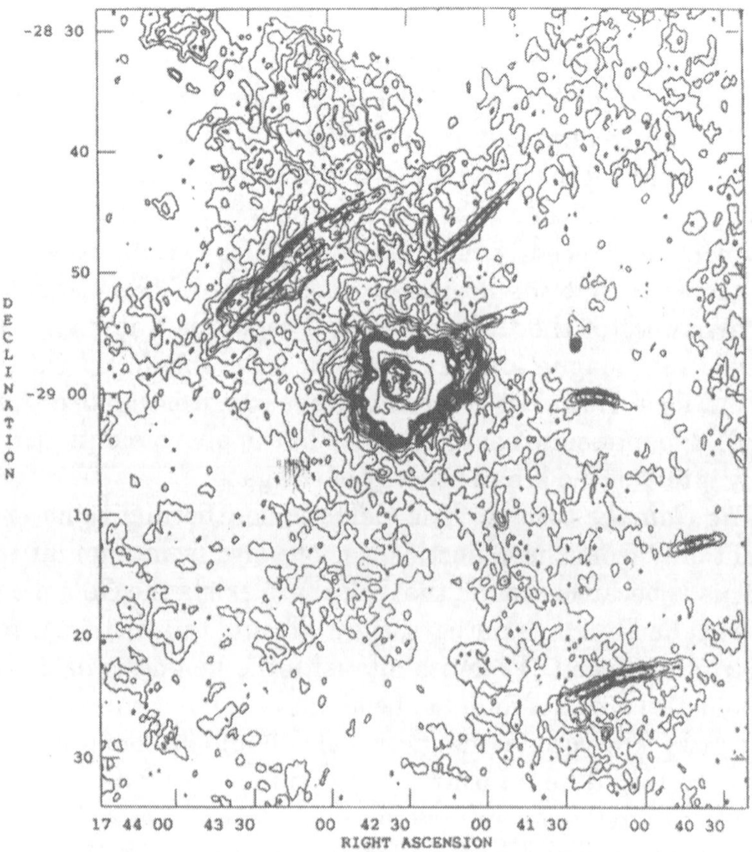

Fig. 10.5. A 90 cm image around the Galactic center.

fields of many external galaxies also have such a general configuration.

10.2 Magnetic Fields of Extragalactic Nebulae

In the foregoing section we have given a brief account of the Galactic magnetic field. Since ours is merely one of billions of galaxies in the grand universe, we may expect that the countless extragalactic nebulae as well as the intergalactic space are also permeated by magnetic fields. Due to the enormous distances their observational data are scarce. The existing results of investigation are far from being complete and what follows is merely a very rough sketch.

10.2.1 THE MAGELLANIC CLOUDS

Both these extragalactic objects nearest to us were discovered in 1518–1520 by F. Magellan and his subordinate sailors during their navigation in the southern hemisphere. Some polarimetric observations have been carried out on these nebulae since 1966. For instance, Schmidt (1970) measured the degree of polarization and got the upper and lower limits of their magnetic field strengths as 10^{-4} and $10^{-4.5}$ G. Besides, in the gaseous connection between the Clouds, discovered by Hindman (1963) through observation of the 21 cm hydrogen line, there is a magnetic field too. Hence he suggested that the two clouds and the intermediate connection form the "pan-Magellanic system".

Later on, the Clouds began to be studied separately. According to the radio polarization observation of Haynes *et al.* (1990), the magnetic field of the *Large Magellanic Cloud* (LMC) seems to possess a series of filaments originating in the 30 Doradus nebula, which may be thought to be the nucleus of the LMC. Costa *et al.* (1991) measured the RM for the pulsar PSR 0529−66 in the LMC and found a value of 4 ± 5 m^{-2} rad, the DM being 100 cm^{-3} pc. From these values they derived the magnetic field intensity of the LMC to be 0–5 μG and oriented away from the Sun. As for the *Small Magellanic Cloud* (SMC), which is a dwarf galaxy, the magnetic field strength is probably less than 3 μG (Wielebinski, 1990).

The magnetic fields in the Magellanic Clouds can also be studied with the polarization of both radio waves and starlight, and the field directions given by these two kinds of radiation are consistent with each other. It is interesting to note that in the vicinity of 30 Doradus the pan-Magellanic field smoothly blends in with the field around this nebula, and a similar situation exists for the SMC. This possibly implies that the magnetic fields in the pan-Magellanic system have to be studied together (Haynes *et al.*, 1990).

10.2.2 SPIRAL GALAXIES

For the measurements of the magnetic fields of extragalactic objects, one may apply the methods described in Section 10.1. Now we like to add that the *principle of equipartition of energy densities* of the magnetic field and cosmic rays can also be utilized. The amount of energy of relativistic electrons in cosmic rays may be taken to be $\int_{E_1}^{E_2} EN_e \, dE$ where N_e is the number of such electrons defined by Eq. (10.4) and E_1 and E_2 are, respectively, the lower and upper limits of the energy spectrum. These limits correspond to the cutoff frequencies ν_1 and ν_2 of the synchrotron radio spectrum. If a is taken to be the ratio of the energy of cosmic rays to the electron energy, then the condition of equipartition may be expressed as

$$\frac{B^2}{8\pi} = a \int_{E_1}^{E_2} EN_e \,\mathrm{d}E. \tag{10.10}$$

According to Moffet (1973), the magnetic field strength can be evaluated with the formula

$$B = 2.3(aA\varepsilon)^{2/7}, \tag{10.11}$$

where ε is the volume emissivity and A is defined by the expression

$$A = C\left(\frac{\alpha+1}{\alpha+\frac{1}{2}}\right)\left(\frac{\nu_2^{\alpha+1/2}-\nu_1^{\alpha+1/2}}{\nu_2^{\alpha+1}-\nu_1^{\alpha+1}}\right). \tag{10.12}$$

Here $C(= 1.057 \times 10^{12})$ is a constant, $\alpha(= -(\gamma-1)/2)$ is the radio spectral index and γ is the index in Eq. (10.4). When some adequate values are assigned to a and ν_1, e.g., Moffet (1973) took $a = 100$ and $\nu_1 = 107$ Hz, one may use the observational data of cosmic rays to obtain B from Eq. (10.11). This method has become widespread in the study of galactic magnetic fields, but, strictly speaking, the equipartition is still an assumption. Recently Duric (1990) has demonstrated that it is not far from reality.

The first extragalactic object whose magnetic field has been measured by the optical polarization is the Andromeda Nebula or M 31 (Öhman, 1942). The measurement through the synchrotron radiation was first realized also for M 51 (Mathewson *et al.*, 1972). Up to the present, the magnetic fields of more than 20 spiral galaxies have been observed and a complete list has been published by Krause (1990). Their average total field strengths range from several to more than 10 μG. As for the orientation, the fields are generally parallel to spiral arms. It is meaningful to mention that the magnetic field configuration of galaxies is helpful to the classification of galaxies according to the origin of their magnetism. To be precise, in the case of the dynamo mechanism the magnetic field is expected to display a circular distribution, while the hypothesis of primordial origin requires a bisymmetric spiral structure, open to the intergalactic space. These two configurations and the corresponding *RM* variations along the major axis and with the azimuthal angle are shown in Figure 10.6. A more detailed account of the magnetic field structure of spiral galaxies can be found in Sofue *et al.* (1986). Individual brief descriptions of the magnetic fields of 18 extragalactic objects are given by Wielebinski (1990).

10.2.3 CLUSTERS OF GALAXIES

Many galaxies may gather into groups and a group, generally consisting of more than 100 galaxies, is often called a *cluster*. Our own Galaxy is a member of the Local Group, which is a small and irregular cluster. Within a cluster there may be intergalactic or intracluster magnetic fields. For instance, such fields exist in the Coma cluster, a rich cluster above the north Galactic pole. As the evidence, its radio halo has a large diameter as well as a steep spectrum, which can hardly be interpreted as the integrated emission of many discrete sources. Kim and Kronberg (1990) studied the magnetism of this cluster with the following methods: (1) the direct polarization observation of the radio halo, (2) the equipartition estimate, and (3) the Faraday rotation measurements of sample sources in or through the intracluster medium. For the strength of the intracluster magnetic field these authors found a value of about 2 μG. Furthermore, it is interesting to mention that the Coma cluster may be a member of a super-cluster of galaxies and that its plane appears to be defined by the Coma cluster itself and another cluster called Abell 1367. As suggested by Giovannini *et al.* (1990), in this super-cluster there is a large-scale magnetic field of a few tenths of one μG. This may again remind us that there is magnetism everywhere in the immense universe.

As we have stated, the magnetic fields of extragalactic objects are measured mainly by the Faraday rotation. In recent years it has become clear that with the widening of the scope of investigation the number of available RMs increases and at the same time the measured field strength in the deep space decreases. Up to the present, the lowest value has been found to be 10^{-11} G or even less. In the limiting case one may have a cosmological magnetic field. This field seems to be too weak to provide a basis for the hypothesis of the relic origin of galactic magnetic fields. A detailed theoretical discussion of these problems can be found in Ruzmaikin *et al.* (1988).

Before closing this chapter the author would like to say that in the realm of galactic and extragalactic magnetic fields there are many other interesting topics, such as the magnetism of black holes, accretion

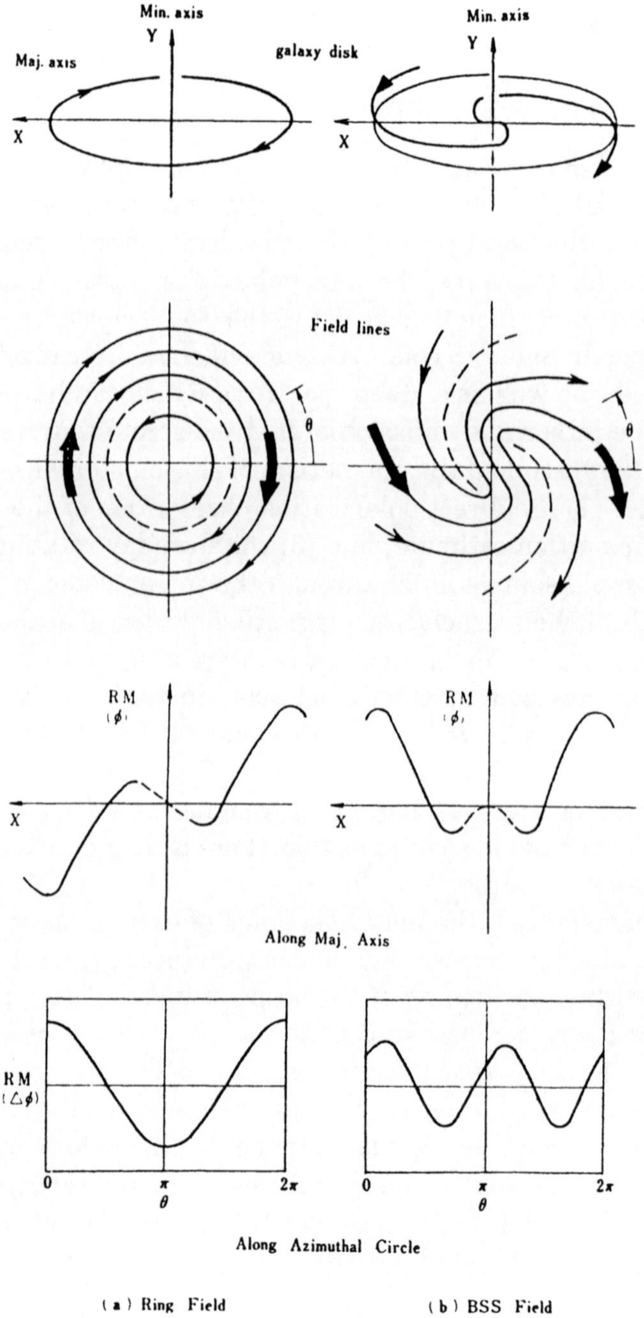

Fig. 10.6. Ring and bisymmetric (BSS) magnetic field configurations in disk galaxies.

disks, active galactic nuclei and supernova remnants. In order to limit the size of this book these topics are not discussed. It is suggested that the interested reader may consult, e.g., Beck *et al.* (1990), Ruzmaikin *et al.* (1988), Belvedere (1989) and Rees (1984).

CHAPTER 11

Some Theoretical Problems of Cosmic Magnetic Fields

The Sun's 22-year cycle is the best known example of a rapidly changing cosmic magnetic field. Its explanation provides the clue for a proper understanding of the origin of cosmic magnetism.
Ya.B. Zeldovich, A.A. Ruzmaikin and D.D. Sokoloff,
Magnetic Fields in Astrophysics (1983).

Magnetic fields are produced by fluid motions.
E.N. Parker,
Cosmical Magnetic Fields (1979).

In Chapters 4–10 we have described in order the magnetic fields of the Sun, the Moon, planets, stars, the Milky Way, interstellar space and extragalactic nebulae. The reader may have been deeply impressed by the ubiquitous permeation of the magnetic field into every corner of the cosmos and by its important role in various astrophysical phenomena, such as solar flares, prominences, sunspots, the solar wind, cosmic rays and the polarization of starlight. Abundant and colorful observational data naturally lead to deepgoing theoretical research. However, our rational knowledge is in general backward in comparison with our perceptional one. Until the present, reliable and mature theoretical recognition has been achieved only for some individual topics, and several of them will be introduced in this paper.

11.1 Theory of Spectral Line Formation in the Magnetic Field

The fundamental knowledge of the theory was stated in Section 2.2. What follows are some supplements.

11.1.1 A BRIEF HISTORICAL RETROSPECT

The classical theories were set up independently by Unno (1956) and Stepanov (1958a, b). Because Unno adopted the Stokes parameters which can fully express the characteristics of polarized radiation, his system of equations is more complete and convenient to use. It is a pity that Unno did not take account of the magneto-optical effect, scattering, fine structures of magnetic fields etc. and that he adopted some assumptions which are far from reality. However, in spite of all this, his work has laid a sound foundation for the theory of spectral line formation and has become its mainstream. In the last thirty more years Unno's theory has been inherited and improved by many authors and the following is a chronological sketch.

(1) Stepanov's theory has its merits. It implicitly includes the magneto-optical effects. The process of scattering is also taken into account. Afterwards, he derived the matrices of absorption and scattering (Stepanov, 1960, 1962).

(2) Rachkovskii (1961a) demonstrated the equivalence of Unno's and Stepanov's systems of equations. Besides, he (Rachkovskii, 1962) explicitly treated the magneto-optical effect as well as the incoherent scattering.

(3) The Landi Degl'Innocenti's (1972) derived the equations of transfer of polarized radiation in magnetic fields in the quantum-mechanical framework.

(4) The non-LTE theory of polarized radiative transfer was developed successively by Domke and Staude (1973), Dolginov and Pavlov (1973), Šidlichovský (1974) and House and Steinitz (1975). In their study of the non-LTE transfer, Stenholm and Stenflo (1978) took the scattering in multi-dimensional media into consideration.

(5) The Hanle effect was investigated by House (1970, 1971), Lamb and ter Haar (1971), Omont *et al.* (1973) and Bommier and Sahal-Bréchot (1978).

(6) Landi Degl'Innocenti (1983) achieved the unified derivation of the transfer equations from first principles and with consideration of both the Zeeman and Hanle effects. This work was done within the framework of quantum electrodynamics.

(7) Recently Stenflo (1991b) did the same but within the classical framework. Starting from Maxwell's equations and the Hamiltonian for an electron in a central potential, he developed the theory of radiative transfer in the magnetic field. His theory includes the magneto-optical effect, scattering and partial redistribution.

(8) Apart from the commonly used system of coupled transfer equations of the Stokes parameters, the normal wave transfer equation has also been set up. Nagendra and Peraiah (1985) performed a numerical solution of such an equation.

Besides what has been stated above, we have to mention that many methods of solving the transfer equations of polarized radiation in magnetic fields have been devised, and they will be systematically introduced in Section 11.1.3. So we can say that at present the theory of spectral line formation in magnetic fields has developed to an advanced stage. In principle one may determine the effect of the magnetic field on any spectral line in an arbitrary atmosphere in spite of the absence of LTE and uniformity of the medium and for any mechanism of reemission. Nevertheless, in the application of the theory, especially in the extraction of the information on the magnetic vector from observational data, many problems still remain to be overcome. They will be briefly discussed later in this section.

11.1.2 MATRIX REPRESENTATION OF THE TRANSFER EQUATIONS

As we have explained, the transfer of polarized radiation in a magnetized atmosphere is a very complicated process. Unlike the transfer equation of unpolarized radiation, which is represented merely by the first Stokes parameter (I), now we have to deal with all four parameters. Moreover, quite a number of physical factors should be considered. In order to meet these requirements, many solar astronomers prefer to the *matrix representation* of the transfer equations in magnetic fields and an example (Obridko, 1985) is as follows:

$$\mu \frac{\mathrm{d}\mathbf{S}_i}{\mathrm{d}\tau} = (E_{ik} + \eta_{ik})\mathbf{S}_k - $$
$$-(1-\varepsilon)\int\int D_{ik}(\mu,\mu')\mathbf{S}_k \frac{\mathrm{d}\omega'}{4\pi}\,\mathrm{d}\nu' - (E_{ik} + \varepsilon\eta_{ik})B_\nu. \qquad (11.1)$$

Here S_i is the matrix (S_1, S_2, S_3, S_4) and its elements S_1–S_4 represent the intensity and polarization of radiation. In general we take them to be the Stokes parameters, i.e., $S_i = (I, Q, U, V)$. The lower and upper levels of the spectral line transition are represented by i and k, respectively. η_{ik} is the *matrix of absorption*. In Unno's theory it is equal to $(\eta_I, \eta_Q, 0, \eta_V)$ and the expressions of the matrix elements are given by Eq. (2.16). In Beckers' (1969) formulation we have $\eta_{ik} = (\eta_I, \eta_Q, \eta_U, \eta_V)$ defined by (4.18). $\mathbf{B}_\nu = (B_\nu, 0, 0, 0)$ and B_ν is the source function in the case of true absorption, viz., Planck's function. ε determines the relative proportion of true absorption ($\varepsilon = 1$) and scattering ($\varepsilon = 0$). μ and μ' are, respectively, the cosines of position angles of the scattered and absorbed light rays. E_{ik} is a unit matrix. τ is the optical depth of the continuous spectrum close to the magneto-sensitive line under consideration. The integration with respect to ν' implies the redistribution of frequencies within the line. D_{ik} is the *matrix of scattering*. For a transition between the upper level with quantum number $j_u = 1$ and the lower level with $j_l = 0$ (this is the case for Fe I λ5250), Obridko (1965) derived the following expression:

$$D_{ik} = \begin{pmatrix} \frac{1+\cos^2\gamma}{\sqrt{2}} & \frac{\sin^2\gamma}{\sqrt{2}} & 0 \\ -\frac{\sin^2\gamma}{\sqrt{2}} & \frac{\sin^2\gamma}{\sqrt{2}} & 0 \\ 0 & 0 & \sqrt{2}\cos\gamma \end{pmatrix} \begin{pmatrix} \frac{\eta_l+\eta_r}{4} & 0 & \frac{\eta_l-\eta_r}{4} \\ 0 & \eta_p & 0 \\ \frac{\eta_l-\eta_r}{4} & 0 & \frac{\eta_l+\eta_r}{4} \end{pmatrix} \times$$
$$\times \begin{pmatrix} \frac{1+\cos^2\gamma'}{\sqrt{2}} & -\frac{\sin^2\gamma'}{\sqrt{2}} & 0 \\ \frac{\sin^2\gamma'}{\sqrt{2}} & \frac{\sin^2\gamma'}{\sqrt{2}} & 0 \\ 0 & 0 & \sqrt{2}\cos\gamma' \end{pmatrix} \qquad (11.2)$$

Another frequently used magneto-sensitive line, Fe I λ6302, has $j_u = 0$ and $j_l = 1$. Its matrix of scattering is

$$D_{ik} = \frac{3}{\eta_l + \eta_r + \eta_p} \begin{pmatrix} \eta_I\eta_I' & \eta_I\eta_Q' & \eta_I\eta_V' \\ \eta_Q\eta_I' & \eta_Q\eta_Q' & \eta_Q\eta_V' \\ \eta_V\eta_I' & \eta_V\eta_Q' & \eta_V\eta_V' \end{pmatrix}, \qquad (11.3)$$

where η_I', η_Q' and η_V' have expressions similar to those in Eq. (2.16), but the angle γ is changed to γ'. More accurate expressions of the matrix of scattering can be found in Rachkovskii (1967, 1971).

Landi Degl'Innocenti (1987) treated the problem of polarized radiation transfer with 4×4 matrices. In analogy with the common transfer equation for unpolarized radiation, viz.,

$$\frac{dI}{ds} = -k(I - S), \qquad (11.4)$$

he wrote the equation of transfer of polarized radiation to be

$$\frac{d\mathbf{I}}{ds} = -\mathbf{K}(\mathbf{I} - \mathbf{S}). \qquad (11.5)$$

Here $\mathbf{I} = (I, Q, U, V)$ is the Stokes vector and $\mathbf{K}$ is a 4×4 matrix which describes the variations of the Stokes parameters along the path s. Besides $\mathbf{S} = (S_I, S_Q, S_U, S_V)$ is the source function vector in the four Stokes parameters. The concrete expressions for $\mathbf{K}$ and $\mathbf{S}$ can be found in his paper just cited.

11.1.3 Various Solutions of Equations of Transfer of Polarized Radiation

The equations of transfer of polarized radiation in a magnetized atmosphere are a system of linear differential equations of the first order. In appearance they look rather simple, but, due to the following reasons, it is a formidable task to obtain their exact solutions. A series of physical factors have to be considered. The transfer equations and some quantities are strongly coupled together. There are certain integrals, like the Faraday–Voigt function, that must be calculated numerically. In the last three decades many authors tried to solve these equations and various methods have been established. The solutions yielded by the existing methods may be classified into the following categories:

(1) If the solutions are explicit expressions of the form $S_i = f(\eta_i, \tau, \ldots)$, they are called *algebraic* or *analytical* solutions.

(2) If certain functions still await integration, we have only *formal* solutions.

(3) When the solution must be calculated in an iterative way, such as $S_i = f(\eta_i, \tau - \Delta\tau, \ldots)$, it is a *numerical* solution.

(4) If a perturbation method similar to that of quantum mechanics is applied, one may obtain a *perturbative solution*. Besides, under the assumption of WFA, some approximate solutions can also be obtained. In the following we present a very brief introduction to a series of the existing solutions.

(1) *Algebraic solutions*. The classical solutions of Unno and Beckers were described, respectively, in Section 2.2.3 and Section 6.1.3. Now we want to make a few supplements.

(i) Staude's solution: Staude (1970a) adopted a source function of the following form

$$B(\tau) = B_0[1 + \beta_0\tau + A\exp(-\alpha T)], \qquad (11.6)$$

where A and α are constants. Besides this, he retained all the other assumptions of Unno and obtained the following algebraic solution:

$$
\begin{aligned}
I(0) &= B_0\left[1 + A + \mu B_0\frac{1-\eta_I}{n} - A\frac{1+\eta_I+\alpha\mu}{n/(\alpha\mu)+\alpha\mu+2(1+\eta_I)}\right],\\
Q(0) &= B_0\left[A\frac{\eta_Q}{n/(\alpha\mu)+\alpha\mu+2(1+\eta_I)} - B_0\mu\frac{\eta_Q}{n}\right],\\
V(0) &= B_0\left[A\frac{\eta_V}{n/(\alpha\mu)+\alpha\mu+2(1+\eta_I)} - B_0\mu\frac{\eta_V}{n}\right],
\end{aligned}
\tag{11.7}
$$

where $n = (1+\eta_I)^2 - \eta_Q^2 - \eta_V^2$. Furthermore, for the sake of the magneto-optical effect he (Staude, 1971) took

$$
\frac{\partial B}{\partial s} = 0, \qquad \frac{\partial \gamma}{\partial s} = 0, \qquad \frac{\partial \chi}{\partial s} = \text{const}, \tag{11.8}
$$

and divided his solution into two parts which correspond, respectively, to the linear and nonlinear terms in his expression of the source function (11.6).

(ii) The solution of Landolfi and Landi Degl'Innocenti (1982): These authors reserved all the other assumptions of Unno only except the absence of the magneto-optical effect, i.e. $U \equiv 0$. Then they obtained the following set of expressions:

$$
\begin{aligned}
I &= B_0 + \mu B_1\Delta^{-1}\left\{(1+\eta_I)[(1+\eta_I)^2 + \rho_Q^2 + \rho_U^2 + \rho_V^2]\right\},\\
Q &= -\mu B_1\Delta^{-1}[(1+\eta_I)^2\eta_Q + (1+\eta_I)(\eta_V\rho_U - \eta_U\rho_V) +\\
&\quad + \rho_Q(\eta_Q\rho_Q + \eta_U\rho_U + \eta_V\rho_V)],\\
U &= -\mu B_1\Delta^{-1}[(1+\eta_I)^2\eta_U + (1+\eta_I)(\eta_Q\rho_V - \eta_V\rho_Q) +\\
&\quad + \rho_U(\eta_Q\rho_Q + \eta_U\rho_U + \eta_V\rho_V)],\\
V &= -\mu B_1\Delta^{-1}[(1+\eta_I)^2\eta_V +\\
&\quad + \rho_V(\eta_Q\rho_Q + \eta_U\rho_U + \eta_V\rho_V)],
\end{aligned}
\tag{11.9}
$$

where $\Delta = (1+\eta_I)^2[(1+\eta_I)^2 - \eta_Q^2 - \eta_U^2 - \eta_V^2 + \rho_Q^2 + \rho_U^2 + \rho_V^2] - (\eta_Q\rho_Q + \eta_U\rho_U + \eta_V\rho_V)^2$.

This solution is an improvement on that of Auer *et al.* (1977) and may be useful for the *inversion technique*, i.e., the determination of vector magnetic fields from observed Stokes parameters.

(iii) Lites' *et al.* (1988) solution: In order to account for departures from LTE, these authors adopted a linear form for the continuum source function i.e.,

$$
S_c = B_0 + B_1\tau_0, \tag{11.10}
$$

and a nonlinear expression for the line source function,

$$S_l = B_0 + B_1\tau_0 - \sum_{i=1}^{2} A_i \exp(-\varepsilon_i\tau_0), \tag{11.11}$$

where A_1, A_2, B_0, B_1, ε_1, ε_2 are all constants. Then they found the following analytical solution:

$$\begin{aligned}\mathbf{I}(0,\mu) &= \mathbf{B}_0\mathbf{e}_0 + \mu B_1[r_0\mathbf{1} + \boldsymbol{\phi}]^{-1}\mathbf{e}_0 - \\ &\quad - \sum_{i=1}^{2} A_i\mathbf{e}_0 + \sum_{i=1}^{2} A_i(r_0 + \varepsilon_i\mu)[(r_0+\varepsilon_i\mu)\mathbf{1} + \boldsymbol{\phi}]^{-1}\mathbf{e}_0, \\ &\quad (r_0 = 1/\eta_0).\end{aligned} \tag{11.12}$$

This solution permitted the authors to generalize the existing least-square inversion procedures for Stokes profiles.

As a short summary, we would like to say that all the algebraic solutions briefly described above are basically within the framework of Unno's pioneering work. Although some improvements have been made, they are still not accurate enough. Nevertheless, these solutions are of practical importance, because they can be conveniently used in the inversion of magnetic field information from observed Stokes profiles.

(2) *Formal solutions.* Such solutions are not so clear and simple as algebraic ones, but with some further hypotheses they may give rise to more accurate algebraic and numerical solutions. In what follows we want only to state the main ideas of several formal solutions. For lack of space, few mathematical expressions will be given. The definitions of notations can be found in the respective references.

(i) Moe (1968) transformed Unno's three coupled equations into a system of uncoupled ones, which are similar to the common transfer equation of unpolarized radiation and so are easier to solve. He defined the following quantities:

$$\mathbf{Z}_i = \mathbf{I} + b_i\mathbf{Q} + c_i\mathbf{V} \quad (i = 1,2,3). \tag{11.13}$$

If $\mathbf{Z}_i$, b_i and c_i can be determined, then by solving a system of linear algebraic equations with three unknowns, one may get I, Q and V.

(ii) Staude (1969) wrote the transfer equation in the matrix form,

$$\mu\frac{\mathrm{d}}{\mathrm{d}\tau}\mathbf{I}(\tau) = \mathbf{A}(\tau)\mathbf{I}(\tau) + \mathbf{B}(\tau)\boldsymbol{\eta}(\tau), \tag{11.14}$$

and transformed it into a Volterra integral equation. Through iteration he obtained the formal solution,

$$\mathbf{I}(0,\mu) = \lim_{\substack{\tau\to\infty\\ N\to\infty}} \left[1 + \sum_{n=1}^{N} \left(\frac{1}{\mu}\right)^n \int_0^\tau \mathbf{K}_n(\tau,x)\,\mathrm{d}x\right]^{-1} \times$$
$$\times \frac{1}{\mu}\int_0^\tau \left[1 + \sum_{n=1}^{N} \left(\frac{1}{\mu}\right)^n \int_x^\tau \mathbf{K}_n(\tau,y)\,\mathrm{d}y\right] \times$$
$$\times \exp\left[-\frac{1}{\mu}\int_0^x (1+\eta_I(x')\,\mathrm{d}x'\right] \mathbf{B}(x)\boldsymbol{\eta}(x)\,\mathrm{d}x. \tag{11.15}$$

(iii) The transfer equation for polarized light was written by van Ballegooijen (1985, 1987) in terms of complex 2×2 matrices, and the following solution similar to the integral expression for the intensity of unpolarized radiation was obtained,

$$\overline{J} = \frac{1}{4\pi} \int_{-\infty}^{\infty} \oint (\phi_I I + \phi_Q Q + \phi_U U + \phi_V V)\,\mathrm{d}\Omega\,\mathrm{d}(\Delta\nu). \tag{11.16}$$

This formal solution can be used together with the statistical equilibrium equations to deal with the non-LTE problems.

(iv) Landi Degl'Innocenti (1987) solved the transfer equation of matrix form (11.5) with a linear evolution operator $O(s, s')$ defined in the following way. When it acts on the Stokes vector at the point s', the Stokes vector at s is obtained. In mathematical language we have

$$\mathbf{I}(s) = \mathbf{O}(s,s')\mathbf{I}(s'). \tag{11.17}$$

Using such an operator, he found the following formal solution:

$$\mathbf{I}(s) = \int_{\infty}^{\infty} \mathbf{O}(s,s')\mathbf{K}(s')\mathbf{S}(s')\,\mathrm{d}s'. \tag{11.18}$$

More details can be found in his paper.

The four formal solutions introduced above have their respective merits and shortcomings. The first one can be conveniently developed into an algebraic solution, but it ignores the parameter U and so is not a complete solution. Both the second and fourth solutions are expressed in series and the one of the former converges rather slowly. The third and fourth solutions can be used to treat non-LTE problems and, furthermore, with some assumptions may lead to numerical solutions that will be described below.

(3) *Numerical solutions.* With such solutions as well as an atmospheric model, we may calculate theoretical or synthetic Stokes profiles. By comparing them with the observed ones, some magnetic field information can be secured. In recent years, thanks to the rapid increase of the power of electronic computers, the numerical solutions are being more frequently adopted.

Except for some common methods, such as those of Runge–Kutta and Feautrier, the following ones have also been applied.

(i) The *diagnostic element lambda operator (DELO) method.* This technique, developed by Rees and Murphy (1987) and Rees *et al.* (1989), involves the lambda operator associated with the diagonal element of the absorption matrix. These authors adopted a modified absorption matrix with zeros on its diagonal,

$$\tilde{\mathbf{K}} = \mathbf{K}/\kappa_I - 1, \tag{11.19}$$

and a modified total source function,

$$\tilde{S} = j/\kappa_I. \tag{11.20}$$

Eq. (11.5) can then be rewritten as

$$\frac{\mathrm{d}}{\mathrm{d}\tau}\mathbf{I} - \mathbf{I} = \tilde{\mathbf{K}}\mathbf{I} - \tilde{\mathbf{S}}. \tag{11.21}$$

The formal solution of this equation involves the lambda transform with respect to τ and has the following form,

$$\mathbf{I}(\tau) = \int_{\tau}^{\infty} \exp(\tau' - \tau)[\tilde{\mathbf{S}}(\tau') - \tilde{\mathbf{K}}(T')\mathbf{I}(\tau')]\,\mathrm{d}\tau'. \tag{11.22}$$

The method of lambda iteration can be applied to this formal solution.

(ii) Landi Degl'Innocenti's (1987) numerical solution. On the basis of his formal solution introduced above, one may proceed to a numerical solution by means of the following algorithm:

(a) The path of integration ($-\infty \leq s' \leq s$) is divided into n intervals.

(b) For each interval, the operator $\mathbf{0}_i$ is defined to be

$$\mathbf{0}_i = \exp[(-(s_i - s_{i-1})\mathbf{K}_i]. \tag{11.23}$$

(c) The Stokes parameter $I(s)$ can be evaluated with the following equality:

$$\mathbf{I}(s) = \sum_{i=1}^{n} \mathbf{O}_n \mathbf{O}_{n-1} \cdots \mathbf{O}_{i+1} \mathbf{V}_i. \tag{11.24}$$

Here the vector $\mathbf{V}_i$ is defined by

$$\mathbf{V}_i = \int_{S_{i-1}}^{S_i} \exp[-(s_i - s')\mathbf{K}_i]\mathbf{K}_i \mathbf{S}(s')\, \mathrm{d}s'. \tag{11.25}$$

(iii) Van Ballegooijen 's (1987) method. In order to get a numerical solution for the equations of non-LTE line transfer in magnetic fields, he adopted the model of two-level atoms and used the following equality

$$(1+\varepsilon)S_L = \overline{J} + \varepsilon B_\nu(T_e). \tag{11.26}$$

Here S_L is the line source function, ε is the ratio of collision and spontaneous de-excitation rates and $\overline{J}$ is a generalized mean integrated intensity with contributions from all Stokes parameters. $\overline{J}$ and S_L, are related by a lambda operator, i.e.,

$$\overline{J} = \Lambda_B S_L. \tag{11.27}$$

The subscript B implies that the operator is concerned with the magnetic vector. From the preceding equalities the following integral equation for S_L can be derived

$$[(1+\varepsilon) - \Lambda_B]S_L = \varepsilon B_\nu(T_e). \tag{11.28}$$

Van Ballegooijen recast Eq. (11.5) in terms of the 2×2 complex coherency matrix,

$$\frac{1}{2}\begin{pmatrix} I+Q & U+\mathrm{i}V \\ U-\mathrm{i}V & I-Q \end{pmatrix}.$$

Then Eq. (11.28) can be replaced by a system of linear equations and this may be solved by a standard matrix method.

Among the above-mentioned numerical solutions, that of van Ballegooijen is the most complete and strict one, yet it is too complicated. In comparison, the DELO method is simpler and easier to handle.

(4) *The perturbative solution.* By using a perturbative method the Landi Degl'Innocentis (1973) reduced the system of four coupled linear differential equations to an iterative series of common differential equations. The perturbation parameter has the same order of magnitude as that of the ratio of the Larmor frequency to the Doppler half-width. Hence the method is particularly suitable for weak magnetic fields. Besides, by means of the perturbative solution it is easy to learn the degrees of influence of various factors on the Stokes profiles.

(5) *Approximate solutions in the case of WFA.* The system of equalities (3.23) is the solution found by Stenflo (1985a). Now we want to make a supplement. Jefferies *et al.* (1989) derived the following expressions for the three components of the magnetic vector:

$$\begin{aligned} B\cos\gamma &= \frac{4\pi mc^2}{eg\lambda^2}\frac{V}{\mathrm{d}I/\mathrm{d}(\Delta\lambda)}, \\ B\sin\gamma &= \frac{8\pi mc^2}{eg\lambda^2}\left(\frac{H'}{\nu H''}\right)^{1/2}\left[\frac{(Q^2+U^2)^{1/2}\Delta\lambda}{\mathrm{d}I/\mathrm{d}(\Delta\lambda)}\right]^{1/2}, \\ \varphi &= \frac{1}{2}\tan^{-1}\frac{U}{Q}. \end{aligned} \tag{11.29}$$

Here H' and H'' are, respectively, the first and second derivatives of the Voigt profile with respect to v.

Although this subsection is very long, our descriptions of individual solutions are quite short. For detailed accounts of the methods of various solutions, the interested reader may consult Rees (1987) and Qu Zhong-quan (1992).

11.2 Theory of the Sunspot Magnetic Field

Although sunspots have been observed for many centuries and studied with modern instruments and physical theories for many decades, our knowledge about their nature is still poor. It is no wonder that Parker (1979a) made the remark quoted at the beginning of Chapter 4. Nevertheless, it has been established after all that the two basic characteristics of spots pointed out by him — the *reduced temperature* and the *intense magnetic field* — are not equal in importance and that the former is subordinate to the latter. In other words, it is the strong magnetism that gives rise to the comparatively low temperature. Therefore, in order to achieve a deep understanding of sunspots, we have to study their magnetic fields.

11.2.1 BASIC EQUATIONS

Due to the intense magnetism, the dynamic processes in spots belong to the realm of *magnetohydrodynamics* (MHD). To be precise, we take $\rho = 1.5 \times 10^{-7}$ g cm^{-3} and $v = 2$ km s^{-1}, then we have $\frac{1}{2}\rho v^2 = 3 \times 10^3$ erg cm^{-3}. On the other hand, for $B = 1000$ G, $B^2/8\pi$ may attain 4×10^4 erg cm^{-3}. This shows that the criterion (1.3) is valid for spots. So in the study of spot magnetic fields the fundamental equations of MHD are applicable and what follows is a list of them.

(1) The equation of conservation of mass,

$$\frac{\mathrm{d}\rho}{\mathrm{d}t} + \rho\nabla \cdot \mathbf{v} = 0. \tag{11.30}$$

(2) The equation of conservation of momentum,

$$\rho\frac{\mathrm{d}\mathbf{v}}{\mathrm{d}t} = \rho\mathbf{g} + \frac{1}{c}\mathbf{j} \times \mathbf{B} - \nabla p. \tag{11.31}$$

(3) The equation of conservation of heat energy,

$$\frac{\mathrm{d}U}{\mathrm{d}t} + U\nabla \cdot \mathbf{v} = \mathbf{j} \cdot (\mathbf{E} + \frac{1}{c}\mathbf{v} \times \mathbf{B}) - p\nabla \cdot \mathbf{v} - \nabla \cdot \mathbf{F}. \tag{11.32}$$

(4) The equation of state,

$$p = \frac{k\rho T}{\mu}. \tag{11.33}$$

(5) The equation of currents,

$$\frac{m_e}{n_e e^2}\frac{\partial \mathbf{j}}{\partial t} + \frac{1}{n_e e c}\mathbf{j} \times \mathbf{B} + \frac{1}{\sigma}\mathbf{j} = \mathbf{E} + \frac{1}{c}\mathbf{v} \times \mathbf{B} + \frac{1}{n_e e}\nabla p_e. \tag{11.34}$$

(6) Maxwell's equations,

$$\begin{aligned}
&\frac{\partial \mathbf{B}}{\partial t} = -c\nabla \times \mathbf{E},\\
&\mathrm{j} = \frac{c}{4\pi}\nabla \times \mathbf{B},\\
&\nabla \cdot \mathbf{E} = 4\pi q,\\
&\nabla \cdot \mathbf{B} = 0.
\end{aligned} \tag{11.35}$$

The symbols are commonly used ones, and their meanings are not explained here. On the basis of these equations we now proceed to discuss some problems concerned with spot magnetic fields.

11.2.2 THE ORIGIN AND DECAY OF THE SUNSPOT MAGNETIC FIELD

The basic conception of the formation of sunspots is due to Babcock (1961), who suggested that under the action of the *magnetic buoyancy* a flux rope arises from below the solar atmosphere and the two cross sections with the surface become spots with opposite polarities. This idea has been refined and developed by many authors. For instance, according to Piddington (1975a, b; 1981), the magnetic rope consists of many long lived ($\gtrsim$ 1 y) and helically twisted flux tubes which originate beneath the convective zone. Thus the spot magnetic field comes from the solar interior and, as will be stated in the next section, its behavior is closely related to the solar activity cycle. In the following we continue to talk about the question of its decay.

By use of the first, second and fourth Maxwell equations, the simplified current equation, viz.,

$$\mathrm{j} = \sigma(\mathbf{E} + \frac{1}{c}\mathbf{v} \times \mathbf{B}), \tag{11.36}$$

as well as the vectorial formula,

$$\nabla \times (\nabla \times \mathbf{B}) = \nabla\nabla \cdot \mathbf{B} - \nabla^2\mathbf{B},$$

one may get the rate of change of **B** as:

$$\frac{\partial \mathbf{B}}{\partial t} = \nabla \times (\mathbf{v} \times \mathbf{B}) + \frac{c^2}{4\pi\sigma}\nabla^2\mathbf{B}. \tag{11.37}$$

If the matter is stationary, i.e., $v = 0$, then the above equality becomes the *equation of diffusion*,

$$\frac{\partial \mathbf{B}}{\partial t} = \frac{c^2}{4\pi\sigma}\nabla^2\mathbf{B}. \tag{11.38}$$

This is the rate of *natural decay* of the magnetic field due to the *ohmic diffusion*. In order of magnitude we have

$$\nabla^2\mathbf{B} = \frac{\partial^2 B}{\partial x^2} + \frac{\partial^2 B}{\partial y^2} + \frac{\partial^2 B}{\partial z^2} \simeq \frac{B}{l^2}, \tag{11.39}$$

where l is the scale of the object under consideration. It follows from the two equations (11.38) and (11.39) that

$$\frac{B}{t_0} = \frac{c^2}{4\pi\sigma}\frac{B}{l^2}. \tag{11.40}$$

Thus the time constant of decay of the magnetic field is

$$t_0 = \frac{4\pi\sigma l^2}{c^2}. \tag{11.41}$$

The time constant of decay can also be defined to be $t_D = \sigma l^2/c^2$.

If we take for sunspots $l = 1500$ km (this is approximately the scale of fine structures of the spot magnetic field) and $\sigma = 1 \times 10^{11}$ esu, then t_D is about one month. This time interval is longer than the lives of most spots. This tells us that the disintegration of the magnetic field of spots is mainly caused by the mass motion, and not by the ohmic diffusion. In other words, on the right side of Eq. (11.37) the first term is much more important than the second one. In fact, observations in recent years have illustrated that the motions of supergranules can erode the spot magnetic field and lead to the disappearance of the field, piece by piece.

11.2.3 THE COOLING OF SUNSPOTS

As measured by Bray (1981), the radiation flux deficit of the umbra of a large spot is 4.9×10^{10} erg cm^{-2} s^{-1} or 65% of the photospheric flux. Wilson (1981) pointed out that the deficit may be due to the following reasons: (1) a diversion of energy flux away from the spot into the neighboring photosphere, (2) a conversion of thermal flux to either mechanical or magnetic energy, (3) a temporal storage of energy and subsequent release.

In the twenties of this century some astronomers suggested that sunspots could be formed by adiabatic cooling in the ascending motion of gas and that the strong magnetic field were only a byproduct of this process. (For a detailed account see Cowling, 1953.) This hypothesis has long ago been negated, because the life of the spot magnetic field is much longer than that of the spot itself. The first plausible theory of the cooling of spots was proposed by Biermann (1941) and it belongs to the first category mentioned above. He argued that the spot magnetic field can suppress the upward transport of the heat energy carried by convective gas streams and this leads to the decrease of spot temperature. This idea is still popular today, but the question is not so simple. First, it follows from Biermann's model that there should be a bright ring around the spot, yet this is not justified by observation. Second, high resolution heliograms reveal that granules or grains can be discerned even in the umbra (e.g., Sobotka *et al.*, 1993). So we may believe that the convection is at least not completely suppressed in spots.

Parker (1974a, b; 1975) set up a theory that pertains to the second category. He suggested that the spot magnetic field does not suppress but, on the contrary, promotes the heat transfer. This is because energy can be transported away by *Alfvén waves* along magnetic lines of force and this also gives rise to the energy deficit and cooling. In other words, the heat energy is converted into magnetic energy. Parker's theory was criticized by Cowling (1976), who said that the efficiency of Parker's "heat engine" of cooling is too high. However, Parker (1977) did not agree with this comment. Since the theoretical dispute is difficult to clarify, it seems better to carry out observational tests. Beckers (1976) studied the influence of Alfvén waves on the spot spectral lines. He claimed that the observed umbral profiles of Fe I λ5691.505 and Ti I λ5713.896 completely coincide with his theoretical ones calculated with consideration of the effect of Alfvén waves. Nevertheless, we (Ye Shi-hui and Jin Jie-hai, 1983b) found some inexactitudes in Beckers' work and made more accurate computations of the theoretical profiles of Fe I $\lambda\lambda$5691.505 and 5434.534. By comparison with observation we probably achieved a preliminary justification of Parker's theory.

According to the measurements of the cavity radiometer aboard the SMM spacecraftt the solar constant diminishes during the presence of spots and the amount of decrease is proportional to their total area. Since the radiometer measures *all* the solar radiant energy, it is improbable that the missing energy in spots might be transferred from the visible to other spectral regions. As suggested by Zirin (1988), this energy may be temporally stored in the magnetic field of plages. This idea can be considered to belong to the third category.

In conclusion, we have to admit that the problem of sunspot cooling has not been resolved definitively. However, it is important to note that in all the existing theories, no matter which categories they belong to, the cooling mechanism is closely concerned with magnetic fields.

11.2.4 THE STABILITY OF SUNSPOTS

Spots are relatively stable magnetic configurations, and the stability itself is a challenge to solar physicists. As remarked by Piddington (1975a, b), a flux tube with $B_\varphi = 0$ is subject to *flute instability*[1] and will rapidly subdivide into smaller tubes. Parker (1975) also pointed out that spots should be unstable and immediately after formation might break apart. Hence it is meaningful to learn with what configurations the spot magnetic field can remain stable.

The equations of MHD equilibrium are

$$\frac{1}{4\pi}(\nabla \times \mathbf{B}) \times \mathbf{B} - \nabla p + \rho \mathbf{g} = 0, \tag{11.42}$$

$$\nabla \cdot B = 0. \tag{11.43}$$

In addition, we have the equation of state (11.33). Altogether we have five scalar equations, but six unknowns: B_z, B_r, B_φ, p, ρ and T. So the solutions may contain certain arbitrary constants, which remain to be determined with the geometrical characteristics of the spot magnetic field.

As shown by Low (1975; 1980a, b), in the case of axial symmetry the above two equations (11.42) and (11.43) can be reduced to the following scalar one:

$$\frac{\partial^2 \psi}{\partial z^2} + r\frac{\partial}{\partial r}\left(\frac{1}{r}\frac{\partial \psi}{\partial r}\right) + I_A \frac{\mathrm{d}I_A}{\mathrm{d}\psi} = -4\pi r^2 \frac{\partial}{\partial \psi} p(\psi, z), \tag{11.44}$$

where $p(\psi, z)$ is the gas pressure. The magnetic field components and I_A have the following expressions:

$$B_z = \frac{\partial \psi}{\partial r}/r, \qquad B_r = -\frac{\partial \psi}{\partial z}/r, \qquad B_\varphi = I_A/r. \tag{11.45}$$

The function ψ and the magnetic flux Φ are related in the following way:

$$\Phi = 2\pi \int_0^R B_z r \, \mathrm{d}r = 2\pi[\psi(R, z) - \psi(0, z)]. \tag{11.46}$$

[1]This is a kind of macroscopic electrostatic instability in plasmas.

Here R is the radius of the spot. If a certain geometry is chosen for the spot, an exact solution of Eq. (11.44) can be secured.

The simplest geometry was proposed by Schlüter and Temesvary (1958) They assumed that the relative distribution of vertical component of the magnetic field is the same across the flux tube at all depths. This is called the *similarity assumption* and its mathematical expression is

$$\frac{B_z(r,z)}{B_z(0,z)} = \frac{D(\alpha)}{D(0)}, \tag{11.47}$$

where $\alpha = r\xi(z)$. The function $\xi(z)$ describes the dependence on depth of the diameter of the flux tube and $D(\alpha)$ determines the shape. By use of this condition and Eqs. (11.42) and (11.43), we have the following equation for $B_\varphi = 0$,

$$fyy'' - y^4 + 8\pi\Delta p = 0, \tag{11.48}$$

where

$$\Delta p = p(\infty, z) - p(0, z) \tag{11.49}$$

and f is a constant which will be determined below.

Yun (1971a, b; 1972) suggested that between the vertical and azimuthal components there is also a relation of similarity, which gives rise to another term in Eq. (11.48). Then we have

$$fyy'' - y^4 + y^2k^2f + 8\pi\Delta p = 0, \tag{11.50}$$

where k is a constant. If we assume

$$\psi = \psi_0 \exp(-\alpha^2), \tag{11.51}$$

then we have $f = \Phi/2\pi$.

Either Eq. (11.48) or (11.50) was solved successively by Schlüter and Temesvary (1958), Yun (1971 a, b; 1972), Osherovich (1980, 1982, 1984), Low (1980b) *et al.* These authors established some models, which contain reasonable values of the magnetic field strength, temperature etc. and can be used to interpret the Wilson depression. Moreover, with the return-flux model Osherovich (1982, 1984) explained the observed inclination of magnetic field lines and its variation across the spot. (For more details the reader may consult Obridko, 1985.)

11.3 Theory of the Solar Activity Cycle

After briefly introducing the sunspot theory we would now like to emphasize that spots and all other solar active features have a common trend of rise and fall of the activity level and that they all exhibit the well-known *11-year cycle*. (When the interchange of magnetic polarities of spots in both hemispheres is taken into account, we have the *22-year magnetic cycle*.) For a long time now the nature of the *solar activity cycle* has become one of the basic problems of solar physics. Although the puzzle is still far from being completely solved, the existence and variation of the magnetic field is believed to be the main prerequisite. It is nowadays generally accepted that spots and entire active regions are controlled by the weak general magnetic field beneath the solar surface. This field is thought to exist in comparatively shallow layers instead of the deep interior of the Sun. This is because the conductivity of solar plasma is very high and so the natural decay of its magnetic field is extremely slow. Therefore, if the general field permeated the entire Sun, it would be impossible that the polarity reversal takes place in every 11 years.

Walén (1941) and Alfvén (1942) suggested that sunspot magnetic fields might originate in the solar core. By the reasoning just stated, their theories have long ago been negated. The *theory of meridional flow* (Bullard, 1955) was also unsuccessful. Only the *theory of differential rotation* of Babcock (1961) and Leighton (1964, 1969) soon achieved general recognition and is still popular nowadays. The differential or nonuniform rotation of the Sun[2] is caused by the coupling of rotation and convection. Due to its action the lines of the general magnetic field are wound up around the axis of rotation and so, besides the *poloidal* field, there appears a *toroidal* one, which lies below the photosphere. Furthermore, under the action of the magnetic buoyancy the horizontal flux tubes belonging to the toroidal field can float up to the solar surface and then bipolar magnetic regions are formed. The Babcock–Leighton model can satisfactorily interpret the magnetic cycle of solar activity, the butterfly-like distribution of spots in latitudes, the regular reversal of the polarity of the general field, etc. So it is not an exaggeration to say that the Babcock–Leighton theory has become the mainstream in the study of the solar activity cycle. This section is mainly devoted to this theory.

[2]The equatorial region rotates faster than the high-latitude and polar ones, but the distributions of rotational periods with latitudes determined with various methods may differ significantly (see e.g., Stenflo, 1989).

11.3.1 MAGNETIC BUOYANCY AND BIRTH OF SUNSPOTS

Eq. (11.42) can be applied to a horizontal magnetic flux tube, which is in the state of magnetohydrostatic equilibrium. Let us adopt a Cartesian system of coordinates ($OXYZ$) with the axis OZ in the vertical direction and OX parallel to the magnetic field. Then from this equation the following equality can be derived:

$$\frac{\partial}{\partial y}\left(p + \frac{B^2}{8\pi}\right) = 0. \tag{11.52}$$

Integration with respect to y gives

$$p_0 = p_i + \frac{B^2}{8\pi}\,, \tag{11.53}$$

where p_0 and p_i are the pressures outside and inside the flux tube. Because the second term on the right side of the last equation is always positive, we have $p_0 > p_i$. If a thermal equilibrium exists between the matter in the tube and that in the surroundings, then by use of the equation of state (11.33) we have

$$\rho_0 = \rho_i + \frac{\mu}{kT}\frac{B^2}{8\pi}, \tag{11.54}$$

or

$$(\rho_0 - \rho_i)g = \frac{\mu g}{kT}\frac{B^2}{8\pi}. \tag{11.55}$$

The right side represents the force exerted on the magnetic flux tube in a unit volume, and this is called the *magnetic buoyancy*. It can be readily seen in the above equality that the magnetic field plays a much more important role than temperature. If the field is intense enough, flux tubes automatically float upward. It is commonly believed that sunspot magnetic fields originate in the upper part of the convection zone, and the field strength there is about 10^2 G.

11.3.2 DRIFT OF SUNSPOT AREAS

At the beginning of an activity cycle spots of both hemispheres appear in high-latitude regions ($\varphi \simeq 25°$–$30°$). Afterwards spots occur in lower and lower latitudes. Around the activity minimum, the mean latitude becomes 10°–15°. At the end of the 11-year cycle, most spots appear near the equator ($\varphi \simeq 5°$–$15°$). This regularity is called *Spörer's law*.

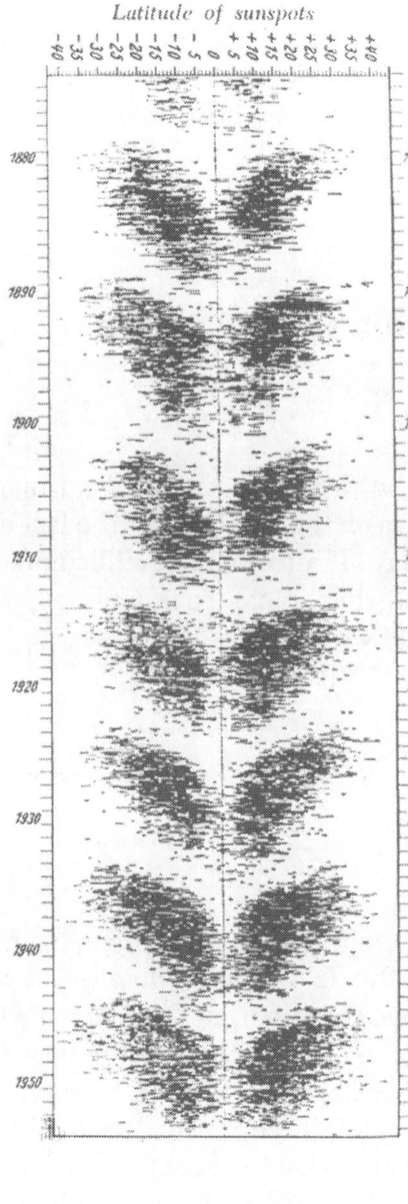

Fig. 11.1. Butterfly diagram.

The change of the mean latitudes of spots with time can be represented by the *butterfly diagram* (Figure 11.1). The interpretation of this diagram may be taken to be a touchstone of the theory of solar activity cycle. Now let us see how this is done by the differential rotation theory.

Babcock suggested that before the differential rotation comes into effect the original solar magnetic field possesses a certain distribution. To be precise, the field strength at latitude φ (B_φ) and that on the equator (B_0) have the following relation:

$$B_\varphi = B_0 f(\varphi), \tag{11.56}$$

where the function of latitude was assumed by Babcock to be

$$f(\varphi) = \sec\varphi. \tag{11.57}$$

Besides, the dependence of the angular velocity of rotation (ξ) of the solar surface on the latitude is

$$\xi = \xi_0 - b\sin^2\varphi. \tag{11.58}$$

Here b is a constant that can be determined from observation. If the winding of magnetic field lines is started up by the differential rotation in the mth year before the beginning of the activity cycle, then in the nth year of the cycle the difference between the angular velocities at the latitude φ and on the equator, i.e. $\Theta(\equiv \xi_0 - \xi)$, is given by

$$\Theta = b(n+m)\sin^2\varphi. \tag{11.59}$$

Under the influence of differential rotation, a magnetic field line, which was originally parallel to the meridian, gradually deviates from this direction and forms an angle ψ with it. Babcock took

$$\tan\psi = \frac{d\Theta}{d\varphi}. \tag{11.60}$$

Then the magnetic field strength becomes

$$B = B_\varphi \sec\psi = B_0 \sec\varphi \sec\psi. \tag{11.61}$$

When ψ is sufficiently large, we may take

$$B = B_0 \sec\varphi \tan\psi. \tag{11.62}$$

From Eqs. (11.59) and (11.60) we get

$$B = 2b(n+m)B_0 \sin\varphi. \tag{11.63}$$

Let us take the value of B at $\varphi = 30°$ and at the beginning of the activity cycle to be the critical field strength B_c (viz., the field intensity that can cause flux tubes to float up to the solar surface and to produce spots), then we have

$$B_c = bmB_0. \tag{11.64}$$

From the preceding two equalities we obtain the relation between the time of appearance of a spot n and its latitude φ to be

$$\sin\varphi = \frac{m}{2(m+n)}\,. \tag{11.65}$$

Taking $m = 3$, we get

$$\sin\varphi = \frac{1.5}{n+3}\,. \tag{11.66}$$

This is just the mathematical expression of Spörer's law. In this way Babcock interpreted the butterfly diagram.

Babcock's work has been improved by some authors. For instance, Kopecky (1963) remarked that Eq. (11.60) is inaccurate, because it does not take into consideration that the longitudinal difference caused by differential rotation depends on latitude. So this equality should be changed to be

$$\tan\psi = \frac{d\Theta}{d\varphi}\cos\varphi. \tag{11.67}$$

Then Eq. (11.65) becomes

$$\sin\varphi\cos\varphi = \frac{\sqrt{3}\,m}{4(n+m)}\,. \tag{11.68}$$

A still more precise formula derived by him (Kopecky, 1970) is

$$B_\varphi \sin\varphi\cos^2\varphi = \frac{m}{n+m} B_{\varphi_c}\sin\varphi_c\cos^2\varphi_c, \tag{11.69}$$

where the index "c" indicates the critical condition.

As suggested by Godoli (1966), if we can determine the drift velocity of spots $v(y) = d\varphi/dn$ by observation, then from the last equality the following expression can be derived:

$$\frac{v(\varphi)}{B_\varphi\cos\varphi}\left[\frac{2(\cos 2\varphi - sin^2\varphi)}{\sin^2 2\varphi} + \frac{1}{\sin 2\varphi}\frac{1}{B_\varphi}\frac{dB_\varphi}{d\varphi}\right] = -\frac{b}{B_c} = \text{const.} \tag{11.70}$$

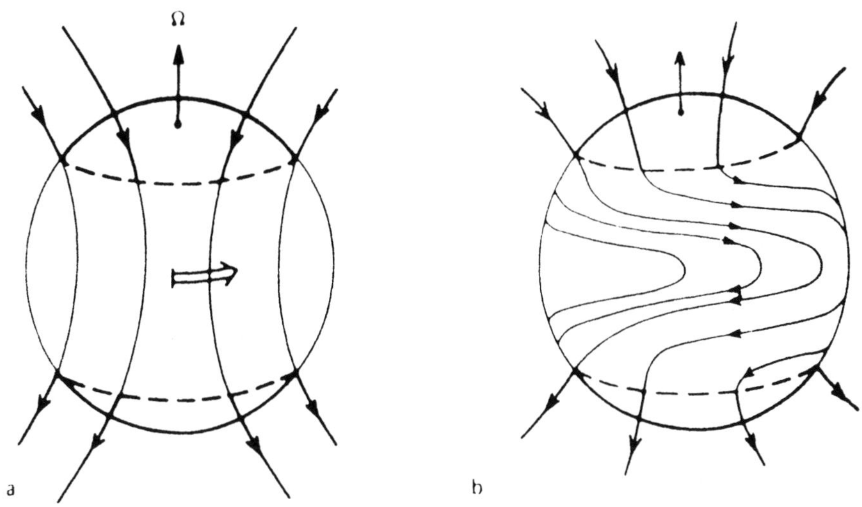

Fig. 11.2. Illustration of Babcock and Leighton's model. (a) Initial poloidal field. (b) Predominantly toroidal field generated by differential rotation.

Otherwise, by use of Eq. (11.56) we have

$$\frac{v(\varphi)}{f(\varphi)\cos\varphi}\left[\frac{1}{2\sin^2\varphi}-\frac{1}{\cos^2\varphi}+\frac{1}{\sin 2\varphi}\frac{1}{f(\varphi)}\frac{\mathrm{d}f(\varphi)}{\mathrm{d}\varphi}\right]$$
$$=-b\frac{B_0}{B_c}=\text{const.} \tag{11.71}$$

These two equations can be utilized to determine the functions B_φ and $f(\varphi)$, i.e., to investigate the distribution of magnetic fields on the solar surface. These equations are called the *Godoli criteria*.

Kopecky (1970) took $B_\varphi = B_0\cos^\alpha\varphi$ and obtained the following expression of Spörer's law:

$$\sin\varphi\cos^{\alpha+2}\varphi=\frac{m}{n+m}\frac{(\alpha+2)^{(\alpha+2)/2}}{(\alpha+3)^{(\alpha+3)/2}}\,. \tag{11.72}$$

From the comparison of theoretical calculation and observation it was found that $\alpha = +1$.

In short, we can state that the interpretation of Spörer's law or the butterfly diagram is an outstanding success of Babcock's theory.

11.3.3 Interpretation of the Solar Magnetic Cycle

(1) *Babcock and Leighton's theory*. The starting point of this theory is that the Sun has a general and poloidal magnetic field (see Figure 11.2a). The differential rotation causes the lines of force of this

field to be more and more densely wound up and then a toroidal field is formed (Figure 11.2b). Although the poloidal field has a strength of only 1–2 G, the toroidal field may be rather strong. Concretely speaking, if the observed value of b is substituted into Eq. (11.63), and letting $m = 3$, then

$$B = 35.2B_0(n + 3)\sin\varphi. \quad (11.73)$$

If we take $B_0 = 5$ G, $n = 0$ and $\varphi = 30°$, then $B = 264$ G $\gg B_0$. The increased magnetic energy comes from a part of the kinetic energy involved in differential rotation.

Being acted upon by the strong buoyancy of the toroidal field, magnetic flux tubes rise to the solar surface and become loops. The two cross-sections of a loop on the surface have oppositely directed lines of force and so different magnetic polarities. They constitute a bipolar group of spots (Figures 11.3). The senses of convolution of magnetic field lines in the northern and southern hemispheres are different, and this can explain the opposite dispositions of spot field polarities in the two hemispheres. The wound up field lines are not exactly parallel to the solar equator. To be precise, the preceding spot is nearer to the equator than the following one. Furthermore, with the elapse of time the magnetic fields of a large number of spots expand, diffuse and neutralize the original poloidal field. Then a new general magnetic field with polarities opposite to those of the initial one is formed. The 22-year magnetic cycle can be interpreted in this way.

Although the Babcock-Leighton model can explain a series of phenomena, it has a vital defect. According to Babcock's (1961) own estimate, the magnetic energy needed for a solar activity cycle is about 10^{36} ergs, and the differential rotation can provide no more than 6×10^{38} ergs. Thus the solar activity could be maintained for only several thousand years. Compared with the age of the Sun (approximately 10^9 years), this is too short.

After Babcock and Leighton, many astronomers have been engaged in the study of the solar cycle, and many notable achievements have been secured. Due to the limitation of space, only a few of them are briefly described in the following.

(2) *Numerical simulation of the evolution of solar magnetic fields.* The continuous and cyclic variations of the toroidal and poloidal fields can be simulated by numerically solving MHD equations with a prescribed velocity field **v**. This work has been done systematically by Yoshimura. In one of his papers (Yoshimura, 1983), he solved

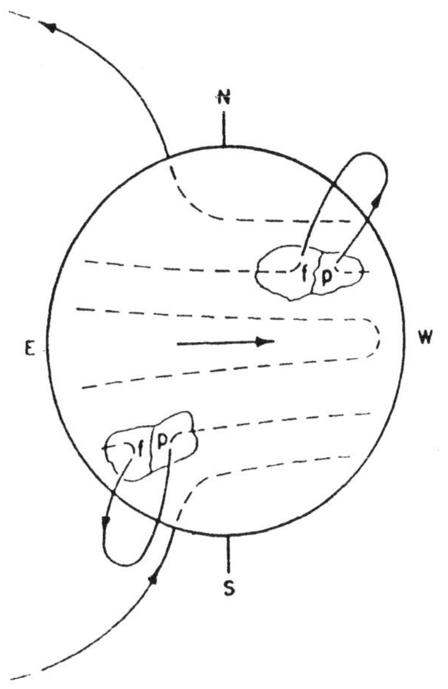

Fig. 11.3. Formation of bipolar magnetic regions.

Eqs. (11.37) and (11.43) with a v representing flows of the differential rotation and global convection in a spherical shell modeled for the convective zone. Then the vivid picture of the perpetual creation and reversal of magnetic flux tubes encircling the Sun may be obtained, and Figure 11.4 is an example. The solid and dotted lines represent, respectively, positive and negative fields. The ordinate is $\sin\varphi$, while the abscissa denotes the time step. The total time span of 7000 steps corresponds to 73 years.

(3) *Dynamo waves*. Gudzenko and Chertoprud (1964) analyzed the Wolf sunspot number W and its rate of change with time W and found that an 11-year cycle can be represented in the phase space $(W, \dot{W})$ by an ellipse-like, two-parameter cycle (Figure 11.5). This led Zeldovich *et al.* (1983) and Ruzmaikin (1990) to interpret the solar cycle with two parameters, viz., the azimuthal field strength B_ϕ, and the azimuthal component A of the vector potential of the poloidal field. These two quantities can be obtained by solving the following equations:

$$\frac{\partial A}{\partial t} = \alpha B_\phi + \nu_T \nabla^2 A, \tag{11.74}$$

$$\frac{\partial B_\phi}{\partial t} = r\sin\theta(\nabla\Omega \times \nabla A)_\phi + \nu_T \nabla^2 B_\phi, \tag{11.75}$$

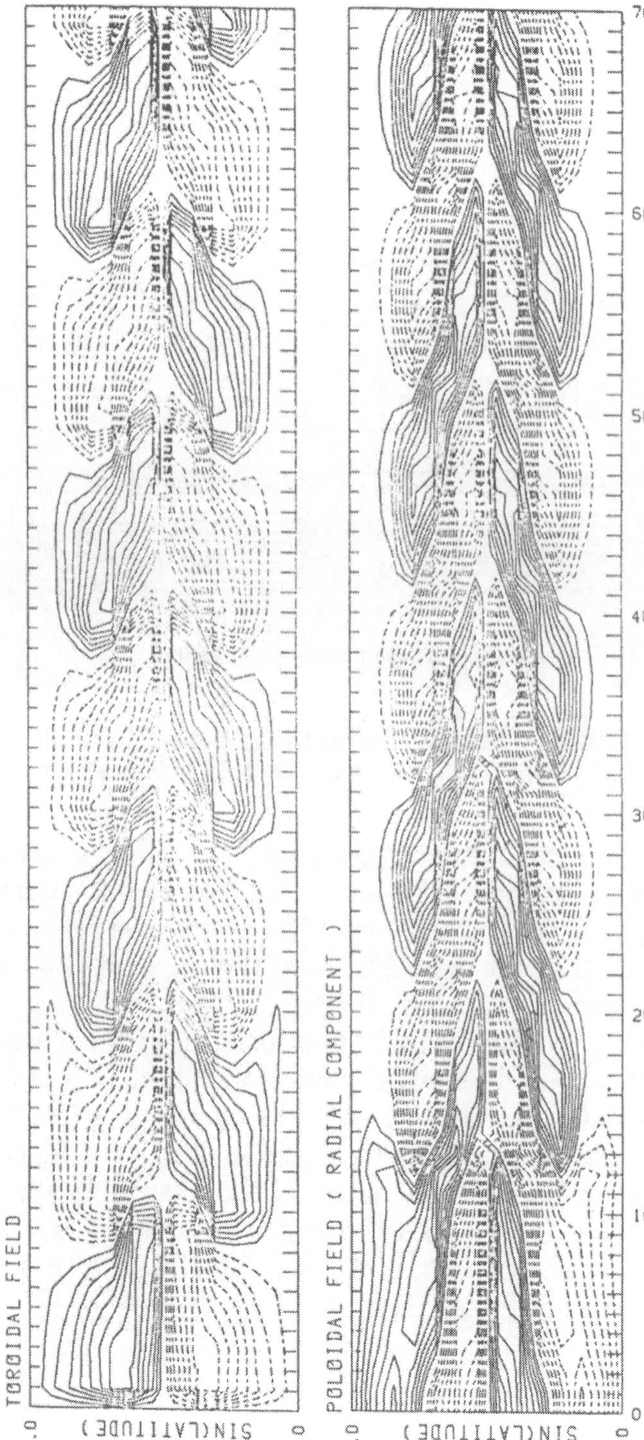

Fig. 11.4. Numerical simulation of evolution of solar toroidal (upper panel) and poloidal (lower panel) magnetic fields.

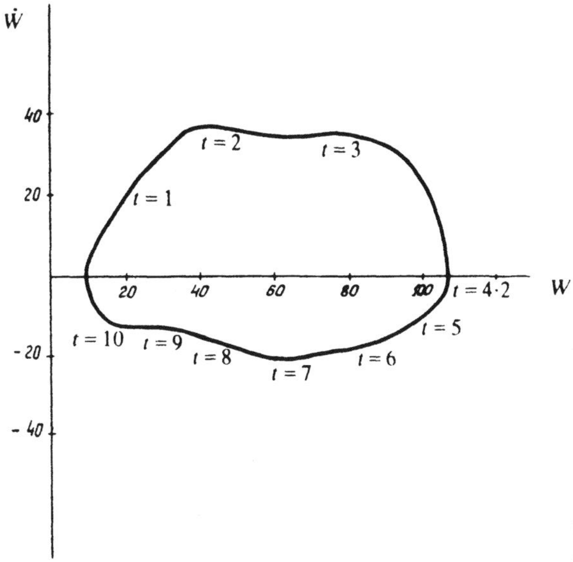

Fig. 11.5. Limit cycle of the observed sunspot number and its rate of change.

where $\nabla\Omega$, α and ν_T represent, respectively, the differential rotation, mean helicity and turbulent diffusion. These equations have wave-like solutions, i.e.,

$$A = A(\omega t - k\theta), \tag{11.76}$$

$$B_\phi = B_\phi(\omega t - k\theta + \delta). \tag{11.77}$$

Such solutions may be thought to represent *dynamo waves*. It is interesting to note that the 22-year magnetic cycle can be interpreted with the dynamo waves from the poles and the equator (see Figure 11.6).

In passing, it is worth mentioning that the investigation of stellar activity cycle has already a good beginning (see, e.g., Baliunas and Vaughan, 1985). The chromospheric and coronal activity of about 100 lower main sequence stars may be magnetic in nature. The RS CVn and BY Dra variables can also provide examples of stellar magnetic activity. Besides, the magnetic activity of stars can be studied through its analogues with the solar activity (Beasley and Cram, 1990).

(4) *High-latitude magnetic fields*. As stated before, in the Babcock–Leighton model of the solar cycle all the flux emergence from the solar interior takes place at low latitudes ($\varphi < 40°$) and the evolution of the high-latitude fields, including the reversal of the polar fields, is exclusively due to the dispersal and surface transport of active-region fluxes. However, Stenflo (1992b) remarks that this model as well as the

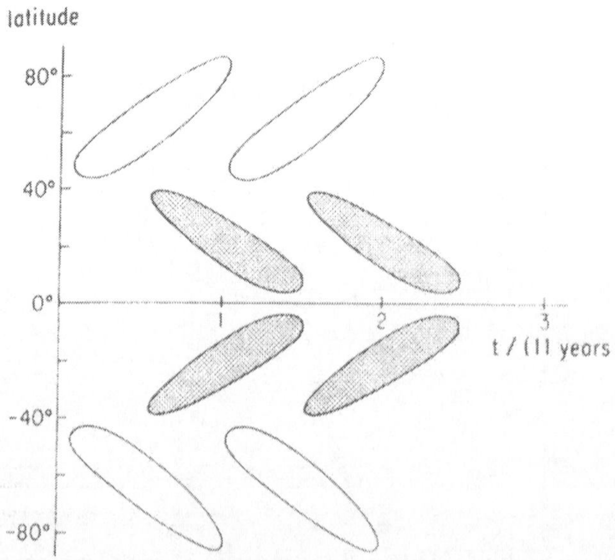

Fig. 11.6. Polar and equatorial dynamo waves produced by mean helicity and differential rotation.

flux transport model (Sheeley *et al.*, 1987; Sheeley, 1992) are unable to reproduce the rotational properties of the high-latitude fields. So he suggests that the high-latitude fluxes must be continuously replenished from sources in the solar interior. This may be regarded as an improvement on the Babcock–Leighton theory.

11.4 Magnetic Fields and the Structure and Evolution of Stars

Stellar magnetic fields are expected to play an important role in the formation, configuration and evolution of stars. This idea is based on the following two reasons. First, as stated in Chapter 9, for magnetic fields "frozen" in plasmas the strength is proportional to the 2/3 power of the density. Since the interior of stars is very dense, the internal magnetic field should be quite intense and may produce high magnetic pressure. This can exert a large influence on the structure of stars. Second, due to the high electric conductivity the natural decay time of the primary stellar magnetism is extremely long. Hence it is possible that the present stellar magnetic field existed in pre-stellar matter. This is the basic concept of the fossil theory. Because magnetism has accompanied the stellar matter for billions of years, it may have certain important effect on the formation and evolution of stars. In order to clarify such an effect, the following problems have been studied.

(1) *The mechanical equilibrium of molecular clouds.* Stars may originate from molecular clouds. In a quasi-stable cloud the self-gravitation should be balanced by some force, such as the thermal pressure, turbulence, centrifugal force and magnetic pressure. Due to the low temperature ($\simeq$ 10 K), the thermal pressure is trivial in comparison with the magnetic one. For a typical case, their ratio is $8\pi\rho v_s^2/B^2 = (v_s/v_A)^2/2 \leq 1/40$ where v_s and v_A are, respectively, the velocity of sound and the Alfvén velocity. Owing to the energy dissipation, the time scale of the existence of turbulence should be much shorter than the age of the cloud. As for the centrifugal force, it can counteract the self-gravitation only for small clouds. Then the unique plausible candidate seems to be the *magnetic force*. Because it is not so easy to dissipate like turbulence, it may be believed to play a major role in the equilibrium of molecular clouds. As argued by Spitzer (1978), the cloud, which is supported by a penetrating magnetic flux Φ_B, has an upper limit of mass given by

$$M_c = \frac{c_1}{\pi}\left(\frac{5}{9G}\right)^{1/2}\Phi_B, \tag{11.78}$$

where c_1 is a correction factor. After its evaluation, the above equality becomes

$$M_c = 1.9 \times 10^4 \frac{(B/1\ \mu\mathrm{G})^3}{(n/1\ cm^{-3})^2} M_\odot. \tag{11.79}$$

For a typical molecular cloud $B_0 \simeq 2.5\ \mu$G and $n = 20$ cm^{-3}, then we have $M_c \simeq 740\ M_\odot$. This value is fairly large. Therefore, it may be thought to be certain that the magnetic field plays a vital role in the equilibrium of molecular clouds. Concretely speaking, in the direction perpendicular to magnetic lines of force, the self-gravitation is overcome by the magnetic pressure, while in the parallel direction by nonlinear Alfvén waves (Shu *et al.*, 1987).

(2) *The stability of stars.* The stable state of a star is maintained mainly by its gravity. But this state may be destroyed by strong magnetic fields. This possibility was probed by Chandrasekhar and Fermi (1953). Taking the magnetic field into consideration, they extended the *virial theorem* to

$$2T + 3(\gamma - 1)U + M + \Omega = 0, \tag{11.80}$$

where T, U and M are, respectively, the kinetic, thermal and magnetic energy, Ω is the potential energy of the gravitational field, and γ is the ratio of two specific heats c_p/c_v. For stationary stars, T can be taken to be equal to zero and the above equality becomes

$$3(\gamma - 1)U + M + \Omega = 0. \tag{11.81}$$

On the other hand, the total energy of the star is

$$E = U + M + \Omega. \tag{11.82}$$

After eliminating U with these two equalities, we get

$$E = -\frac{3\gamma - 4}{3(\gamma - 1)}(|\Omega| - M). \tag{11.83}$$

It then follows that for $\gamma > 4/3$ the condition for the existence of the dynamical stability is

$$M = \frac{1}{8\pi}\iiint |B|^2 \, dx \, dy \, dz = \frac{1}{6}R^3(\overline{B^2}) < |\Omega|, \tag{11.84}$$

where R is the radius of the star. For a spherical star with homogeneous density the potential energy is $\Omega = -\frac{3}{5}(Gm/R)$, where m is the star's mass. When the solar radius and mass are taken to be the units of R and m, respectively, the last equation becomes

$$\sqrt{(\overline{B^2})} < 2.0 \times 10^8 \frac{m}{R^2}. \tag{11.85}$$

This is a criterion for the stability of stars. Once it is not satisfied, viz., when the magnetic energy exceeds the potential one, the star becomes unstable. This is the case for magnetic stars, which are in general not stable and oscillate incessantly (Ledoux, 1967).

(3) *The magnetic braking and transfer of angular momentum.* The angular momentum of molecular clouds is quite large. For a cloud with a mass of 1 $M_\odot$ and a number density of 1 cm^{-3}, the angular momentum gained through its participation in the Galactic rotation may attain 10^{55} $\text{g cm}^2\,\text{s}^{-1}$. This is far greater than that of an ordinary star (which is about 10^{49} $\text{g cm}^2\,\text{s}^{-1}$.) Even for a binary star system with members of mass of 1 $M_\odot$ each and with a period of 100 y, the angular momentum is only about 10^{53} $\text{g cm}^2\,\text{s}^{-1}$. Therefore, in the process of star formation there should be a certain mechanism, which can violate the principle of conservation of angular momentum and causes a heavy loss of this quantity. Now it is generally agreed that this is *magnetic*

braking. Like the Galaxy, the rotation of clouds is also differential and this leads to the distortion of magnetic field lines frozen in plasmas and then to the appearance of Alfvén waves. The propagation of these waves may transfer a part of the internal angular momentum of the cloud to the external medium. The time scale of loss of angular momentum through the magnetic braking is

$$\tau_B = -\frac{\Omega}{\mathrm{d}\Omega/\mathrm{d}t} = \frac{4M}{5(\pi\rho)^{1/2}BR^2} = \frac{8}{15}\frac{\rho_{\rm cl}}{\rho_{\rm ext}}\frac{R}{V_{A,\rm ext}}, \tag{11.86}$$

where $\rho_{\rm cl}$ is the mean density of the cloud, $\rho_{\rm ext}$ is that of the external medium and $v_{A,\rm ext}$ is the Alfvén speed of the external medium. For detailed accounts of this problem the reader may consult Mouschovias (1981) and Deng Shi-bing and Yue Zeng-yuan (1991).

(4) *The ambipolar diffusion and loss of magnetic flux*. For a molecular cloud having a mass equal to that of a common star, the magnetic flux is in 2–5 orders of magnitude larger than that of a magnetic star. To be precise, let us take for the cloud $B_0 = 10^{-6}$ G, $M = 1\ M_\odot$ and $R = 0.1$ pc, then under the condition of conservation of magnetic flux the star with a radius of 1 $R_\odot$ formed by this cloud would have a magnetic field strength of $B \simeq 10^7$ G. However, the average field intensity of magnetic stars is only about 10^3 G. Needless to say, for common stars like the Sun ($B \simeq 1$ G) the difference is still larger. Hence there should be a mechanism that leads to the loss of a large amount of magnetic flux. Now it is generally believed that this is the so-called *ambipolar diffusion*. Its significance can be explained as follows.

The degree of ionization of the interstellar gas is usually very low, 10^{-8}–10^{-6}. In this gas the charged particles may roam randomly in the ocean of neutral ones. The Lorentz force $[(\nabla \times \mathbf{B}) \times \mathbf{B}]/(4\pi)$ can act only upon charged particles, while the neutral ones tend to shrink under the influence of self-gravitation. Then relative displacements arise between the two constituents. In other words, relative to the contracting neutral particles, the charged ones diffuse outwards, and this is called ambipolar diffusion. The molecular clouds, mainly consisting of neutral particles, may continue to contract, but the magnetic fields, permeating ions and electrons, do not shrink in the same proportion. This can lead to a breakdown in the freezing of magnetic fields at a relatively early stage of star formation. In reality, there is no loss of magnetic flux or any violation of its conservation. The essence of the question is that, due to the ambipolar diffusion, only a small portion of the magnetic flux originally contained in a molecular cloud is reserved in the star finally formed.

The time scale of the ambipolar diffusion (Spitzer, 1978) is

$$\tau_{\rm AD} = 5 \times 10^{13} \frac{n_i}{n_{\rm H}} \text{ y.} \tag{11.87}$$

In the expression of $\tau_{\rm AD}$ derived by Mouschovias (1981) there is an additional factor $\xi(r)$. This implies that the rate of ambipolar diffusion varies with the location in the cloud.

(5) *The magnetic field and efficiency of star formation.* The efficiency of formation of stars in our Galaxy is very low. The total mass of Galactic CO clouds is $2.5 \times 10^9\ M_\odot$. The time scale of collapse due to selfgravitation (Deng Shi-bing and Yue Zheng-yuan, 1991) is

$$\begin{aligned}\tau_{\rm ff} &= \left[\frac{3\pi}{32G\rho(0)}\right]^{1/2} \\ &= \frac{4.3 \times 10^7}{[n_{\rm H}(0)]^{1/2}} < 10^6 \text{ y.}\end{aligned} \tag{11.88}$$

Hence the rate of formation of stars should be about $10^3 M_\odot\ {\rm y}^{-1}$. However, the observed rate is merely 3–5 $M_\odot\ {\rm y}^{-1}$ (Shu *et al.*, 1987). The early explanations of the low efficiency with stellar winds and supernova explosions have been found to be unsuccessful (Mouschovias, 1987). Now it is generally believed that the riddle can be solved, again with the magnetic field.

The collapse of a cloud is determined by the gravitation f_g and the magnetic force f_m. As shown by Den Shi-bing and Yue Zeng-yuan (1991), their ratio is

$$f_m/f_g \propto 1/R_e^2 R_p, \tag{11.89}$$

where R_e and R_p are, respectively, the equatorial and polar radii. At a certain stage of collapse both R_e and R_p are so small that the magnetic force can overcome the self-gravitation. Although the collapse ceases at this time, stars can still be formed. This is because in the cloud there may be a nucleus, whose comparatively small dimensions and large density can give rise to ambipolar diffusion. The nucleus gradually contracts to form stars, while the envelope, which contains the major part of the cloud mass, is thrown to the outside. Since only a small portion of the mass of the cloud takes part in star formation, the efficiency is naturally very low.

Space forbids us discussing some additional problems, but the five subjects described above are sufficient to show that the magnetic field plays a vital role in the structure and evolution of stars.

11.5 The Origin of Cosmic Magnetic Fields

Where does the cosmic magnetism, including the geomagnetic field, come from? This question of paramount importance has been studied for a very long time. More than 400 years ago, W. Gilbert, the physician to Queen Elizabeth I of England, asserted that the Earth is a giant lodestone. This idea was disproved near the end of the last century, when it became clear that the interior of the Earth has a temperature higher than the Curie point and so cannot be ferromagnetic. After this the piezoelectric effect, the thermoelectric effect, etc. were applied to explain the terrestrial magnetic field, yet all these attempts failed. A brief historical retrospect on this can be found in Parker (1979a). In what follows we shall introduce some theories which appeared since the turn of the century, and our attention will be turned mainly to the *dynamo theory*.

11.5.1 ROTATOR THEORY

Just 100 years ago Schuster (1892) suggested that there might be a certain physical connection between the magnetic field and the rotation of heavenly bodies. The representative figure of this idea is Blackett (1947), who made a comparison of the Earth, the Sun and 78 Vir.[3] It is interesting to note that in spite of the large differences of their masses, sizes and magnetic field intensities their respective ratios of the magnetic moment (P) and the angular momentum (U) are almost the same. Furthermore, he found that these quantities satisfy the following relation

$$P = \beta \frac{\sqrt{G}}{c} U, \tag{11.90}$$

where β is a constant close to unity. According to his opinion, this equation might reveal the relation between gravitational and electromagnetic fields. Nevertheless, soon afterwards it became clear that this is not true. A large amount of astronomical observations disprove the above relation. In fact, all the magnetic stars are magnetic variables, both in strength and polarity, while their state of rotation can hardly change much. It is still more important to learn that, according to direct physical experiment, the rapid rotation of bodies does not give rise to any significant magnetic effects (Blackett, 1952). Although a

[3]At that time this was the unique star with a known magnetic field strength.

fast rotation may cause the separation of electrons and ions in astrophysical plasmas and then produce magnetic fields, this effect is quite trivial. It is estimated that the strength of fields created in this way might be even smaller than 10^{-13} G (Raychaudhuri, 1972). For these reasons, the rotator theory has long ago been negated and has been almost completely forgotten.

11.5.2 Battery Theory

When charged particles revolve around a radiating massive object, they can be slowed down. This is called the *Poynting–Robertson* effect. Electrons and ions have opposite charges and they are decelerated in opposing directions. Because the Thompson cross section (σ_T) of the electron is $(m_p/m_e)^2$ ($\simeq 3 \times 10^6$) times larger than that of the proton, the deceleration of electrons is very much larger than that of ions. Therefore, on a fast rotating star the ions remain almost unaffected, while the electrons flow in currents and then give rise to magnetic fields (e.g., Cattani and Sacchi, 1966). Because, just like in a battery, the magnetic energy comes from the thermal energy, this process is called the *battery effect*.

Under the action of the Poynting–Robertson effect, the rate of change of the angular momentum L of an electron is given by the equation,

$$\frac{1}{L}\frac{\mathrm{d}L}{\mathrm{d}t} = \frac{L_s}{4\pi R^2}\frac{\sigma_T}{m_e c^2}\,, \tag{11.91}$$

where R is the distance from the source and L_s is its luminosity. Then the work done on the electron in unit time is

$$\frac{\mathrm{d}W}{\mathrm{d}t} = \left(\frac{1}{R}\frac{\mathrm{d}L}{\mathrm{d}t}\right)\frac{L}{Rm_e} = \frac{L_s\sigma_T L^2}{4\pi R^4 m_e^2 c^2}\,. \tag{11.92}$$

The total work that can be done on the cloud of electrons is

$$N\frac{\mathrm{d}W}{\mathrm{d}t} \simeq \frac{L_s}{R^2}\frac{L^2}{m_e^2 c^2}\,, \tag{11.93}$$

where $N \simeq 4\pi R^2/\sigma_T$ is the largest number of electrons that may be slowed down. According to the argument of Harwit (1973), this amount of work is sufficient for the building up of the Galactic magnetic field.

Biermann (1950) first pointed out that in a nonuniformly rotating star quite strong magnetic fields could be generated by the battery effect: about 10^3 G for the Sun and up to 10^7–10^8 G for rapidly rotating stars. However, this theory has some difficulties. For example, Mestel (1967) remarked that a primeval poloidal field of strength 10^{-3} G or even less can suppress the battery effect. Hence it is doubtful that the existing stellar and solar magnetic fields might have been produced by this effect (Parker, 1979a).

11.5.3 FOSSIL THEORY

It can be inferred from Eq. (11.41) that due to the high electrical conductivity of astrophysical plasmas the natural decay time of cosmic magnetic fields should be extremely long. For instance, according to the estimate of Cowling (1945), the decay time of the solar field is about 10^{10} y. Although, by an argument of Wrubel (1952), this value has to be reduced by one order of magnitude, it is still comparable to the Sun's age. The magnetic decay time of the Galactic disk may still be longer: 10^{10}–10^{12} y! Therefore, it is natural to conceive that the existing magnetic fields might be the "fossil" or "relic" of the primordial magnetism of the matter which formed the celestial bodies.

The basic idea of the fossil theory seems to be easily accessible to astronomers, but it has been reproached in some aspects. First, in the process of contraction and before arriving at the main sequence, a star has to pass through a completely *convective phase* (Hayashi, 1961). If the magnetic field is not strong enough, it can be destroyed by the convective motion. This is a critical moment for the survival of the fossil magnetic field. Although some way out of this dilemma may be found (see, e.g. Spitzer, 1957), this is still an unsettled question. Second, the field formed from the fossil magnetism should be basically stationary. However, it is well known that the fields of magnetic stars often display drastic changes and even the weak general field of the Sun reverses its polarity at regular intervals. These facts can hardly be interpreted with the fossil theory. Third, according to the observation of the *3 K background radiation*, the universe is homogeneous and isotropic at large spatial scales. So, if there were an intergalactic fossil magnetic field, it should be also homogeneous. As estimated by Zeldovich *et al.* (1983), the strength of such a field might be 10^{-9} G, which after all could be detected through the Faraday rotation of the radio emission of remote galaxies and quasars. Nevertheless, this has not been justified by observation.

Now, it has to be stressed that in spite of the difficulties described above, the fossil theory seems to be plausible on some occasions. For instance, Dudorov *et al.* (1990) calculated the internal magnetic field of the Sun with this theory and got reasonable results. Dolginov (1988) remarked that the magnetic fields of neutron stars, white dwarfs and chemically peculiar stars might all be fossil ones. Now let us say a few words about the neutron stars. Such a star is generally believed to be the product of a supernova explosion, which leads to the fast compression of the remnant core and the neutralization of nuclear charges. This process may be too short for the magnetic field of the parent star to dissipate and so the remnant remains as a fossil.

11.5.4 DYNAMO THEORY

According to Ampère's law, electric currents can produce magnetic fields. Since the overwhelming part of matter in the universe is in a state of ionization, its motion is equivalent to electric currents and may give rise to magnetic fields. This is like a dynamo, which converts the kinetic energy into electromagnetic energy. Thus all the theories of the origin of magnetic fields based on plasma motions belong to the category of the *dynamo theory*. The earliest theory of this kind was proposed by Larmor (1919), who asserted that the sunspot magnetic field could be generated by the swirling motions around spots. It later became clear that such motions do not exist. Moreover, this argument contradicts the *Cowling* (1934) *theorem*, which will be introduced immediately below. Although Larmor's theory was long ago negated, a lot of other versions of the dynamo theory appeared, one after another, and many books and review papers are devoted to them. Here we would like to recommend the monographs of Moffatt (1978), Parker (1979a), Krause and Rädler (1980) and Zeldovich *et al.* (1983), as well as the reviews of Cowling (1981), Dolginov (1988) and Belvedere (1990). What follows is merely a few essential points of the dynamo theory.

(1) *Non-existence theorems.* One should never think that any kind of plasma motion is able to play the role of a dynamo. Cowling (1934) pointed out that *a magnetic field with axial symmetry cannot be generated or maintained by fluid motions*. This theorem was further extended to some specific circumstances. For example, Bullard and Gellman (1954) and Backus (1958) proved that the dynamo action is impossible in the spherical case with $v_r = 0$. Moffatt (1978) and Zeldovich and Ruzmaikin (1980) arrived at a similar conclusion for the planar case with $v_z = 0$. Hide and Palmer (1982) concluded that an axisymmetrical field cannot be maintained by motion of a compressible fluid. These non-existence theorems are useful for the construction of dynamo models.

(2) *α–ω dynamo*. The basic principle of the dynamo theory can be expressed by the *induction equation*, viz.

$$\frac{\partial \mathbf{B}}{\partial t} = \nabla \times (\mathbf{v} \times \mathbf{B}) - \nabla \times (\eta \nabla \times \mathbf{B}). \tag{11.94}$$

This equality is another form of Eq. (11.37) and here $\eta = c^2/4\pi\sigma$ is the *magnetic diffusivity*. This equation tells us that the change of the magnetic field intensity **B** is due to both its interaction with the velocity field **v** as well as the natural diffusion or decay. In the system of spherical coordinates (r, θ, ϕ) the ϕ-component of the above equation is

$$\frac{\partial B_\phi}{\partial t} - \eta \left(\nabla^2 - \frac{1}{r^2 \sin^2 \theta} \right) B_\phi = B_r \left(\frac{\partial v_\phi}{\partial r} - \frac{v_\phi}{r} \right). \tag{11.95}$$

It can be inferred from this equation that the differential rotation is very important. As a concrete example, let us discuss the *geomagnetic field*. Taking for the Earth's core $\eta \simeq 10^5$ cm^2 s^{-1} and $B_r = 5$ G, Elsasser (1946) solved this equation and obtained $B_\phi \simeq 10^2$ G. This implies that the main component of the terrestrial magnetic field should be recognized to be the toroidal field (B_ϕ) in the interior of the Earth and not the dipolar field (B_p) on its surface. It is a pity that, although we live on the Earth, we cannot directly measure the chief constituent of its magnetic field.

It follows from the above that due to the differential rotation a weak B_p can give rise to a strong B_ϕ. This is called the *ω-process*. Now we have to stress that the inverse process $B_\phi \to B_p$ is also needed, unless the dipolar field would be completely exhausted. This is the *α-process* caused by turbulence. These two reactions constitute the dynamo cycle $B_p \rightleftarrows B_\phi$ and so the magnetic field can exist for a long time.

(3) *Mean field electrodynamics (MFE)*. In this framework the induction equation is solved by adopting

$$\begin{aligned} \mathbf{B} &= \langle \mathbf{B} \rangle + \mathbf{B}', \\ \mathbf{v} &= \langle \mathbf{v} \rangle + \mathbf{v}', \end{aligned} \tag{11.96}$$

where $\langle \mathbf{B} \rangle$ and $\langle \mathbf{v} \rangle$ are, respectively, the mean magnetic and velocity fields and $\mathbf{B}'$ and $\mathbf{v}'$ are the fluctuating ones. Substituting these equalities into Eq. (11.94), we obtain the respective induction equations for the two kinds of fields, i.e.,

$$\frac{\partial \langle \mathbf{B} \rangle}{\partial t} = \nabla \times (\langle \mathbf{v} \rangle \times \langle \mathbf{B} \rangle + \langle \mathbf{E} \rangle) - \nabla \times (\eta \nabla \times \langle \mathbf{B} \rangle), \tag{11.97}$$

$$\frac{\partial \mathbf{B}'}{\partial t} = \nabla \times (\langle \mathbf{v} \rangle \times \mathbf{B}' + \mathbf{v}' \times \langle \mathbf{B} \rangle) + \mathbf{G}) - \nabla \times (\eta \nabla \times \mathbf{B}'), \tag{11.98}$$

where $\langle \mathbf{E} \rangle = \langle \mathbf{v}' \times \mathbf{B}' \rangle$ is a mean electromotive force (EMF) generated by the interaction between the fluctuating velocity and magnetic fields and G is defined to be $\mathbf{v}' \times \mathbf{B}' - \langle \mathbf{v}' \times \mathbf{B}' \rangle$. As a first approximation, we take $\langle \mathbf{v} \rangle = 0$ and $G = 0$, then Eq. (11.98) becomes

$$\frac{\partial \mathbf{B}'}{\partial t} = \nabla \times (\mathbf{v}' \times \langle \mathbf{B} \rangle) - \nabla \times (\eta \nabla \times \mathbf{B}'). \tag{11.99}$$

This is the simplified induction equation for $\mathbf{B}'$. On the other hand, in the case of isotropic turbulence, we have

$$\mathbf{E} = \alpha \langle \mathbf{B} \rangle - \beta \nabla \times \langle \mathbf{B} \rangle, \tag{11.100}$$

and also the MFE induction equation for $\langle \mathbf{B} \rangle$, i.e.,

$$\frac{\partial \langle \mathbf{B} \rangle}{\partial t} = \nabla \times (\langle \mathbf{v} \rangle \times \langle \mathbf{B} \rangle + \alpha \langle \mathbf{B} \rangle) - \nabla \times [(\eta + \beta) \nabla \times \langle \mathbf{B} \rangle]. \tag{11.101}$$

Here β is the turbulent magnetic diffusivity and $\alpha\langle \mathbf{B} \rangle$ is a mean EMF generated by turbulence. This EMF produces an electric current which flows in the direction of $\langle \mathbf{B} \rangle$. This is called the *$\alpha$-effect*. Due to its action, the mean toroidal magnetic field yields a toroidal current and this current can produce a mean poloidal field. This is the α-process $B_\phi \to B_p$ mentioned at the end of the last subsection.

The MFE $\alpha\omega$-theory has proved to be successful in simulating the main characteristics of the solar cycle, such as the butterfly diagram, the polarity reversal of the polar fields, etc. This is the interesting contents of, e.g., Yoshimura (1975, 1983), Stix (1976) and Schüssler (1983).

(4) *Nonlinear dynamo theory*. Dynamo theories can be divided into two categories. What was stated above belongs to the kinematic and linear theory, in which the velocity field v is not affected by the magnetic field B. The theories of this kind are not sufficiently rigorous and have some limitations (Belvedere, 1990). Of another kind is the *hydromagnetic* and *nonlinear* theory, which takes into consideration the back-reaction of B on v via the Lorentz force. The whole system of MHD equations is solved simultaneously with the requirement of internal consistency. As formulated by Meneguzzi *et al.* (1981), the equations are

$$\begin{aligned} &\left(\frac{\partial}{\partial t} - \nu \nabla^2\right) \mathbf{v} = -(\mathbf{v} \cdot \nabla)\mathbf{v} + (\mathbf{B} \cdot \nabla)\mathbf{B} - \nabla p + \mathbf{f}, \\ &\left(\frac{\partial}{\partial t} - \eta \nabla^2\right) \mathbf{B} = -(\mathbf{v} \cdot \nabla)\mathbf{B} + (\mathbf{B} \cdot \nabla)\mathbf{v}, \\ &\nabla \cdot \mathbf{v} = \nabla \cdot \mathbf{B} = 0. \end{aligned} \tag{11.102}$$

Here p is the total pressure (kinetic plus magnetic), ν and η are, respectively, the coefficients of viscosity and magnetic diffusivity, and f is a force that drives the flow. The fluid is assumed to be incompressible. These authors performed direct numerical simulation of the 3-D MHD turbulence with the kinetic and magnetic Reynolds numbers up to 100. They found that the small-scale helical driving can create strong large-scale magnetic fields.

The nonlinear problem has been investigated for several configurations. For instance, a convectively driven dynamo in a spherical shell was studied by Gilman and Miller (1981) and this can be used to model some aspects of the solar dynamo.

In passing, it is worth mentioning that a quasi-steady magnetic field may be generated by nonlinear wave–wave and wave–particle interactions in plasmas (Xiao-qing Li and Yue-hua Ma, 1993)

Before ending this section we want to bring the following three points to the reader's attention:

(1) The solar dynamo is believed to lie below the surface and so cannot be observed directly. However, as suggested by Stenflo (1991), it is feasible to diagnose the dynamo with the observed pattern of surface magnetic fields.

(2) Although the dynamo theory may be said to have become the mainstream in the study of the origin of cosmic magnetism, there are still objections and doubts about its correctness. For example, Piddington (e.g., 1971, 1976) criticized the solar dynamo theory and among his comments we mention only the following. (i) The turbulence may mix magnetic elements, but it does not destroy large-scale fields. So the whole concept of the turbulent diffusion of solar fields is unsound. (ii) The generation of poloidal field from toroidal field by the twisting of field lines in rising eddies would leave the new poloidal field at a higher level than that which it replaced. Hence the field would tend to be lifted out of the Sun. On the basis of such arguments, he and also Layzer *et al.* (1979) proposed their respective alternative theories. (For a detailed account see Cowling, 1981.) Therefore, we agree with Belvedere's (1990) remark, i.e. the "dynamo theory has still some obscure and bright aspects. The next decade will probably decide on its validity."

(3) It is quite probable that there is no universal mechanism of magnetic field generation, and different theories may be applied to various celestial bodies (Dolginov, 1988).

11.6 Magnetic Fields and Cosmic Rays

The cosmic rays are highly isotropic fluxes of relativistic particles coming from the depth of the universe. Their highest energy is about 10^{20} eV per nucleon. This is far larger than the uppermost energy which can be attained nowadays in the physical laboratory. The origin, propagation and mechanism of acceleration of the cosmic radiation are of great importance for the physics of elementary particles, space physics and astrophysics. What we narrate in this book is merely the relation between the cosmic rays and the magnetic field. It was stated in Section 10.1 that the isotropy of cosmic rays is caused by the interstellar magnetic field in the Galaxy. Now it has to be added that the movement and acceleration of cosmic ray particles are also closely concerned with magnetism.

11.6.1 MOTION OF COSMIC RAY PARTICLES

In the ideal case the trajectory of motion of a particle with charge Ze and mass m_i in a magnetic field of strength B is a spiral around the line of force. The angular velocity ω and radius r of the circle projected by the trajectory on a plane perpendicular to the magnetic line of force can be calculated with the following expressions:

$$\begin{aligned} \omega &= \frac{ZeB}{m_i c}, \\ r &= \frac{v_\perp}{\omega} = \frac{m_i c v_\perp}{ZeB} = \frac{m_i c v}{ZeB}\sin\theta. \end{aligned} \tag{11.103}$$

Here θ is the angle formed by the line of force and the direction of motion. These formulae are valid for nonrelativistic particles. In the relativistic case, i.e., for particles moving with velocities close to that of light, the above equalities should be modified to be

$$\begin{aligned} \omega &= \frac{ZeB}{m_i c}\sqrt{1-\frac{v^2}{c^2}} = \frac{ZeBc}{E}, \\ r &= \frac{c\sin\theta}{\omega} = \frac{E\sin\theta}{300ZB}. \end{aligned} \tag{11.104}$$

The unit of the particle's energy is the electron volt. Now it should be pointed out that the real situation is much more complicated. When the magnetic field is inhomogeneous, the charged particles not only move along spirals, but also drift at large angles with magnetic lines of force. Generally speaking, the amount of drift is small. If we take the size of the inhomogeneous structure of the magnetic field to be $l = 30$ pc and let $B = 10^{-5}$ G and $E = 10^{10}$ eV, then in the long period of 10^{17} s ($\simeq 3 \times 10^9$ y) a proton can travel through a total distance of 10^9 pc, but the drift is merely 30 pc (Kaplan and Pikel'ner, 1963).

11.6.2 Acceleration of Cosmic Ray Particles

How can elementary particles be accelerated to energies as high as 10^{20} eV? This is a problem of vital importance for the physics of cosmic rays. For a long time now, numerous theories have been set up, and they can be divided into the following two categories.

(1) *Concomitant acceleration and propagation* (CAP). The cosmic ray particles are continuously accelerated during their propagation through the interstellar medium. To this kind belong the following theories.

(i) Fermi statistical mechanism. In the turbulent motion of the interstellar plasma, charged particles may move to and fro in the inhomogeneous magnetic fields frozen in the plasma. In other words, particles undergo multiple reflections between *magnetic mirrors*. In this process they can gain or lose energy through mutual collisions. So part of the particles are accelerated step by step (Fermi, 1949; Jokippi, 1977). For relativistic particles, the rate of change of energy is

$$\frac{\mathrm{d}E}{\mathrm{d}t} = \frac{U^2}{lc}E, \tag{11.105}$$

where $U = (E \sin^2\theta / ZBl) \times 10^8$ is the drift velocity and l is the scale of interstellar inhomogeneous structures.

(ii) Stochastic particle acceleration. The cosmic ray particles may also be accelerated by the stochastic resonant scattering of hydromagnetic waves and by shock waves. On the other hand, particles lose momentum in some interactions with the ambient matter. All these processes can be treated quantitatively. For instance, due to the interaction with an isolated shock wave the momentum of a charged particle changes with the rate

$$\dot{p} = \frac{1}{3}(U_1 - U_2)\left(\frac{K_1}{U_1} + \frac{K_2}{U_2}\right)^{-1} p, \tag{11.106}$$

where U_1 (U_2) and K_1 (K_2) are, respectively, the flow velocities and spatial diffusion coefficients upstream (downstream) of the shock. When all the relevant processes are taken into consideration, a transport equation can be set up. The particle spectra obtained from solutions of this equation agree with observation, and this is in favor of the continuous acceleration models (Schlickeiser, 1986).

(2) *Sequential acceleration and propagation* (SAP). The cosmic ray particles can at first be rapidly accelerated by discrete sources and then they escape into interstellar space. Such discrete sources may be taken to be young supernova remnants, binary neutron stars, pulsars, flare stars, etc. For lack of space, we now only briefly mention the acceleration by binary neutron stars. If there is a sufficiently strong coupling between the corotating magnetosphere and the impinging plasma, a large amount of the rotational energy of neutron stars can be expended in the acceleration of beams of cosmic ray particles. As described by Kundt (1986), this is like the sparking of a grindstone. The highest energy could not much exceed 10^{20} eV. Besides, the observed ionic cosmic-ray spectrum might be formed from a superposition of the contributions of fewer than 10^6 young binary neutron stars.

As a compromise between these two categories, there may be a hybrid scenario: the injection and acceleration at localized sites could be followed by additional acceleration in interstellar space (Bell, 1978; Blandford and Ostriker, 1980).

At present it is still unclear which kind of theory is closer to reality. According to the argument of Fransson and Epstein (1980), the SAP model is better. This is because, in the framework of the CAP model, these authors calculated the abundance ratio of predominantly secondary nuclei (Li + Be + B + N) to predominantly primary nuclei (C + O) and found that the result disagrees with observation. Similar conclusions were obtained by Cowsik (1980) and Eichler (1980). However, Schlickeiser (1986) has the following point of view. If the momentum dependence of escape lifetimes and the different source distributions of primary and secondary nuclei are adequately considered, the observed ratio can also be interpreted with the CAP theories. Although much still remains to be done in the study of the acceleration of cosmic ray particles, it may be said to be certain that the acceleration mechanisms are closely connected with the cosmic magnetic field.

References

Abdusamatov, H. I. and Krat, V. A.: 1970, Solar Phys. 14, 132.
Abetti, G.: 1932, Publ. R. Osser. Arceti 50, 47.
Abramenkov, E. A. and Krymkin, V. V.: 1990, in: Galactic and Intergalactic Magnetic Fields, ed. R. Beck, P. P. Kronberg and R. Wielebinski, Kluwer Acad. Publ., Dordrecht, p. 49.
Adam, M. G.: 1985, Solar Phys. 96, 27.
Adam, M. G.: 1990, Solar Phys. 125, 37.
Ai Guoxiang and Hu Yuefeng: 1986, Acta Astron. Sin. 27, 173.
Aitken, D. K., Roche, P. F., Bailey, J. A., Briggs, G. P., Hough, J. H. and Thomas, J. A.: 1986, Mon. Not. R. Astron. Soc. 218, 363.
Akasofu, S.-I., Lee, L.-H. and Saito, T.: 1991, Planet. Space Sci. 39, 1259.
Alfvén, H.: 1937, Z. Phys. 105, 579; 107, 319.
Alfvén, H.: 1942, Ark. f. Mat. Astron. och. Fys. 29B, No. 2.
Alfvén, H.: 1950, Cosmical Electrodynamics, Clarendon Press, Oxford.
Aller, M. F.: 1971, Sky Telesc. 41, 220.
Altschuler, M. D. and Newkirk, G., Jr.: 1969, Solar Phys. 9, 131.
Ambrož, P.: 1989, in: Solar Magnetic Fields and Corona, ed. R. B. Teplitskaja, Nauka, Novosibirsk, p. 138.
Anantharamaiah, K. R. and Pedlar, A.: 1990, in: Galactic and Intergalactic Magnetic Fields, ed. R. Beck, P. P. Kronberg and R. Wielebinski, Kluwer Acad. Publ., Dordrecht, p. 375.
Anderson, K. A., Lin, R. P., McGuire, R. E. and McCoy, J. E.: 1975, Space Sci. Instrum. 1, 439.
Andreassian, R. R. and Makarov, A. N.: 1990, in: Galactic and Intergalactic Magnetic Fields, ed. R. Beck, P. P. Kronberg and R. Wielebinski, Kluwer Acad. Publ., Dordrecht, p. 54.
Angel, J. R. P. and Landstreet, J. D.: 1971, in: White Dwarfs, ed. W. J. Luyten, D. Reidel Publ. Co., Dordrecht, p. 79.
Antalova, V.: 1961, Bull. Astron. Inst. Czech. 12, 108.
Arena, P., Landi Degl'Innocenti, E. and Noci, G.:1990, Solar Phys. 129, 259.
Athay, R. G.: 1976, The Solar Chromosphere and Corona: Quiet Sun, D. Reidel Publ. Co., Dordrecht.
Athay, R. G., House, L. L. and Newkirk, G., Jr. (eds.): 1972, Line Formation in the Presence of Magnetic Fields, NCAR, Boulder, Colorado.
Auer, L. H., Heasley, J. H. and House, L. L.: 1977, Solar Phys. 55, 47.
Baade, W. and Zwicky, F.: 1934, Phys. Rev. 45, 138.
Babcock, H. D.: 1959, Astrophys. J. 130, 364.
Babcock, H. W.: 1947, Astrophys. J. 105, 105.

Babcock, H. W.: 1953, Astrophys. J. 118, 387.
Babcock, H. W.: 1958a, Astrophs. J. Suppl. Ser. 3, No. 30.
Babcock, H. W.: 1958b, Astrophys. J. 128, 228.
Babcock, H. W.: 1960a, Astrophys. J. 132, 521.
Babcock, H. W.: 1960b, in: Stellar Atmospheres, ed. J. L. Greenstein, Univ. Chicago Press, p. 282.
Babcock, H. W.: 1961, Astrophys. J. 133, 572.
Babcock, H. W. and Babcock, H. D.: 1955, Astrophys. J. 121, 349.
Babin, A. N., Kartashova, L. G., Koval, A. N., Ogir, M. B. and Stepanian, N. N.:1974, Izv. Krym. Astrofiz. Obs. 52, 92,
Backus, G. E.: 1958, Ann. Phys. 4, 372.
Bahcall, J. N. and Ulrich, R. K.: 1971, Astrophys. J. 170, 593.
Baliunas, S. L. and Vaughan, A. H.: 1985, Ann. Rev. Astron. Astroph. 23, 379.
Ballester, J. L.: 1984, Solar Phys. 94, 151.
Ballester, J. L. and Kleczek, J.: 1984, Solar Phys. 90, 37.
Baranovskij, E. A.: 1960, Izv. Krym. Astrofiz. Obs. 23, 304.
Baranovskij, E. A. and Stepanov, V. E.: 1959, Izv. Krym. Astrofiz. Obs. 21, 180.
Baur, T., Curtis, G. W., Hull, H. and Rush, J.: 1974, in: Planets, Stars and Nebulae Studied with Photopolarimetry, ed. T. Gehrels, Univ. of Arizona Press, Tucson, p. 246.
Baur, T. G., Elmore, D. E., Lee, R. H., Querfeld, C. W. and Rogers, S. R.: 1981, Solar Phys. 70, 395.
Baur, T. G., House, L. L. and Hull, H. K.: 1980, Solar Phys. 65, 111.
Beasley, A. J. and Cram, L. E.:1990, Solar Phys. 125, 191.
Beck, R.: 1987, in: Magnetic Fields and Extragalactic Objects, ed. E. Asseo and D. Grésillon, Editions de Physique, France, p. 113.
Beck, R. and Gräve, R. (eds.): 1987, Interstellar Magnetic Fields, Springer-Verlag.
Beck, R., Kronberg, P. P. and Wielebinski, R. (eds.): 1990, Galactic and Intergalactic Magnetic Fields, Kluwer Acad. Publ., Dordrecht.
Beckers, J. M.: 1968a, Solar Phys . 3, 258.
Beckers, J. M.: 1968b, Solar Phys. 3, 367.
Beckers, J. M.: 1968c, Solar Phys. 5, 15.
Beckers, J. M.: 1969, Solar Phys. 9, 372.
Beckers, J. M.: 1972, Ann. Rev. Astron. Astrophys. 10, 73.
Beckers, J. M.: 1976, Astrophys. J. 203, 739.
Beckers, J. M. and Schröter, E. H.: 1968, Mitt. Astron. Ges. No. 25, 197.
Beckers J. M. and Schröter, E. H.: 1969a, Solar Phys. 7, 22.
Beckers, J. M. and Schröter, E. H.: 1969b, Solar Phys. 10, 384.
Behannon, K. W. and Neubauer, F. M.: 1981, in: Solar Wind Four, Report MPAE-W100-81-31, p. 179.
Bell, A. R.: 1978, Mon. Not. R. Astron. Soc. 187, 142.
Belvedere, G.: 1989, Accretion Disks and Magnetic Fields in Astrophysics, Kluwer Acad. Publ., Dordrecht.
Belvedere, G.: 1990, Mem. Soc. Astron. Ital. 61, 273.
Biermann, L.: 1941, Vierteljahr. Astron. Ges. 76, 194.
Biermann, L.: 1950, Z. Naturforsch. 5a, 65.
Biermann, L.: 1951, Z. Astrophys. 29, 274.
Bigelow, F. H.: 1889, Contributions to Knowledge No. 691, Smithsonian Institution.
Bioschot, A. and Clavelier, B.: 1967, Astrophys. Lett. 1, 7.
Blackett, P. M. S.: 1947, Nature 159, 658.
Blackett, P. M. S.: 1952, Phil. Trans. R. Soc. London A245, 309.

Blandford, R. D. and Ostriker, J. P.:1980, Astrophys. J. 237, 793.
Bogod, V. I. and Gelfreich, G. B.: 1980, Solar Phys. 67, 29.
Bogod, V. I., Vial'shin, G. F., Gelfreich, G. B. and Petrova, N. S.: 1982, Solnechnye Dannye Byull. No. 1, 104.
Bolton, J. G. and Wild, J. P.: 1957, Astrophys. J. 125, 296.
Bommier, V.: 1986, in: Coronal and Prominence Plasmas, ed. A. I. Poland, NASA CP-2442, p. 209.
Bommier, V. and Sahal-Bréchot, S.: 1978, Astron. Astrophys. 69, 57.
Börner, P. and Kneer, F.: 1992, Astron. Astrophys. 259, 307.
Borra, E. F.: 1980, Astrophys. J. 235, 915.
Borra, E. F. and Vaughan, A. H.:1976, Astrophys. J. 210, L145.
Borra, E. F., Landstreet, J. D. and Mestel, L.: 1982, Ann. Rev. Astron. Astrophys. 20, 191.
Boyer, D. W. and Levy, E. H.: 1984, Astrophys. J. 277, 848.
Bray, R. J.: 1981, Solar Phys. 69, 3.
Bray, R. J. and Loughhead, R. E.: 1964, Sunspots, Chapman and Hall, London.
Bray, R. J. and Loughhead, R. E.: 1974, The Solar Chromosphere, Chapman and Hall, London.
Bray, R. J., Loughhead, R. E. and Durrant, C. J.: 1984, The Solar Granulation, Cambridge Univ. Press.
Broxon, J. W.: 1942, Phys. Rev. 65, 521.
Bruzek, A.: 1960, Z. Astrophys. 50, 110.
Bugoslavskaja, E. Ya.: 1950, Works of Sternberg State Astron. Inst. 19, 144.
Bullard, E. C.: 1955, in: Vistas in Astronomy, vol. 1, ed. A. Beer, Pergamon Press, p. 685.
Bullard, E. C. and Gellman, H.: 1954, Phil. Trans. R. Soc. (London) A247, 213.
Bumba, V.: 1958, Izv. Krym. Astrofiz. Obs. 19, 105.
Bumba, V.: 1960a, Izv. Krym. Astrofiz. Obs. 23, 212.
Bumba, V.: 1960b, Izv. Krym. Astrofiz. Obs. 23, 253.
Bumba, V.: 1960c, Izv. Krym. Astrofiz. Obs. 23, 277.
Bumba, V.: 1961, Bull. Astron. Inst. Czech. 12, 82.
Bumba, V. and Howard, R.: 1969, Solar Phys. 7, 28.
Bumba, V. and Topolova-Ruzickova, B.: 1962, Bull. Astron. Inst. Czech. 13, 35.
Bumba, V., Howard, R. and Smith, S. F.:1967, in: The Magnetic and Related Stars, ed. R. C. Cameron, Mono Book Corp. , Baltimore, p. 131.
Cameron, A. G. W. and Canuto, V.: 1974, in: Astrophysics and Gravitation, Univ. de Bruxelles, p. 221.
Cameron, R. C. (ed.): 1967, The Magnetic and Related Stars, Mono Book Corp. , Baltimore.
Campos, L. M. B. C.: 1984, Mon. Not. R. Astron. Soc. 207, 547.
Canfield, R. C. and Fisher, R. R.: 1976, Astrophys. J. 210, L149.
Canfield, R. C., Priest, E. R. and Rust, D. M.: 1974, in: Flare-Related Magnetic Field Dynamics, ed. Y. Nakagawa and D. M. Rust, NCAR, Boulder, Colorado, p. 361.
Cantu, A. M., Godoli, G. and Poleto, G.:1968, Mem. Soc. Astron. Ital. 38, 367.
Canuto, V. (ed.): 1975, Role of Magnetic Fields in Physics and Astrophysics, New York Academy of Sciences.
Cattani, D. and Sacchi, C.: 1966, Nuovo Cimento 46B, 8046.
Chandrasekhar, S.: 1950, Radiative Transfer, Clarendon Press, Oxford.
Chandrasekhar, S. and Fermi, E.: 1953, Astrophys. J. 118, 116.
Chapman, G. A. and Sheeley, N. R.: 1968, Solar Phys. 5, 442.

Chapman, S.: 1957, Smithsonian Contrib. Astrophys. 2, 1.
Chapman, S. and Bartels, J.: 1940, Geomagnetism, Oxford Univ. Press.
Chapman, S. and Ferraro, V. C. A.: 1931, Terr. Magn. 36, 77.
Cheng, Chung-Chieh, Phillips, K. J. H. and Wilson, A. M.: 1973, Solar Phys. 29, 383.
Chitre, S. M., Ezer, E. and Stothers, R.: 1973 , Astrophys. Lett. 14, 37.
Cimino, M. (ed.): 1966, Proceedings of the Meeting on Solar Magnetic Field and High Resolution Spectroscopy, Osser. Astron. di Monte Mario, Rome.
Cisowski, S. M. and Hood, L. L.: 1991, in: The Sun in Time, ed. C. P. Sonett, M. S. Giampapa and M. S. Mathews, Univ. of Arizona Press, Tucson, p. 761.
Clark, A. and Johnson, H. K.: 1967, Solar Phys. 2, 433.
Costa, M. E., McCullock, P. M. and Hamilton, P. A.: 1991, in: The Magellanic Clouds, ed. R. Haynes and D. Milne, Kluwer Acad. Publ., Dordrecht, p. 101.
Couturier, P. and Leblanc, Y.: 1970, Astron. Astrophys. 7, 254.
Cowling, T. G.: 1934, Mon. Not. R. Astron. Soc. 94, 39.
Cowling, T. G.: 1945, Mon. Not. R. Astron. Soc. 105, 166.
Cowling, T. G.: 1946, Mon. Not. R. Astron. Soc. 106, 218.
Cowling, T. G.: 1952, Mon. Not. R. Astron. Soc. 112, 527.
Cowling, T. G.: 1953, in: The Sun, ed. G. P. Kuiper, Univ. of Chicago Press, p. 532.
Cowling, T. G.: 1957, Magnetohydrodynamics, Wiley Interscience Publ.
Cowling, T. G.: 1976, Mon. Not. R. Astron. Soc. 177, 409.
Cowling, T. G.: 1981, Ann. Rev. Astron. Astrophys. 19, 115.
Cowsik, R.: 1980, Astrophys. J. 241, 1195.
Davies, R. D., Slater, C. H., Shuter, W. L. H. and Wild, P. A. T.: 1960, Nature, 187, 1088.
Davis, L. and Smith, E. J.: 1990, J. Geophys. Res. 95, 15257.
de Jager. C.: 1957, Bull. Astron. Inst. Netherlands 13, No. 474, 133.
de Jager, C.: 1985, Solar Phys. 98, 267.
Demidov, M. L., Kotov, V. A. and Grigoryev, V. M.: 1990, Izv. Krym. Astrofiz. Obs. 82, 147.
Deming, D.: 1992, in: Programme and Abstracts of IAU Colloquium No. 141, The Magetic and Velocity Fields of Solar Active Regions, Beijing Astron. Obs., p. 46.
Deng Shi-bing and Yue Zeng-yuan: 1991, Progress in Astron. 9, 41.
Deutsch, A. J.: 1954, Trans. Intern. Astron. Union 8, 801.
Deutsch, A. J.: 1958, in: Electromagnetic Phenomena in Cosmical Physics, ed. B. Lehnert, Cambridge Univ. Press, p. 441.
Deutsch, A. J.: 1970, Astrophys. J. 159, 985.
De Vore, C. R.: 1987, Solar Phys. 112, 17.
De Vore, C. R. and Sheeley, N. R.: 1987, Solar Phys. 108, 47.
Dicke, R. H.: 1977, Astrophys. J. 218, 547.
Dicke, R. H.: 1 979, Astrophys. J . 228, 898 .
Doell, R. R., Grommé, C. S. and Thorpe, A. N.: 1970, Science 167, 695.
Dolginov, A. Z.: 1988, Phys. Reports 162, 337.
Dolginov, A. Z.: 1989, Pis'ma Astron. Zh. 15, 89.
Dolginov, A. Z. and Pavlov, G. G.: 1973, Astron. Zh. 50, 762.
Dolginov, Sh. Sh., Eroshenko, E. G. and Zhuzgov, L. N.: 1972, Dokl. Akad. Nauk SSSR, Ser. Mat. Fiz. 207, 1296.
Dolginov, Sh. Sh., Eroshenko, E. G., Zhuzgov, L. N. and Pushkov, N. V.: 1966, Dokl. Alkad. Nauk SSSR, Ser. Mat. Fiz. 170, 574.
Dombrovskii;, V. A.: 1973, in: A Course of Astrophysics and Stellar Astronomy, ed. A. A. MIihailov, 3rd ed., vol. 1, ch. 26.
Domke, H. and Staude, J.: 1973, Solar Phys. 31, 279.

Dryer, M.: 1987, in: The Solar Wind and the Earth, ed. S.-I. Akasofu and Y. Kamide, Terra Sci. and D. Reidel, p. 19.
Dudorov, A. E., Krivodulskij;, V. N., Ruzmaikin, A. A. and Ruzmaikin, T. V.: 1990, in: Solar Photosphere: Structure, Convection and Magnetic Fields, ed. J. O. Stenflo, Kluwer Acad. Publ., Dordrecht, p. 391.
Dudorov, A. E., Krivodulskij, V. N. Ruzmaikin, T. V. and Ruzmaikin, A. A.: 1989, Astron. Zh. 66, 809.
Duric, N.: 1990, in: Galactic and Intergalactic Magnetic Fields, ed. R. Beck, P. P. Kronberg and R. Wielebinski, Kluwer Acad. Publ., Dordrecht, p. 235.
Dziembowski, W. A. and Goode, P. R.: 1989, Astrophys. J. 347, 540.
Dziembowski, W. A. and Goode, P. R.: 1990, in: Inside the Sun, ed. G. Berthomieu and M. Cribier, Kluwer Acad. Publ., Dordrecht, p. 341.
Eichler, D.: 1980, Astrophys. J. 237, 809.
Elsasser, W. M.: 1946, Phys. Rev. 69, 106.
Encrenaz, T., Bibring, J.-P. and Blanc, M.: 1990, The Solar System, Springer-Verlag.
Evershed, J.: 1909, Kodaikanal Obs. Bull. 2, 63.
Evershed, J.: 1910, Mon. Not. R. Astr. Soc. 70, 217.
Evershed, J.: 1944, Observatory 65, 190.
Fermi, E.: 1949, Phys. Rev. 75, 1169.
Ferrario, L. and Wickramasinghe, D. T.: 1989, in: White Dwarfs, ed. G. Wegner, Springer-Verlag, p. 324.
Ferris, G. A. and Goldsworthy, F. A. (eds.): 1971, Quart. J. R. Astron. Soc. 12, No. 4.
Fransson, C. and Epstein, R. I.: 1980, Astrophys. J. 242, 411.
Frazier, E. N.: 1970, Solar Phys. 14, 89.
Fuller, M.: 1974, Rev. Geoph. Space Phys. 12, 23.
Gehrels, T.: 1974, Planets, Stars and Nebula Studied with Photopolarimetry, Univ. of Arizona Press, Tucson.
Gelfreich, G. B.: 1972, Astron. Tsirkyliar No. 699.
Gelfreich, G. B., Koval, A. N. and Stepanian, N. N.: 1981, in: Solar Maximum Year, 2nd ed., IZMIRAN, p. 193.
Gilman, P. A. and Miller, J.: 1981, Astrophys. J. Suppl. 46, 211.
Ginzburg, V. L.: 1985, Physics and Astrophysics: A Selection of Key Problems, Pergamon Press.
Ginzburg, V. L. and Syrovatskii, S. I.: 1965, Ann. Rev. Astron. Astrophys. 3, 297.
Ginzburg, V. L. and Usov, V. V.: 1972, Pis'ma Zh. Eksp. Teor. Fiz. 15, 280.
Giovanelli, R. G.: 1980, Solar Phys. 68, 49.
Giovannini, G., Kim, E.-T., Kronberg, P. P. and Venturi, T.: 1990, in: Galactic and Intergalactic Magnetic Fields, ed. R. Beck, P. P. Kronberg and R. Wielebinski, Kluwer Acad. Publ., Dordrecht, p. 492.
Glagolevsky, Yu. V. and Kopylov, I. M.: 1988, Magnetic Stars, Nauka, Leningrad.
Godoli, G.: 1966, Observatory 86, 243.
Gold, T.: 1968, Nature 218, 731.
Gold, T. and Hoyle, F.: 1960, Mon. Not. R. Astron. Soc. 120, 89.
Goldreich, P. and Kylafis, N. D.: 1981, Astrophys, J. 243, L75.
Goldreich, P. and Kylafis, N. D.: 1982, Astrophys. J 253, 606.
Golnev, V. Z., Pariisky, Y. N. and Soboleva, N. S.: 1969, Izv. Glavnoi Astron. Obs. in Pulkovo 23, 22.
Goncharskii, A. V., Ryabchikova, T. A., Stepanov, V. V., Khokhlova, V. L. and Yagola, A. G.: 1983, Astron. Zh. 60, 83.
Gopasyuk, S. I., Ogir, M. B., Severny, A. B. and Shaposhnikova, E. F.: 1963, Izv. Krym.

Astrofiz. Obs. 29, 15.
Grebinskij, A. S.: 1989, in: Solar Magnetic Fields and Corona, ed. R. B. Teplitskaya, Nauka, Novosibirsk, p. 132.
Grigoriev, V. M.: 1991, Doctoral Thesis, SibIZMIR, Irkutsk.
Grigoriev, V. M. and Katz, I. M.: 1972, Solar Phys. 22, 119.
Grigoriev, V. M., Osak, E. F., Pflyng, K. and Selivanov, V. L.: 1980, Phys. Solariterr. (Potsdam) No. 14, 81.
Grigoriev, V. M. and Demidov, M. L.: 1991, Solar Phys. 133, 103.
Groth, E. J.: 1975, Astrophys. J., Suppl. Ser. 29, 453.
Gudzenko, L. I. and Chertoprud, V. E.: 1964, Astron. Zh. 41, 697.
Gullahorn, G. E., Isaacman R., Rankin, J. M. and Payne, R. R.: 1977, Astron. J. 82, 309.
Gunn, J. E. and Ostriker, J. P.: 1969, Nature 221, 454.
Gurman, J. B. and House, L. L.: 1981, Solar Phys. 71, 5.
Guseinov, M. J.: 1974a, Izv. Krym. Astrofiz. Obs. 49, 15.
Guseinov, M. J.: 1974b, Izv. Krym. Astrofiz. Obs. 50, 168.
Guthrie, B. N. G.: 1968, Publ. R. Obs. Edinburgh 6, 145.
Hagyard, M. J.: 1 990, Mem. Soc. Astron. Ital. 61, 337.
Hagyard, M. J. and Cumings, N. P.: 1975, NASA TM X-64922.
Hagyard, M. J., West, E. A. and Cumings, N. P.: 1977, Solar Phys. 53, 3.
Hale, G. E.: 1908, Astrophys. J. 28, 100, 315.
Hale, G. E.: 1913, Astrophys. J. 38, 27.
Hale, G. E.: 1914, Publ. Astron. Soc. Pacific 26, 146.
Hale, G. E. and Nicholson, S. B.: 1938, Magnetic Observations of Sunspots 1917–1924, Carnegie Institute of Washington.
Hale, G. E., Ellerman, F., Nicholson, S. B. and Joy, A. H.: 1919, Astrophys. J. 49, 153.
Hale, G. E., Seares, F. H., von Maanen, A. and Ellerman, F.: 1918, Astrophys. J. 47, 206.
Hall, J. S.: 1949, Science 109, 166.
Harvey, J. W.:1973, Solar Phys. 28, 9.
Harvey, J. W. and Livingston, W.: 1969, Solar Phys. 10, 283.
Harvey, J. W. and Tandberg-Hanssen, E.: 1968, Solar Phys. 3, 316.
Harvey, K. L. and Martin, S. F.: 1973, Solar Phys. 32, 389.
Harvey, K. L., Harvey, J. W. and Martin, S.F.: 1975, Solar Phys. 40, 87.
Harvey, K. L. . Livingston, W. C., Harvey, J.W. and Slaughter, C.D.: 1970, in: Solar Magnetic Fields, ed. R. Howard, D. Reidel Publ. Co., Dordrecht, p. 422.
Harwit, M.: 1973, Astrophysical Concepts, John Wiley and Sons, N.Y.
Hatzes, A. P.: 1990, Mon. Not. R. Astron. Soc. 245, 56.
Hatzes, A. P.: 1991, Sky Telesc. 82, 120.
Hatzes, A. P., Penrod, G. D. and Vogt, S. S.: 1989, Astrophys. J. 341, 456.
Hayashi, C.: 1961, Publ. Astron. Soc. Japan 13, 450.
Haynes, R. F., Wayte, S., Harnett, J. I., Klein, U., Wielebinski, R., Buczilowski, U., Bajaja, E., Malin, D., Murray, J. D., Hunt, A. J., Wark, R. and Sciacca, L.: 1990, in: Galactic and Intergalactic Magnetic Fields, ed. R. Beck, P. P. Kronberg and R. Wielebinski, Kluwer Acad. Publ., Dordrecht, p. 205.
Heiles, C.: 1976, Ann. Rev. Astron. Astrophys. 14, 1.
Heiles, C. and Stevens, M.: 1986, Astrophys. J. 301, 331.
Hénoux, J.-C.: 1983, in: Proceedings of Japan-France Seminar on Active Phenomena in the Outer Atmospheres of the Sun and Stars, ed. J.-C. Pecker and Y. Uchida, Obs. Paris-Meudon, p. 200.

Henze, W., Tandberg-Hanssen, E., Hagyard, M. J., Woodgate, B. E., Shine, R. A., Beckers, J. M., Bruner, M., Gurman, J. B., Hyder, C. L. and West, E. A. 1982, Solar Phys. 81, 231.
Hewish, A.: 1962, Trans. Intern. Astron. Union XIB, 431.
Hewish, A., Bell, S. J., Pilkington, J. D. H., Scott, P. F. and Collins, R. A.: 1968, Nature 217, 709.
Hide, R. and Palmer, T. N.: 1982, Geophys. Astrophys. Fluid Dyn. 19, 301.
Hillebrandt, W. and Müller, E.: 1976, Astrophys. J. 207, 589.
Hiltner, W. A.: 1949, Astrophys. J. 109, 471.
Hindman, J. V.: 1963, Aust. J. Phys. 16, 570.
Hirayama, T.: 1985, Solar Phys. . 100, 414.
Högbom, J. A.: 1960, Mon. Not. R. Astron. Soc. 126, 469.
Holzer, T. E.: 1979, in: Solar System Plasma Physics, ed. C. E. Kennel, L. F. Lanzerott and E. N. Parker, North-Holland Publ. Co., vol. 1, p. 101.
Hood, L. L. and Williams, C. R.: 1988, in: Lunar and Planetary Science XIX, Johnson Space Flight Center, NASA, p. 503.
House, L. L.: 1970, J. Quant. Spectrosc. Radiat. Transfer 10, 909, 1171.
House, L. L.: 1971, J. Quant. Spectrosc. Radiat. Transfer 11, 367.
House, L. L. and Steinitz, R.: 1975, Astrophys. J. 195, 235.
Houtgast, T. and van Sluiters, A.: 1948, Bull. Astron. Inst. Netherlands 10, 325.
Howard, R.: 1959, Astrophys. J. 130, 193.
Howard, R.: 1962, Astrophys. J. 136, 211.
Howard, R. (ed.): 1971, Solar Magnetic Fields, D. Reidel Publ. Co., Dordrecht.
Howard, R.: 1989, Solar Phys. 123, 271.
Howard, R.: 1990, Solar Phys. 126, 299.
Howard, R.: 1991a. Solar Phys. 131, 239.
Howard, R.: 1991b, Solar Phys. 131, 259.
Howard, R.: 1991c, Solar Phys. 132, 49.
Howard, R.: 1991d, Solar Phys. 132, 257.
Howard, R.: 1991e, Solar Phys. 134, 233.
Howard, R.: 1991f, Solar Phys. 135, 43.
Howard, R. and Babcock, H. W.: 1960, Astrophys. J. 132, 218.
Howard, R. and LaBonnte, B. J.: 1981, Solar Phys. 74, 137.
Hubenet, H.: 1954, Z. Astrophys. 34, 110.
Hyder, C. L.: 1966, Z. Astrophys. 63, 78.
Ioshpa. B. A. and Obridko, V. N.: 1966, Solnechnye Dannye Byull. No. 5, 68.
Jefferies, J. T. and Mickey, D. L.: 1991, in: Solar Polarimetry, ed. L. J. November, NSO/SacPeak, p. 373.
Jefferies, J. T., Lites, B. W. and Skumanich, A. P.: 1989, Astrophys. J. 343, 920.
Jensen, E.: 1983, Solar Phys. 89, 275.
Jokippi, J. R.: 1977, Proceedings 15th Intern. Cosmic Ray Conf. (Plovdiv), vol. 1, p. 429.
Jones, H. P.: 1985, in: Chromospheric Diagnostics and Modelling, ed. B. W. Lites, NSO/SacPeal, p. 175.
Jones, H. P. and Giovanelli, R. G.: 1982, Solar Phys. 79, 247.
Kahler. S. W. . Cliver, E. W., Cane, H. K., McCuire, R. E., Stone, R.G. and Sheeley, N. R., Jr.: 1986, Astrophys. J . 302, 504.
Kamide, Y. and Slavin, J. A.: 1986, Solar Wind-Magnetosphere Coupling, Terra Sci. and D. Reidel Publ. Co.
Kaplan, S. A. and Pikel'ner, S. B.: 1963, The Interstellar Medium, Nauka, Moscow.
Kawakami, M.: 1983, Publ. Astron. Soc. Japan 35, 459.

Kazes, I. and Crutcher, R. M.: 1986, Astron. Astrophys. 164, 328.
Kemp, J. C.: 1970, Astrophys. J. 162, 169.
Kemp, J. C., Swedlund, J. B. and Evans, B. D.: 1970a, Phys. Rev. Lett. 24, 1211.
Kemp, J. C., Swedlund, J. B., Landstreet, J. D. and Angel, J. R. P.: 1970b, Astrophys. J. 161, L77.
Khokhlova, V. L.: 1985, in: Astrophys. Space Phys. Rev. (Soviet Sci. Rev., Sect. E), ed. R. A. Synyaev, Harwood Acad. Publ., Montreux, p. 99.
Khokhlova, V. L.: 1986, in: Upper Main Sequence Stars with Anomalous Abundances, ed. C. R. Cowley, M. M. Dworetsky and C. Mégessier, D. Reidel Publ. Co., Dordrecht, p. 125.
Kiepenheuer, K. O.: 1953a, in: The Sun, ed. G. P. Kuiper, Univ. of Chicago Press, p. 322.
Kiepenheuer, K. O.: 1953b, Astrophys. J. 117, 447.
Kim, I. S.: 1990, Preprint 11, Middle Asia High Altitude Obs., Sternberg State Astron. Inst., Moscow State Univ.
Kim, I. S., Klepikov, V. Yu., Koutchmy, S., Stepanov, A. I. and Stellmacher, G.: 1988a, Solnechnye Dannye Byull. No. 1, 75.
Kim, I. S., Klepikov, V. Yu., Koutchmy, S., Stepanov, A. I. and Stellmacher, G.: 1988b, Solchnye Dannye Byull. No. 5, 77.
Kim, I. S., Koutchmy, S., Stellmacher, G. and Stepanov, A. I.: 1987, in: The Role of Fine-Scale Magnetic Fields on the Structure of the Solar Atmosphere, ed. E.-H. Schröter, M. Vázquez and A. A. Wyller, Cambridge Univ. Press, p. 289.
Kim, K.-T. and Kronberg, P. P.: 1990, in: Galactic and Intergalactic Magnetic Fields, ed. R. Beck, P. P. Kronberg and R. Wielebinski, Kluwer Acad. Publ., Dordrecht, p. 483.
King, R. B.: 1934, Astrophys. J. 80, 136.
Kinman, T. D.: 1952, Mon. Not. R. Astron. Soc. 112, 425.
Kinman, T. D.: 1953, Mon. Not. R. Astron. Soc. 113, 613.
Kippenhahn, R. and Schlüter, A.: 1957, Z. Astrophys. 43, 36.
Kitai, R., Funakoshi, Y., Nakai, Y., Kurokawa, H. and Makita, M.: 1993, Proceedings of 1st China–Japan Seminar on Solar Physics, eds. G.X. Ai, E. Hiei, Y.J. Ding and Z.X. Ling, Kunming Tondar Institute, p. 84.
Kleczek, J. and Kuperun, M.: 1969, Solar Phys. 6, 72.
Klepikov, V. Yu.: 1989, in: Solar Atmosphere, Interplanetary Medium and Planetary Atmospheres, ed. R. A. Gulyaev, IZMIRAN, Moscow, p. 12.
Kneer, F.: 1973, Solar Phys. 28, 361.
Kodaira, K. and Unno, W.: 1969, Astrophys. J. 157, 769.
Kopecky, M.: 1963, Bull. Astron. Inst. Czech. 14, 231.
Kopecky, M.: 1970, Adv. Astron. Astrophys. 7, 57.
Kotov, B. A.: 1970, Izv. Krym. Astrofiz. Obs. 41–42, 67.
Kotov, B. A.: 1985, Solar Phys. 100, 101.
Koval, A. N. and Stepanian, N. N.: 1983, Izv. Krym. Astrofiz. Obs. 68, 3.
Krall, K. R., Smith Jr., J. B., Hagyard, M. J., West., E. A. and Cumings, N. P.: 1982, Solar Phys. 79, 59.
Krause, F. and Rädler, K.-H.: 1980, Mean-Field Electrodynamics and Dynamo Theory, Akademie-Verlag, Berlin.
Krause, M.: 1990, in: Galactic and Intergalactic Magnetic Fields, ed. R. Beck, P. P. Kronberg and R. Wielebinski, Kluwer Acad, Publ., Dordrecht, p. 187.
Krishnamohan, S.: 1987, in: The Origin and Evolution of Neutron Stars, ed. D. J. Helfand and J.-H. Huang, D. Reidel Publ. Co., Dordrecht, p. 377.
Kuchovicz, B.: 1975, in: Physics of Ap-Stars, ed. W. W. Weiss, H. Jenkner and H. J. Wood, D. Reidel Publ. Co., Dordrecht, p. 169.

Kundt, W.: 1986, in: Cosmic Radiation in Contemporary Astrophysics, ed. M. M. Shapiro, D. Reidel Publ. Co., Dordrecht, p. 57.

Kundu, M. R.: 1990, Mem. Soc. Astron. Ital. 61, 431.

Kuperus, M. and Raadu, M. A.: 1974, Astron. Astrophys. 31, 189.

Küveler, G. and Wiehr, E.: 1985, Astron. Astrophys. 142, 205.

Kylafis, N. D.: 1983, Astrophys. J. 275, 135.

Kylaris, N. D. and Shapiro, P. R.:1983, Astrophys. J. 272, L35.

Lamb, F. and ter Haar, D.: 1971, Phys. Rep. 3C, 253.

Landi Degl'Innocenti, E.: 1979, Solar Phys. 63, 237.

Landi Degl'Innocenti, E.: 1983, Solar Phys. 85, 3.

Landi Degl'Innocenti, E.: 1985, in: Measurements of Solar Vector Magnetic Fields, ed. M. J. Hagyard, NASA CP-2374, p. 279.

Landi Degl'Innocenti, E.: 1987, in: Numerical Radiative Transfer, ed. W. Kalkofen, Cambridge Univ. Press, p. 265.

Landi Degl'Innocenti, E. and Landi Degl'Innocenti, M.: 1972, Solar Phys. 27, 319.

Landi Degl'Innocenti, E. and Landi Degl'Innocenti, M.: 1973, Solar Phys. 31, 299.

Landolfi, M. and Landi Degl'Innocenti, E.: 1982, Solar Phys. 78, 355.

Landolfi, M., Landi Degl'Innocenti, E. and Arena, P.: 1984, Solar Phys. 93, 269.

Landstreet, J. D.: 1979, in: White Dwarfs and Variable Degenerate Stars, ed. H. M. Van Horn and V. Weidemann, Univ. of Rochester Press, N. Y., p. 297.

Landstreet, J. D.: 1980, Astron. J. 85, 611.

Landstreet, J. D.: 1992, Astron. Astrophys. Rev. 4, 35.

Landstreet, J. D., Barker, P. K., Bohlender, D. A. and Jewison, M. S.: 1989, Astrophys. J. 344, 876.

Lang, K. R.: 1992, Astrophysical Data: Planets and Stars, Springer-Verlag.

Lang, K. and Willson, R.: 1986, in: Coronal and Prominence Plasmas, ed. A. I. Poland, NASA CP-2442, p. 353.

Langez, R.: 1936, Publ. Astron. Soc. Pacific 48, 208.

Larmor, G.: 1919, Rep. Brit. Assoc. Adv. Sci., 159.

Latter, W. B., Schmidt, G. D. and Green, R. F.: 1987, Astrophys. J. 320, 308.

Layzer, D., Rosner, R. and Doyle, H. T.: 1979, Astrophys. J. 229, 1126.

Ledoux, P.: 1967, in: The Magnetic and Related Stars, ed. R. C. Cameron, Mono Book Corp., Baltimore, p. 65.

Lee, R. H., Rust, D. M. and Zirin, H.: 1965, Appl. Opt. 4, 1081.

Lehnert, B. (ed.): 1958, Electromagnetic Phenomena in Cosmic Physics, Cambridge Univ. Press.

Leighton, R. B.: 1964, Astrophys. J. 140, 1547.

Leighton, R. B.: 1965, in: Stellar and Solar Magnetic Fields, ed. R. Lüst, North-Holland Publ. Co., p. 158.

Leighton, R. B.: 1968, Astrophys. J. 156, 1.

Leroy, J. L.: 1962, Ann. d'Astrophys. 25, 127.

Leroy, J. L.: 1979, in: Physics of Solar Prominences, ed. E. Jensen, P. Maltby and F. Q. Orrall, Blindern, Oslo, p. 56.

Leroy, J. L.: 1988, in: Dynamics and Structure of Quiescent Solar Prominences, ed. E. R. Priest, Kluwer Acad. Publ., Dordrecht, p. 77.

Leroy, J. L., Bommier, V. and Sahal-Brechot, S.: 1983, Solar Phys. 83, 135.

Levine, R. H.: 1975, Solar Phys. 44, 365.

Liamova, G. V.: 1971, Astron. Tsirkyliar No. 622, 5.

Libbrecht, K. G.: 1989, Astrophys. J. 336, 1092.

Li Jing, Ai Guo-xiang, Zhang Hong-qi and Han Feng:1989, Acta Astron. Sin. 30, 188.

Li Xiao-qing and Ma Yue-hua: 1993, Astron. Astrophys. 270, 534.
Lin Yuanzhang and Wu Fei: 1987, Acta Astrophys. Sin. 7, 312.
Lin Yuanzhang, Wei Xiaolei and Zhang Hongqi: 1993, in: The Magnetic and Velocity Fields of Solar Active Regions, eds. H. Zirin, Guoxiang Ai and Haimin Wang, Astron. Soc. Pacific Conf. Ser. vol. 46, p. 446.
Lites, B. W. and Scharmer, G. B.: 1989, in: High Spatial Resolution Solar Observations, ed. O. von der Lühe, NSO/SacPeak.
Lites, B. W. and Skumanich, A.: 1989, in: High Spatial Resolution Solar Observations, ed. O. von der Lühe, NSO/SacPeak, p. 389.
Lites, B. W., Skumanich, A., Rees, D. E. and Murphy, G. A.: 1988, Astrophys. J. 330, 493.
Litvak, M. M.: 1971, Astrophys. J. 170, 71.
Liu Jiangqiang, Cao Ai, Chen Jimin and Ai Guoxiang: 1991, Publ. Beijing Astron. Obs. No. 18, 54.
Livi, S. H. B., Martin. S. F., Haimin Wang and Guoxiang Ai: Solar Phys. 121, 197.
Livi, S.H.B., Wang, J. and Martin, S. F.: 1985, Austr. J. Phys. 38, 855.
Livingston, W. C.: 1968, Astrophys. J. 153, 929.
Livingston, W. C.: 1973, World Data Center for Solar-Terrestrial Phys. Rep., UAG-28, Part I, p. 95.
Livingston, W. C. and Harvey, J.: 1971, in: Solar Magnetic Fields, ed. R. Howard, D. Reidel Publ. Co., Dordrecht, p. 51.
Livingston, W. C., Harvey, J., Pierce, A. K., Schrage, D., Gillespie, B, Simons, J. and Slaughter, C.: 1976a, Appl. Opt. 15, 33.
Livingston. W. C., Harvey, J., Slaughter, C. and Trumbo, D.: 1976b, Appl. Opt. 15, 40.
Low, B. C.: 1975, Astrophys. J. 197, 251.
Low, B. C.: 1980a, Solar Phys. 65, 147.
Low, B. C.: 1980b, Solar Phys. 67, 57.
Lozitskij, V. G. and Tsap, T. T.: 1990, in: Solar Photosphere: Structure, Convection and Magnetic Fields, ed. J. O. Stenflo, Kluwer Acad. Publ., Dordrecht, p. 125.
Lüst, R. (ed.): 1965, Stellar and Solar Magnetic Fields, North-Holland Publ. Co.
Luyten, W. J. (ed.): 1971, White Dwarfs, D. Reidel Publ. Co., Dordrecht.
Makarov, V. I.: 1983, Solnechnye Dannye Byull. No. 4, 100.
Makarov, V. I., Fatianov, M. P. and Sivaraman, K. S.: 1983, Solar Phys. 85, 215, 227.
Makita, M.: 1963, Publ. Astron. Soc. Japan 15, 145.
Makita, M.: 1979, Publ. Astron. Soc. Japan 31, 575.
Makita, M.: 1986a, Astrophys. Space Sci. 118, 163.
Makita, M. . 1986b, Solar Phys. 103, 1.
Makita, M., Funakoshi, Y. and Hanaoka, Y.: 1991, in: Solar Polarimetry, ed. L. J. November, NSO/SacPeak, p. 198.
Maltby, P.: 1971, in: Solar Magnetic Fields, ed. R. Howard, D. Reidel Publ. Co., Dordrecht, p. 141.
Malville, J. M. and Tandberg-Hanssen, E.: 1969, Solar Phys. 6, 278.
Mariana, F. and Neubauer, F. M.: 1990, in: Physics of the Inner Heliosphere, ed. R. S. Schwenn and E. N. Marsch, Springer-Verlag, p. 183.
Martin, S. F., Bentley, R. D., Schadee, A., Antalova, A., Kučera, A., Dezsö, L., Gesztelyi, L., Harvey, K. L., Jones, H., Livi, S. H. B. and Wang, J.: 1984, Adv. Space Res. 4, 61.
Martin, S. F., Dezsö, L., Antalova, A., Kučera, A. and Harvey, K. L.: 1983, Adv. Space Rev. 2, No. 11, 39.
Martin, S. F., Livi, S. H. B., Wang, J. and Shi, Z.: 1985, in: Measurements of Solar Vector Magnetic Fields, ed. M. J. Hagyard, NASA CP-2374, p. 403.

Martres, M. J., Michard, R., Soru-Iscovici, I. and Tsap, T.: 1968a, in: Structure and Development of Solar Active Regions, ed. K. O. Kiepenheuer, D. Reidel Publ. Co., Dordrecht, p. 318.
Martres, M. J., Michard, R., Soru-Iscovici, I. and Tsap, T.: 1968b, Solar Phys. 5, 187.
Mathewson, D. S. and Ford, V. L.: 1970, Mem. R. Astron. Soc. 74, 139.
Mathewson, D. S., van der Kruit, P. C. and Brouw, W. N.: 1972, Astron. Astrophys. 17, 468.
Mathys, G.: 1989, Fund. Cosmic Phys. 13, 143.
Mathys, G. and Stenflo, J. O.: 1986, Astron. Astrophys. 168, 184.
Mathys, G. and Stenflo, J. O.: 1987, Astron. Astrophys. 171, 368.
Mattig, W.: 1953, Z. Astrophys. 31, 273.
Mayfield, E. B. and Chapman, G. A.: 1981, Solar Phys. 70, 351.
McIntosh, P. S.: 1990, Solar Phys. 125, 251.
Meneguzzi, M., Frisch, U. and Pouquet, A.: 1981, Phys. Rev. Lett. 47, 1060.
Mestel, L.: 1967, in: The Magnetic and Related Stars, ed. R. C. Cameron, Mono Book Corp. , Baltimore, p. 101.
Michard, R.: 1953, Ann. d'Astrophys. 16, 217.
Midgley, J. E. and Davis, L., Jr.: 1963, J. Geophys. Res. 68, 5111.
Mikić, Z. Barnes, D. C. and Schnack, D. D.: 1989, Bull. Am. Astron. Soc. 21, 855.
Ming Chang-rong, Han Feng, Zhang Hong-qi, Ai Guo-xiang and Kong Fan-xi: 1989, Acta Astron. Sin. 29, 346.
Moe, O. K.: 1968, Solar Phys. 4, 267.
Moffatt, H. K.: 1978, Magnetic Field Generation in Electrically Conducting Fluids, Cambridge Univ. Press.
Moffet, A. T.: 1973, in: Galaxies and the Universe, ed. A. Sandage, M. Sandage and J. Kristian, Univ. of Chicago Press, vol. 9, ch. 7.
Mogilevskij, E. I., Demkina, L. B., Ioshpa, B. A. and Obridko, V. N.: 1967, in: Structure and Development of Solar Active Regions, ed. K. O. Kiepenheuer, D. Reidel Publ. Co., Dordrecht, p. 215.
Moore, R. L.: 1981, in: The Physics of Sunspots, ed. L. E. Cram and J. H. Thomas, Sac Peak Obs., p. 259.
Morris, M.: 1990, in: Galactic and Intergalactic Magnetic Fields, ed. R. Beck, P. P. Kronberg and R. Wielebinski, Kluwer Acad. Publ., Dordrecht, p. 361.
Mosher, J. M.: 1976, Big Bear Solar Obs. Preprint No. 159.
Moss, D.: 1987, Mon. Not. R. Astron. Soc. 224, 1019.
Mouschovias, T. Ch.: 1981, in: Fundamental Problems in Theory of Stellar Evolution, ed. D. Sugimoto, D. G. Lamb and D. N. Schram, R. Reidel Publ. Co., Dordrecht, p. 27.
Mouschovias, T. Ch.: 1987, in: Physical Processes in Interstellar Clouds, ed. G. E. Morfill and M. J. Scholer, D. Reidel Publ. Co, Dordrecht, p. 453.
Nagendra, K. W. and Peraiah, A.: 1985, Mon. Not. R. Astron. Soc. 214, 203.
Nakagawa, Y. and Rust, D. M.: 1974, in: Flare-Related Magnetic Field Dynamics, ed. R. C. Canfield, E. R. Priest and D. M. Rust, HAO, Boulder, p. 361
Nakagawa, Y., Wu, S. T. and Tandberg-Hanssen, E.: 1978, Astron. Astrophys. 69, 43.
Newkirk, G., Jr.: 1967, Ann. Rev. Astron. Astrophys. 5, 213.
Newkirk, G., Jr.: 1971, in: Physics of the Solar Corona, ed. C. J. Macris, D. Reidel Publ. Co., Dordrecht, p. 66.
Nicholson. S. B.: 1933, Publ. Astron. Soc. Pacific 45, 51.
Nikolsky, G. M., Kim, I. S. and Koutchmy, S.: 1982, Solar Phys. 81, 81.
Nikulin, N. S.: 1960, Izv. Krym. Astrofiz. Obs. 22, 3.
Nikulin, N. S., Severny, A. B. and Stepanov, V. E.: 1958, Izv. Krym. Astrofiz. Obs. 19, 3.

Nishi, K.: 1962, Publ. Astron. Soc. Japan 14, 315.
Nishi, K. and Makita, M.: 1973, Publ. Astron. Soc. Japan 25, 51.
November, L. J. (ed.): 1991, Solar Polarimetry, NSO/SacPeak.
Obridko, V. N.: 1965, Astron. Zh. 42, 102.
Obridko, V. N.: 1968, Bull. Astron. Inst. Czech. 19, 183.
Obridko, V. N.: 1985, Sunspots and Complexes of Activity, Nauka, Moscow.
Öhman, Y.: 1942, Stockholm Obs. Ann. 14(4), 1.
Omont, A., Smith, E. W. and Cooper, J.: 1973, Astrophys. J. 182, 283.
Oort, J. H.: 1977, Ann. Rev. Astron. Astrophys. 15, 295.
Oppenheimer, J. R. and Volkoff, G. M.: 1939, Phys. Rev. 55, 374.
Osherovich, V. A.: 1980, Solar Phys. 68, 297.
Osherovich, V. A.: 1982, Solar Phys. 77, 63.
Osherovich, V. A.: 1984, Solar Phys. 90, 31.
Osherovich, V. A. and Garcia, H. A.: 1989, Astrophys. J. 336, 468.
Pacini, F.: 1967, Nature 216, 567.
Pacini, F.: 1968, Nature 219, 145.
Parker, E. N.: 1958, Astrophys. J. 128, 664.
Parker, E. N.: 1963, Astrophys. J. 138, 552.
Parker, E. N.: 1974a, Solar Phys. 36, 249.
Parker, E. N.: 1974b, Solar Phys. 37, 127.
Parker, E. N.: 1975, Solar Phys. 40, 275, 291.
Parker, E. N.: 1977, Mon. Not. R. Astron. Soc. 179, 93.
Parker, E. N.: 1979a, Cosmical Magnetic Fields – Their Origin and Activity, Clarendon Press, Oxford.
Parker, E. N.: 1979b, Astrophys. J. 230, 905.
Parker, E. N.: 1985, Astrophys. J. 286, 666, 677.
Patty, S. R.: 1981, in: The Physics of Sunspots, ed. L. E. Cram and J. H. Thomas, Sac Peak Obs., p. 64.
Pickel'ner, S. B.: 1966, Principles of Cosmical Electrodynamics, 2nd ed., Nauka, Moscow.
Pickel'ner, S. B.: 1969, Astron. Zh. 46, 328.
Pickel'ner, S. B.: 1971, Comments on Astrophys. Space Sci. 3, 33.
Pickel'ner, S. B. and Khokhlova, V. L.: 1972, Usp. Fiz. Nauk 107, 389.
Piddington, J. H.: 1971, Proc. Astron. Soc. Aust. 2, 7.
Piddington, J. H.: 1975a, Astrophys. Space Sci. 34, 347.
Piddington, J. H.: 1975b, Astrophys. Space Sci. 35, 269.
Piddington, J. H.: 1976, in: Basic Mechanisms of Solar Activity, ed. V. Bumba and J. Kleczek, D. Reidel Publ. Co., Dordrecht, p. 389.
Piddington, J. H.: 1981, Astrophys. Space Sci. 75, 273.
Pillet, V. M. and Vazquez, M.: 1993, in: The Magnetic and Velocity Fields of Solar Active Regions, eds. H. Zirin, Guoxiang Ai and Haimin Wang, Astron. Soc. Pacific Conf. Ser. vol. 46, p. 60.
Pneuman, G. W. and Kopp, R. A.: 1970, Solar Phys. 13, 176.
Podgorny, A. I.: 1989, Solar Phys. 123, 285.
Preston, G. W.: 1967, in: The Magnetic and Related Stars, ed. R. C. Cameron, Mono Book Corp., Baltimore, p. 3.
Preston, G. W.: 1969a, Astrophys. J. 156, 957.
Preston, G. W.: 1969b, Astrophys. J. 157, 247.
Preston, G. W.: 1969c, Astrophys. J. 158, 243.
Preston, G. W.: 1971, Astrophys. J. 164, 309.

Priest, E.: 1982, Solar Magnetohydrodynamics, D. Reidel Publ. Co., Dordrecht.
Priest, E. R.: 1990, Mem. Soc. Astron. Ital. 61, 383.
Pudovkin, M. I. and Benevolenskaja, E. E., 1984, Astron. Zh. 61, 783.
Pyper, D.: 1969, Astrophys. J., Suppl. Ser. 18, 347.
Qu Zhong-quan: 1992, Master Thesis, Yunnan Astron. Obs.
Rabin, D. M. and Graves, J. E.: 1989, Bull. Am. Astron. Soc. 21, 854.
Rachkovskii, D. N.: 1961a, Izv. Krym. Astrofiz. Obs. 25, 277.
Rachkovskii, D. N.: 1961b, Izv. Krym. Astrofiz. Obs. 26, 63.
Rachkovskii, D. N.: 1962, Izv. Krym. Astrofiz. Obs. 27, 148.
Rachkovskii, D. N.: 1967, Izv. Krym. Astrofiz. Obs. 36, 3.
Rachkovskii, D. N.: 1971, Izv. Krym. Astrofiz. Obs. 43, 190.
Rachkovskii, D. N.: 1972, Izv. Krym. Astrofiz. Obs. 44, 64.
Ramsay, J. V., Giovanelli, R. G. and Gillet, H. R.: 1971, in: Solar Magnetic Fields, ed. R. Howard, D. Reidel Publ. Co., Dordrecht, p. 24.
Rand, R. J. and Kulkarni, S. R.: 1990, in: Galactic and Intergalactic Magnetic Fields, ed. R. Beck, P. P. Kronberg and R. Wielebinski, Kluwer Acad. Publ., Dordrecht, p. 44.
Raychaudhuri, P.: 1972, Astrophys. Space Sci. 18, 425.
Rees, D. E.: 1987, in: Numerical Radiative Transfer, ed. W. Kalkofen, Cambridge Univ. Press, p. 213.
Rees, D. E. and Murphy, G. A.: 1987, in: Numerical Radiative Transfer, ed. W. Kalkofen, Cambridge Univ. Press.
Rees, D. E., Murphy, G. A. and Durrant, C. J.: 1989, Astrophys. J. 339, 1093.
Rees, M. J.: 1984, Ann. Rev. Astron. Astrophys. 22, 471.
Rensen, P.:1963a, Ciel et Terre 79, 68.
Rensen, P.: 1963b, Bull. Soc. R. Sci. Liège 32, 664.
Ribes, E.: 1969, Astron. Astrophys. 2, 316.
Ringnes, T. S. and Janssen, E.: 1960, Astrophys. Norvegica 7, 99.
Robinson, R. D.: 1980, Astrophys. J. 239, 961.
Robinson, R. D., Worden, S. P. and Harvey, J.W.: 1980, Astrophys. J. 236, L157.
Romani, R. W.: 1990, Nature 347, 741.
Ronan, R. S., Orrall, F. Q., Mickey, D. L., West, E. A., Hagyard, M. J. and Balasubramanian, K. S.: 1992, Solar Phys. 138, 49.
Rosenberg, H.: 1972, Solar Phys. 25, 188.
Ruderman, M.: 1971, Phys. Rev. Lett. 27, 1306.
Runcorn, S. K.: 1987, Sci. Am. 257, No. 6, 34.
Runcorn, S. E. (ed.): 1988, The Physics of the Planets – Their Origin, Evolution and Structure, John Wiley and Sons, N. Y.
Russell, C. T.: 1987, in: The Solar Wind and the Earth, ed. S.-I. Akasofu and Y. Kamide, Terra Sci. and D. Reidel, p. 71.
Rust, D. M.: 1976a, Solar Phys. 47, 21.
Rust, D. M.: 1976b, Phil. Trans. R. Soc. (London) A281, No. 1304, 427.
Rust, D. M. and O'Byrne, J. W.: 1991, in: Solar Polarimetry, ed. L. J. November, NSO/SacPeak, p. 74.
Ruzmaikin, A. A.: 1990, in: Solar Photosphere: Structure, Convection and Magnetic Fields, ed. J. O. Stenflo, Kluwer Acad. Publ., Dordrecht, p. 343.
Ruzmaikin, A. A.: 1991, Icarus 93, 82.
Ruzmaikin, A. A., Sokolov, D. D. and Shukurov, A. M.: 1988, Magnetic Fields of Galaxies, Nauka, Moscow. (English translation published in 1988 by Kluwer Acad. Publ., Dordrecht.)
Sakurai, K.: 1987, in: The Solar Wind and the Earth, ed. S.-I. Akasofu and Y. Kamide,

Terra Sci. and D. Reidel, p. 37.
Sault, R. T., Killeen, N. E. B., Zmuidzinas, J. and Lauschin, R.: 1990, Astrophys. J. Suppl. 74, 437.
Schatten, K. H., Leighton, R. B., Howard, R. and Wilcox, J. M.: 1972, Solar Phys. 26, 283.
Schlickeiser, R.: 1986, in: Cosmic Radiation in Contemporary Astrophysics, ed. M. M. Shapiro, D. Reidel Publ. Co., Dordrecht, p. 27.
Schlüter, A. and Temesvary, S.: 1958, in: Electromagnetic Phenomena in Cosmic Physics, ed. B. Lehnert, Cambridge Univ. Press, p. 263.
Schmelz, J. T.: 1990, in: Galactic and Intergalactic Magnetic Fields, ed. R. Beck, P. P. Kronberg and R. Wielebinski, Kluwer Acad. Publ., Dordrecht, p. 20.
Schmelz, J. T., Saba, J. L. K., Strong, K. T. and Holman, G. D.: 1988, Adv. Space Res. 8, No. 11, 189.
Schmidt, G. D.: 1987, in: The Second Conference on Faint Blue Stars, ed. A. G. D. Philip, D. S. Hayes and J. W. Liebert, L. Davis Press, N. Y., p. 377
Schmidt, G. D.: 1989, in: White Dwarfs, ed. G. Wegner, Springer-Verlag, p. 305.
Schmidt, G. D., West, S. C., Liebert, J., Green, R. F. and Stockman, H. S.: 1986, Astrophys. J. 309, 218.
Schmidt, Th.: 1970, Astron. Astrophys. 6, 294.
Schröter, E. H.: 1971, in: Solar Magnetic Fields, ed. R. Howard, D. Reidel Publ. Co., Dordrecht, p. 167.
Schröter, E. H., Vázquez, M. and Wyller, A. A. (eds.): 1987, The Role of Fine-Scale Magnetic Fields on the Structure of the Solar Atmosphere, Cambridge Univ. Press.
Schurcliff, A.: 1962, Polarized Light, Harvard Univ. Press.
Schüssler, M.: 1983, in: Solar and Stellar Magnetic Fields: Origin and Coronal Effects, ed. J. O. Stenflo, D. Reidel Publ. Co., Dordrecht, p. 213.
Schuster: 1892, Rep. Brit. Assoc. Adv. sci., 634.
Schwarzschild, M.: 1949, Ann. d'Astrophys. 12, 148.
Schwenn, R. and Marsch, E. (eds.): 1990, Physics of the Inner Heliosphere I. Large-Scale Phenomena, Springer-Verlag.
Seares, F. H.: 1913, Astrophys. J. 38, 99.
Semel, M.: 1962, Compt. Rend. Acad. Sci. 254, 3978.
Severny, A. B.: 1958a, Izv. Krym. Astrofiz. Obs. 20, 22.
Severny, A. B.: 1958b, Astron. Zh. 35, 335.
Severny, A. B.: 1959, Astron. Zh. 36, 208.
Severny, A. B.: 1960a, Izv. Krym. Astrofiz. Obs. 22, 12.
Severny, A. B.: 1960b, Izv. Krym. Astrofiz. Obs. 24, 28.
Severny, A. B.: 1964a, Izv. Krym. Astrofiz. Obs. 31, 126.
Severny, A. B.: 1964b, Space Sci. Rev. 3, 451.
Severny, A. B.: 1965, Izv. Krym. Astrofiz. Obs. 33, 34.
Severny, A. B.: 1966, Astron. Zh. 43, 465.
Severny, A. B.: 1969, N ature 224, 53.
Severny, A. B.: 1970, Astrophys. J. 159, L73.
Severny, A. B.: 1971, in: Solar Magnetic Fields, ed. R. Howard, D. Reidel Publ. Co., Dordrecht, p. 675.
Severny, A. B.: 1988, Some Problems of the Physics of the Sun, Nauka, Moscow.
Severny, A. B. and Bumba, V.: 1958, Observatory 78, 33.
Severny, A. B. and Stepanov, V. E.: 1957, Izv. Krym. Astrofiz. Obs. 16, 3.
Severny, A. B., Kotov, V. A. and Tsap, T. T.: 1976, Nature 259, 87.
Severny, A. B., Kuvshinov, V. M. and Nikulin, N. S.: 1974a, Izv. Krym. Astrofiz. Obs.

50, 3.
Severny, A. B., Moiseev, I. G. and Efanov, V. A.: 1974b, Nature 249, 330.
Shajn, G. A.: 1955, Astron. Zh. 32, 110.
Shapiro, S. L. and Teukolsky, S. A.: 1983, Black Holes, White Dwarfs, and Neutron Stars – The Physics of Compact Objects, John Wiley and Sons, N. Y.
Sheeley, N. R., Jr.: 1992, in: The Solar Cycle, ed. K. L. Harvey, Astron. Soc. Pacific.
Sheeley, N. R., Jr., Nash, A. G. and Wang, Y. M.: 1987, Astrophys. J. 319, 481.
Shu, F. H., Adams, F. C. and Lizamo, S.: 1987, Ann. Rev. Astron. Astrophys. 25, 23.
Sidlichovský, M.: 1974, Bull. Astron. Inst. Czech. 25, 198.
Simon, G. W. and Leighton, R. B.: 1964, Astrophys. J. 140, 1120.
Sistla, G. and Harvey, J. W.: 1970, Solar Phys. 12, 66.
Skumanich, A. and Lites, B. W: 1987, Astrophys. J. 332, 473.
Skumanich, A. and Lites, B. W.: 1991, in: Solar Polarimetry, ed. L. J. November, NSO/SacPeak, p. 307.
Smerd, S. F., Sheridan, K. V. and Stewart, R. T.: 1975, Astrophys. Lett. 16, 23.
Smirnova, T. V.: 1991, Pis'ma Astron. Zh. 17, 159.
Smith, E. J.: 1967, in: Magnetism and the Cosmos, ed. W. R. Hindmarsh, Oliver and Boyd, Edinburgh, p. 271.
Smithson, R. C.: 1975, Solar Phys. 40, 241.
Sobotka, M., Bonet, J. A. and Vázquez, M.: 1993, in: The Magnetic and Velocity Fields of Solar Active Regions, eds. H. Zirin, Guoxiang Ai and Haimin Wang, Astron. Soc. Pacific Conf. Ser. vol. 46, p. 20.
Sofue, Y.: 1987, Publ. Astron. Soc. Japan 39, 547.
Sofue, Y., Fujimoto, M. and Wielebinski, R.: 1986, Ann. Rev. Astron. Astrophys. 24, 458.
Solanki, S. K.: 1986, Astron. Astrophys. 168, 311.
Solanki, S. K.: 1990, in: Solar Photosphere: Structure, Convection and Magnetic Fields, ed. J. O. Stenflo, Kluwer Acad. Publ., Dordrecht, p. 103
Solanki, S. K. and Stenflo, J. O.: 1984, Astron. Astrophys. 140, 185.
Solanki, S. K. and Stenflo, J. O.: 1985, Astron. Astrophys. 148, 123.
Solanki, S. K., Keller, C. and Stenflo, J. O.:1987, Astron. Astrophys. 188, 183,
Solanki, S. K., Rüedi, L. and Livingston, W. C.:1992, Astron. Astrophys. 263, 312
Solanki, S. K., Zayer, I. and Stenflo, J. O.: 1988, in: High Spatial Resolution Solar Observations, ed. O. von der Lühe, Sac Peak Obs.
Spicer, D. S.: 1982, Space Sci. Rev. 35, 349.
Spitzer, L., Jr.: 1957, Astrophys. J. 125, 525.
Spitzer, L., Jr.: 1978, Physical Processes in the Interstellar Medium, Wiley-Interscience, N. Y.
Spruit, H. C.: 1981, in: The Physics of Sunspots, ed. L. E. Cram and J. H. Thomas, Sac Peak Obs., p. 98.
Spruit, H. C. and Zweibel, E. G.: 1979, Solar Phys. 62, 15.
Staude, J.: 1969, Solar Phys. 8, 264.
Staude, J.: 1970a, Solar Phys. 12, 84.
Staude, J.: 1970b, Solar Phys. 15, 102.
Staude, J.: 1971, Solar Phys. 18, 22.
Staude, J.: 1973, Solar Phys. 32, 403.
Staude, J.: 1980, Phys. Solariterr. (Potsdam) 14, 58.
Staude, J. and Hofmann, A.: 1988, ESA SP-285, vol. II, 123.
Staude, J., Hofmann, A. and Bachmann, G.: 1991, in: Solar Polarimetry, ed. L. J. November, NSO/SacPeak, p. 49.

Stellmacher, G. and Wiehr, E.: 1980, Astron. Astrophys. 82, 157.

Stelzried, C. T., Levy, G. S., Sato, T., Rusch, W. V. T., Ohlson, J. E., Schatten, K. H. and Wilcox, J. M.: 1970, Solar Phys. 14, 440.

Stenflo, J. O.: 1968, Acta Univ. Lund II, No. 2. = Medd. Lund Astron. Obs. II, No. 153.

Stenflo, J. O.: 1971, in: Solar Magnetic Fields, ed. R. Howard, D. Reidel Publ. Co., Dordrecht, p. 714.

Stenflo, J. O.: 1973, Solar Phys. 32, 141.

Stenflo, J. O.: 1978a, Astron. Astrophys. 66, 241.

Stenflo, J. O.: 1978b, Rep. Prog. Phys. 41, 865.

Stenflo, J. O. (ed.): 1983, Solar and Stellar Magnetic Fields: Origin and Coronal Effects, D. Reidel Publ. Co., Dordrecht.

Stenflo, J. O.: 1984, Adv. Space Res. 4, 5.

Stenflo. J. O.: 1985a, in: Measurements of Solar Vector Magnetic Fields, ed. M. J. Hagyard, NASA CP-2374, p. 263.

Stenflo, J. O.: 1985b, Solar Phys. 100, 189.

Stenflo, J. O.: 1986a, in: Small-Scale Magnetic Flux Concentrations in the Solar Photosphere, ed. W. Deinzer, M. Knölker and H. H. Voigt, Vandenhoeck and Ruprecht, Göttingen, p. 59.

Stenflo, J. O.: 1986b, Mitt. Astron. Ges. 65, 25.

Stenflo, J. O.: 1988, Astrophys. Space Sci. 144, 321.

Stenflo, J. O.: 1989, Astron. Astrophys. 210, 403.

Stenflo, J. O.: 1991a, in: Hanle Effect and Level-Crossing Spectroscopy, ed. G. Moruzzi and F. Strumia, Plenum Publ. Corp., p. 237.

Stenflo, J. O.: 1991b, in: Solar Polarimetry, ed. L. J. November, NSO/Sac Peak, p. 416.

Stenflo, J. O.: 1992a, in: Electromechanical Coupling of the Solar Atmosphere, ed. D. S. Spicer, Amer. Inst. Phys.

Stenflo, J. O.: 1992b, in: The Solar Cycle, ed. K. L. Harvey, Astron. Soc. Pacific, p. 83.

Stenflo. J. O. and Güdel, M.: 1988, Astron. Astrophys. 191, 137.

Stenflo, J. O. and Harvey, J. W.: 1985, Solar Phys. 95, 99.

Stenflo, J. O. and Vogel, M.: 1986, Nature 319, 285.

Stenflo, J. O. and Weisenhorn, A. L.: 1987, Solar Phys. 108, 205.

Stenholm, Z. G. and Stenflo, J. O.: 1978, Astron. Astrophys. 67, 33.

Stepanov, V. E.: 1958a, Izv. Krym. Astrofiz. Obs. 18, 136.

Stepanov, V. E.: 1958b, Izv. Krym. Astrofiz. Obs. 19, 20.

Stepanov, V. E.: 1960, Izv. Krym. Astrofiz. Obs. 24, 293.

Stepanov, V. E.: 1962, Izv. Krym. Astrofiz. Obs. 27, 140.

Stepanov, V. E. and Gopasyuk, S. I.: 1962, Izv. Krym. Astrofiz. Obs. 28, 194.

Stepanov, V. E. and Severny, A. B.: 1962, Izv. Krym. Astrofiz. Obs. 28, 166.

Stepanov, V. E. et al.: 1962, in: Annals of the International Geophysical Year, vol. XVIII.

Steshenko, N. V.: 1960, Izv. Krym. Astrofiz. Obs. 22, 49.

Stibbs, D. W. N.: 1950, Mon. Not. R. Astron. Soc. 110, 395.

Stift, M. J. and Goossens, M.: 1991, Astron. Astrophys. 251, 139.

Stix, M.: 1976, Astron. Astrophys. 47, 243.

St. John , C. E.: 1913, Astrophys. J. 39, 322.

Strom, R. G.: 1987, Mercury – The Elusive Planet, Cambridge Univ. Press.

Sturrock, P. A.: 1968, in: Structure and Development of Solar Active Regions, ed. K. O. Kiepenheuer, D. Reidel Publ. Co., Dordrecht.

Sturrock, P. A.: 1986, in: Physics of the Sun, eds. P. A. Sturrock, D. Reidel. Publ. Co., Dordrecht, vol. 1, p. 1.

Švestka, Z.: 1976, Solar Flares, D. Reidel Publ. Co., Dordrecht.

Sweet, P.: 1958, in: Electromagnetic Phenomena in Cosmic Physics, ed. B. Lehnert, Cambridge Univ. Press, p. 123.
Syrovatsky, S. I.: 1978, Solar Phys. 58, 89.
Tandberg-Hanssen, E.: 1970, Solar Phys. 15, 359.
Tandberg-Hanssen, E.: 1974, Solar Prominences, D. Reidel Publ. Co., Dordrecht.
Tandberg-Hanssen , E. and Anzer, U.: 1970, Solar Phys. 15, 158.
Tandberg-Hanssen, E. and Emslie, A. E.: 1988, The Physics of Solar Flares, Cambridge Univ. Press.
Tanenbaum , A. S ., Wilcox , J.M. and Howard, R.: 1971, in: Solar Magnetic Fields, ed. R. Howard, D. Reidel Publ. Co., Dordrecht.
Tang, F.: 1982, Solar Phys. 75, 179.
Tang, F. and Haimin Wang: 1991, Solar Phys. 132, 247.
Tarbell, T. D., Title, A. M. and Schoolman, S. A.: 1979, Astrophys. J., 229, 387.
Tatsch, J. H.: 1974, The Moon – Its Past Development and Present Behavior, Tatsch Associates, Mass.
ten Bruggencate, P. and vön Kluber, H.: 1939, Z. Astrophys. 18, 284.
Thiesssen, G.: 1949, Z. Astrophys. 26, 130.
Townes, C. H.: 1989, in: The Center of the Galaxy, ed. M. Morris, Kluwer Acad. Publ., Dordrecht, p. 1.
Treanor. P. J.: 1960, Mon. Not. R. Astron. Soc. 120, 412.
Trimble, V.: 1990, in: Galactic and Intergalactic Magnetic Fields, ed. R. Beck, P. P. Kronberg and R. Wielebinski, Kluwer Acad. Publ., Dordrecht, p. 29.
Troland, T. H. and Heiles, C.: 1986, Astrophys. J. 301, 339.
Trumbo, D., Harvey, J., Livingston, W. and Slaughter, C.: 1974, Bull. Am. Astron. Soc. 6, 296.
Uchida. Y.: 1969, Publ. Astron. Soc. Japan 21, 128.
Unno, W.: 1956, Publ. Astron. Soc. Japan 8, 108.
Urey, H. and Runcorn, S. K.: 1973, Science 180, 636.
Vallée, J. P.: 1988, Astron. J. 95, 750.
van Ballegooijen, A. A.: 1985, in: Measurements of Solar Vector Magnetic Fields, ed. M. J. Hagyard, NASA CP-2374, p. 322,
van Ballegooijen, A. A.: 1987, in: Numerical Radiative Transfer, ed. W. Kalkofen, Cambridge Univ. Press, p. 279.
van Ballegooijen, A. A.: 1990, in: Basic Plasma Proce sses on the Sun, ed. E. R. Priest and V. Krishan, Kluwer Acad. Publ., Dordrecht, p. 303.
van de Hulst, H. C.: 1953, in: The Sun, ed. G. P. Kuiper, Univ. of Chicago Press, p. 207.
van de Hulst, H. C.: 1967, Ann. Rev. Astron. Astrophys. 5, 167.
Vasyliuras, V. M.: 1983, in: Solar-Terrestrial Physics, ed. R. L. Carovillano and J. M. Forbes. D. Reidel Publ. Co., Dordrecht, p. 243.
Veeder, G. J. and Zirin, H.: 1970, Solar Phys. 12, 391.
Verschuur, G. L.: 1968, Phys. Rev. Lett. 21, 775.
Verschuur, G. L.: 1970, in: Interstellar Gas Dynamics, ed. H. J. Habing, D. Reidel Publ. Co., Dordrecht, p. 150.
Verschuur, G. L.: 1971, Astrophys. J. 165, 651.
Verschuur, G. L.: 1979, Fundam. Cosmic Phys. 5, 113.
von Klüber, H.: 1948, Z. Astrophys. 24, 121.
von Klüber, H.: 1954, Mon. Not. R. Astron. Soc. 114, 242.
von Klüber, H.: 1955, Vistas of Astron. 1, 751.
Walén, C.: 1941, Ark. f. Mat. Astron. och. Fys. 30A, No. 15.
Wanders, A. J. M.: 1934, Z. Astrophys. 8, 108.

Wang, Haimin: 1988a, Ph. D. Thesis, Caltech.
Wang, Haimin: 1988b, Solar Phys. 116, 1.
Wang, Haimin: 1993, in: The Magnetic and Velocity Fields of Solar Active Regions, eds. H. Zirin, Guoxiang Ai and Haimin Wang, Astron. Soc. Pacific Conf. Ser. vol. 46, p. 323.
Wang, Haimin and Zirin, H.: 1987, Bull. Am. Astron. Soc. 19, 941.
Wang, Haimin and Zirin, H.: 1988, Solar Phys. 115, 205.
Wang Jing-xiu and Shi Zhong-xian; 1992, Acta Astron. Sin. 33, 95.
Wang Zhengzhi and Lin Yuanchang: 1984, Acta Astrophys. Sin. 4, 222.
Watson, W. D.: 1970, Astrophys. J. 162, L45.
Webb, D. F., Davis, J. M. and McIntosh, P. S.: 1984, Solar Phys. 92, 109.
Wegner, G. (ed.): 1989, White Dwarfs, Springer-Verlag.
Weiler, K. W. and de Pater, I.: 1983, Astrophys. J. Suppl. Ser. 52, 293.
West, S. C. ; 1989, Astrophys. J. 345, 511.
Western, L. R. and Watson, W. D.: 1984, Astrophys. J. 285, 158.
Wickramasinghe, D. T.: 1987, in: The Second Conferene on Faint Blue Stars, ed. A. G. Davis Philip, D. S. Hayes and J. W. Liebert, L. Davis Press, N. Y., p. 389.'
Wiehr, E. and Degenhardt, D.: 1992, Astron. Astrophys. 259, 313.
Wiehr, E. and Stellmacher, G.: 1989, Astron. Astrophys. 225, 528.
Wiehr, E., Koch, A., Knölker, M., Küveler, G. and Stellmacher, G.: 1984, Astron. Astrophys. 140, 352.
Wielebinski, R.: 1990, in: The Interstellar Medium in Galaxies, ed. H. A. Thronson and J. M. Shull, Kluwer Acad. Publ., Dordrecht, p. 349.
Wilcox, J. M.: 1968, Space Sci. Rev. 8, 258.
Wilson. P. R.: 1981, in: The Physics of Sunspots, ed. L. E. Cram and J. H. Thomas, Sac Peak Obs., p. 83.
Wittmann, A.: 1971a, Solar Phys. 20, 78.
Wittmann, A.: 1971b, Solar Phys. 20, 365.
Wittmann, A.: 1972, Mitt. Astron. Ges. No. 31, 150.
Wittmann, A.: 1974, Solar Phys. 36, 29.
Wolff, S. C. and Wolff, R. J.: 1970, Astrophys. J. 160, 1049.
Wrubel, M. H.: 1952, Astrophys. J. 116, 291.
Wu Fei and Wang Zhengzhi: 1984, Acta Astrophys. Sin. 4, 272.
Wu Linxiang and Ai Guoxiang: 1990, Acta Astrophys. Sin. 10, 371,
Yang Wei-min: 1990, Doctoral Thesis, Yunnan Astron. Obs.
Ye Shi-hui: 1991, in: Solar Polarimetry, ed. L. J. November, NSO/SacPeak, p. 403.
Ye Shi-hui: 1992, Publ. Yunnan Astron. Obs. No. 3, 1.
Ye Shi-hui and Jin Jie-hai: 1983a, Acta Astron. Sin. 24, 119.
Ye Shi-hui and Jin Jie-hai: 1983b, Acta Astrophys. Sin. 3, 276.
Ye Shi-hui and Jin Jie-hai: 1985, Publ. Purple Mt. Obs. 4, No. 3, 1.
Ye Shi-hui and Jin Jie-hai: 1986, Solar Phys. 104, 273.
Ye Shi-hui and Jin Jie-hai: 1987, Solar Phys. 112, 305.
Ye Shi-hui and Jin Jie-hai: 1990, Solar Phys. 129, 247.
Ye Shi-hui and Jin Jie-hai: 1993, Solar Phys. 146, 229.
Ye Shi-hui, Wang Zhen-yi and Jin Jie-hai: 1978a, Acta Astron. Sin. 19, 152.
Ye Shi-hui, Wang Zhen-yi and Jin Jie-hai: 1978b, Studia Astron. Sin. 2, 70.
Ye Shi-hui, Wang Zhen-yi and Jin Jie-hai: 1979, Acta Astron. Sin. 20, 275.
Yoshimura, H.: 1975, Astrophys. J. Suppl. Ser. 29, 469.
Yoshimura, H.: 1983, Astrophys. J. Suppl. Ser. 52, 363.
Yun, H. S.: 1971a, Astrophys. J. 162, 975.

Yun, H. S.: 1971b, Solar Phys. 16, 379.
Yun, H. S.: 1972, Solar Phys. 22, 137.
Zeeman, P.: 1897, Phil. Mag. 5, 43, 226.
Zeldovich, Ya. B. and Ruzmaikin, A. A.: 1980, Zh. Eksp. Teop. Fiz. 78, 980.
Zeldovich, Ya. B., Ruzmaikin, A. A. and Sokolov, D. D.: 1983, Magnetic Fields in Astrophysics, Gordon and Breach, Montreux.
Zirin, H.: 1961, Astron. Zh. 38, 5.
Zirin, H.: 1972, Solar Phys. 22, 34.
Zirin, H.: 1986, Aust. J. Phys. 38, 961.
Zirin, H.: 1987, Solar Phys. 110, 101.
Zirin, H.: 1988, Astrophysics of the Sun, Cambridge Univ. Press.
Zirin, H. and Severny, A. B.: 1961, Observatory 81, 155.
Zirin, H., Guoxiang Ai and Haimin Wang (eds.): 1993, The Magnetic and Velocity Fields of Solar Active Regions, Astron. Soc. Pacific Conf. Ser. vol. 46.
Zwaan, C.: 1965, Rech. Astron. Obs. Utrecht 17(4).
Zwaan, C.: 1968, Ann. Rev. Astron. Astrophys. 6, 135.
Zwaan, C.: 1987, Ann. Rev. Astron. Astrophys. 25, 83.
Zwaan, C. and Brants, J. C.: 1981, in: The Physics of Sunspots, ed. L. E. Cram and J. H. Thomas, Sac Peak Obs., p. 210.
Zwaan, C. and Burman, J.: 1971, in: Solar Magnetic Fields, ed. R. Howard, D. Reidel Publ. Co., Dordrecht, p. 220.
Zwaan, C., Brants, J. C. and Cram, L. E.: 1985, Solar Phys. 95, 3.

Name Index

Subject Index

SPRINGER NATURE

GPSR Compliance

The European Union's (EU) General Product Safety Regulation (GPSR) is a set of rules that requires consumer products to be safe and our obligations to ensure this.

If you have any concerns about our products, you can contact us on ProductSafety@springernature.com

In case Publisher is established outside the EU, the EU authorized representative is:

Springer Nature Customer Service Center GmbH
Europaplatz 3
69115 Heidelberg, Germany

Kolibri 360 GmbH
Ferdinand-Jühlke-Straße 7
99095 Erfurt, Deutschland
produktsicherheit@kolibri360.de